Standard Methods
for the Examination of
Dairy Products

15th edition

Standard Methods
for the Examination of
Dairy Products

15th edition

Edited by
Gary H. Richardson, PhD

American Public Health Association
Washington, D.C.

Copyright © 1985
AMERICAN PUBLIC HEALTH ASSOCIATION

All rights reserved. No part of this publication may be reproduced graphically or electronically, including storage and retrieval systems, without prior written permission of the publisher.

3M(p)2M(h) 2/85
ISBN 0-87553-132-6 (paper)
      0-87553-118-0 (hardcover)
ISSN 8755-3554

Printed and bound in the United States of America
  Typography: Monotype Composition, Baltimore, Maryland
  Printing and binding: Port City Press, Inc. Baltimore, Maryland
  Cover design: Donya Melanson Associates, Boston, Massachusetts

TECHNICAL COMMITTEE

Gary H. Richardson, Ph.D., *Chairman*
  Professor of Food Science and Microbiology, Department of Nutrition and Food Sciences, Utah State University, Logan, UT 84322
John C. Bruhn, Ph.D.
  Extension Food Technologist (Dairy Products), Department of Food Science and Technology, 101B Cruess Hall, University of California, Davis, CA 95616
Ronald A. Case, M.S.
  Manager, Laboratory Controls, Kraft Inc., Kraft Court, Glenview, IL 60025
Roy E. Ginn, B.S.
  General Manager, Dairy Quality Control Institute Inc., 2353 Rice St., St. Paul, MN 55113
Robert T. Marshall, Ph.D.
  Professor of Food Science, Department of Food Science, 203 Eckles Hall, University of Missouri, Columbia, MO 65201
Elmer H. Marth, Ph.D.
  Professor of Food Science and Bacteriology, Department of Food Science, University of Wisconsin, Madison, WI 53706
James W. Messer, Ph.D.
  Chief, Laboratory Quality Assurance Branch, Division of Microbiology, Food and Drug Administration, 1090 Tusculum Ave., Cincinnati, OH 45226
H. Michael Wehr, Ph.D.
  Administrator, Laboratory Services Division, Oregon Department of Agriculture, 635 Capitol St. N.E. Salem, OR 97310

Project Director

Howard L. Bodily, Ph.D.
  Staff Associate for Laboratory Programs, American Public Health Association, P.O. Box 69, Midway, UT 84049

Project Officer

Ralston B. Read, Jr., Ph.D.
  Director, Division of Microbiology, Bureau of Foods, Food and Drug Administration, 200 C Street, S.W., Washington, DC 20204

CONTRIBUTORS

William L. Arledge, B.S.
  Director of Quality Control and Related Services, Dairymen Inc., 604 Portland Bldg. 200 W. Broadway, Louisville, KY 40202

Sidney E. Barnard, Ph.D.
  Extension Food Technologist (Dairy Products), Penn State University, 9 Borland Laboratory, University Park, PA 16801

Calvin E. Beckelheimer, M.S.
  Associate Director of Laboratory, Hygienic Laboratory Division, West Virginia Department of Health, 167-11th Ave., South Charleston, WV 25303

Harry M. Behney, Jr., B.S.
  Principal Laboratory Survey Officer, Pennsylvania Department of Agriculture, 2301 N. Cameron St., Harrisburg, PA 17042

Richard H. Bell, Ph.D.
  Manager, Quality Control Laboratory, Difco Laboratories, Inc., Box 1058A, Detroit, MI 48232

Robert L. Bradley, Jr. Ph.D.
  Professor, Department of Food Science, University of Wisconsin, Madison, WI 53706

John C. Bruhn, Ph.D.
  Extension Food Technologist (Dairy Products), Department of Food Science and Technology, 101B Cruess Hall, University of California, Davis, CA 95616

Robert Y. Cannon, Ph.D.
  Professor of Animal Science, Department of Animal Science, Auburn University, Auburn, AL 36830

Ronald A. Case, M.S.
  Manager, Laboratory Controls, Kraft Inc., Kraft Court, Glenview, IL 60025

Roger W. Cochran, B.S.
  Chief, Radiological Health Division, State Hygienic Laboratory, University of Iowa, Iowa City, IA 52242

CONTRIBUTORS

Irvin C. Druchrey, B.S.
Food, Drug and Dairy Laboratory Division, Bureau of Laboratories in Board of Health, 1330 W. Michigan St., Indianapolis, IN 46206

J. E. Edmondson
Professor, Department of Food Science and Nutrition, University of Missouri, Columbia, MO 65201

Robert Fischbeck, B.S.
Director of Client Services and Administration, Hazelton Laboratories America, 3301 Kinsman Blvd, Madison, WI 63704

Joseph F. Frank, Ph.D.
Assistant Professor, Department of Animal and Dairy Sciences, University of Georgia, Athens, GA 30602

Roy E. Ginn, B.S.
General Manager, Dairy Quality Control Institute, Inc., 2353 N. Rice St., St. Paul, MN 55113

Kathryn Glynn, M.S.
Assistant Director (Retired), Laboratory Division, Connecticut State Department of Health, Box 1689, Hartford, CT 06101

Virgil D. Grace, B.S.
Laboratory Quality Assurance Branch, Div. of Microbiology, Food and Drug Administration, 1090 Tusculum Ave., Cincinnati, OH 45226

Lester Hankin, Ph.D.
Biochemist, Connecticut Agricultural Experiment Station, 123 Huntington St., New Haven, CT 06504

Paul A. Hartman, Ph.D.
Distinguished Professor and Chairman, Department of Bacteriology, Iowa State University, Ames, IA 50011

G. A. Houghtby, Ph.D.
Microbiologist, Division of Microbiology, Food and Drug Administration, 1090 Tusculum Ave., Cincinnati, OH 45226

Wesley R. Kelly, Ph.D.
Director, State Dairy Laboratory, Dairy Science Department, South Dakota State University, Brookings, SD 57007

Dick H. Kleyn, Ph.D.
Associate Professor of Food Science, Department of Food Science, Rutgers University, New Brunswick, NJ 08903

John A. Koberger, Ph.D.
Professor, Department of Food Science and Human Nutrition, University of Florida, Gainesville, FL 32611

William S. LaGrange, Ph.D.
Extension Food Technologist, Department of Food Technology, Iowa State University, Ames, IA 50011

CONTRIBUTORS

Lloyd O. Luedecke, Ph.D.
Professor, Department of Food Science and Technology, Washington State University, Pullman, WA 99164

John B. Lindamood, Ph.D.
Professor, Department of Food Science and Nutrition, Ohio State University, Columbus, OH 43210

Robert T. Marshall, Ph.D.
Professor, Department of Food Science and Nutrition, 203 Eckles Hall, University of Missouri, Columbia, MO 65201

Elmer H. Marth, Ph.D.
Professor of Food Science and Bacteriology, Department of Food Science, University of Wisconsin, Madison, WI 53706

R. Burt Maxcy, Ph.D.
Professor, Department of Food Science and Technology, University of Nebraska, Lincoln, NE 68583

Larry J. Maturin, Ph.D.
Bureau of Laboratories, Office of Health Services, State Office Building, 325 Loyolla Ave., New Orleans, LA 70160

James W. Messer, Ph.D.
Chief, Laboratory Quality Assurance Branch, Division of Microbiology, Food and Drug Administration, 1090 Tusculum Ave., Cincinnati, OH 45226

Emil M. Mikolajcik, Ph.D.
Professor, Department of Food Science and Nutrition, Ohio State University, 2121 Fyffe Rd., Columbus, OH 43210

Richard D. Mochrie, Ph.D.
Professor, Department of Animal Science, North Carolina State University, Box 5127, Raleigh, NC 27650

Gopala K. Murthy, Ph.D.
Assistant Chief, Microbial Biochemistry Branch, Division of Microbiology, Food and Drug Administration, 1090 Tusculum Ave., Cincinnati, OH 45226

Vernal S. Packard, Ph.D.
Extension Specialist—Dairy Products, Department of Food Science and Nutrition, University of Minnesota, 1354 Eckles, St. Paul, MN 55108

David A. Power, Ph.D.
Manager of Technical Services, BBL Microbiology Systems, Box 243, Cockeysville, MD 21030

Gary H. Richardson, Ph.D.
Professor, Department of Nutrition and Food Sciences, Utah State University, Logan, UT 84322

L. H. Schultz, Ph.D.
   Professor, Dairy Science Department, University of Wisconsin, Madison, WI 53706
David B. Weddle, Ph.D.
   Director of Product Development, Mid-America Farms, Inc., 800 W. Tampa, Springfield, MO 65805
H. Michael Wehr, Ph.D.
   Administrator, Laboratory Services Division, Oregon Department of Agriculture, 635 Capitol St. NE, Salem, OR 97310
Charles H. White, Ph.D.
   Professor, Department of Dairy Science, Louisiana State University, Baton Rouge, LA 70803
Robert R. Williams, Ph.D.
   Microbiological Supervisor, Laboratory Services Division, Oregon Department of Agriculture, 635 Capitol St. NE, Salem, OR 97310
Earl O. Wright, M.S.
   (Retired), Box 701, 413 Kellogg, Ames, IA 50010

CONSENSUS REVIEWERS

Most contributors were involved in the consensus review and made appropriate suggestions for improvement. Additionally, the following volunteers were provided drafts of one or more chapters in areas of their expertise and invited to suggest modifications to the Technical Committee:

Dan Adams, Henry V. Atherton, Edward B. Aylward, Stan Barlow, Harold K. Bengsch, Floyd W. Bodyfelt, Esther Bone, David M. Barbano, A. Richard Brazis, Stanley E. Charm, Warren S. Clark, Gary M. Dixon, Joseph E. Edmonson, Sai Faraknik, Joseph A. Frank, Ruth Fuqua, Berry Gay, Jr., Roland Golden, Wyatt Gosnell, Earl G. Hammond, Gary D. Hayen, Judy Heady, O. Eldon Henson, Clem Honer, Barbara P. Keogh, Richard M. McKee, William A. Moats, Nancy J. Moon, Dee R. Morgan, Gayle K. Mulberry, Robert Mullen, Norman F. Olson, Joe Peters, Charles F. Piwaronas, George W. Reinbold, Joseph Scharer, James A. Scott, Robert L. Sellars, Phillip E. Shaver, John W. Sherbon, George Sherman, Dennis Westhoff, Faye Feldstein Westhoff, Kenneth W. Whaley, and George York.

PREFACE

A committee was appointed by the American Public Health Association (APHA) in 1905 to standardize methods for the bacteriological examination of milk [1.10,#11]. *Standard Methods for the Examination of Dairy Products (SMEDP)* was an outgrowth of that charge. The 15th edition of *SMEDP* has been produced under the direction of the APHA Committee on Laboratory Standards and Practices (CLASP). Guidance of the project was provided by a Technical Committee. This approach mirrored the Intersociety Council organization used for the previous two editions and encouraged continuity between editions. It also facilitated the process of making changes between editions and improved coordination between chapters, while insuring adequate scientific backgrounds for making decisions on methods and supervising research projects to develop or improve methods. Committee activities were made possible through funds provided by the U.S. Food and Drug Administration (FDA) via contract to the APHA.

The editor was selected by CLASP and charged with nominating the Technical Committee. Eight individuals were named and approved by CLASP. The committee members were chosen to have varied backgrounds and to represent diverse professional societies as well as industrial, academic, and governmental interests. Three member of the 14th edition Intersociety Council were retained to provide continuity. The Technical Committee held one or two meetings annually to coordinate the text writeup activities and research projects. The 15th edition of *SMEDP* represents the culmination of their assignment to provide a statement of current methodology for approximately 9,000 dairy laboratories in this country and, hopefully, guidance for other nations involved in international marketing of dairy foods.

Technical Committee Activities

Reactions to the 14th edition were requested. Most suggestions for improvements came from Dr. James W. Messer and his colleagues who are responsible for coordination of the Laboratory Quality Assurance Program of the FDA. An update was published in the *Journal of Food Protection* [1.10,#12] and Dr. Messer was asked to serve as a Technical Committee member to assure continuation of assistance from that group. A CLASP recommended classification of methods was adopted. A consensus procedure was examined in concert with CLASP and adopted to allow greater participation by users and the scientific community. Committee members made presentations at annual meetings of the International Association of Milk, Food and Environmental Sanitarians, the American Society for Microbiology, the American Dairy Science Association, and the Institute of Food Technologists. In all cases, invitations were extended for reactions to the 14th edition, contributors to help with the 15th edition, and reviewers to serve on the new consensus program.

Members of the Technical Committee each chaired at least one chapter committee and coordinated the efforts of other chapter committee chairpersons. The specific involvements are indicated at the beginning of each chapter.

Chapter texts were circulated between the Technical and Chapter committees until agreement was reached and each chapter's content represented the desired goals. Substantive changes are included in this edition and reactions to these changes were obtained from the technical and scientific communities. Eighty-six participants were mailed chapters in the areas of their expertise and invited to suggest improvements. Each response was considered, appropriate changes were made and the Project Director responded to reviewers with appropriate reasons when suggestions were not accepted.

The FDA contract provided limited funds for mini-research projects. Invitations were published in the *Journal of Food Protection* for interested parties to submit proposals in areas identified by the Technical Committee as likely to contribute to the improvement of this edition. Nineteen proposals were considered; these included offers to conduct studies with no funding from the Technical Committee. The studies supported and guided by the Technical Committee included: development of reference samples for somatic cell counters, an evaluation of a spread vs the Standard Plate Count, replacement of toxic chemicals in some reagents, a collaborative study of evolving antibiotics assays, a survey of sulfa drug methods, an evaluation of media for injured coliforms, a comparison of media for yeast and mold counts, the use of hydrophobic grid membrane technology, the refinement of microwave methods for moisture, the use of reflectance infrared instruments with dry products, the use of near infrared technology for high moisture products, improvement of accelerated psychrotrophic methods, more rapid shelf life tests, and evaluation of interrupted incubation tests. During the development of the 15th edition, the Committee was kept informed of studies involving; the Rurakura Rolling Ball Viscometer, the Bactomatic 120, new sediment tests, the Technicon 400, automated pipets, plastic disposable dilution blanks, air impact samplers, Gerber test bottle specifications, portable conductivity meters, special milk samplers, plastic milk dilution bottles, a hand held conductivity meter, elisa antibiotic assays,

PREFACE

preincubation methods, automated somatic cell counters, barrel cheese sampling, continuous raw milk sampling, and the Catalasemeter.

The Committee considered the problem of approving automated equipment that is subsequently modified by the manufacturer, and chose to develop procedures to cover subsequent improvements that can be shown by the manufacturer to correlate with approved and reference methods. Data from a suitable comparative study were considered adequate to allow such modification. The Committee recognized that this general guideline cannot be followed in all instances and retained the right to refuse classification of a method with unproven correlations. The Committee philosophy reflects that of the Association of Official Analytical Chemists (AOAC) in the Preface of the 1980 *Official Methods of Analysis* by Williams Horwitz [1.10, #7]:

> "A problem that has arisen is how to handle the numerous individual instruments of diverse design and manufacture that have been developed to automate a particular determination. Even if this problem is solved by incorporating all the available instruments into the initial collaborative study, the problem returns with the first "new and improved" modification. Different instrument designs and their subsequent modifications have been handled in the infrared determination of milk constituents by providing performance specifications which must be met by the basic instrument in general, when compared to a reference method or reference sample. In addition, the user must satisfy himself that his particular instrument also meets the performance specifications by frequent comparisons with the reference method or sample. This requirement eliminates the need for repeated collaborative studies every time a manufacturer redesigns or modifies his basic equipment, and in addition provides a continuous quality control technique on the performance of the instrument, method, and laboratory."

Several Technical Committee members joined the United States National Committee (USNAC) of the International Dairy Federation (IDF) and have remained actively aware of developments pertinent to this edition. The APHA approved donation of copies of the 14th edition of SMEDP to participants of the 1984 international seminar, "Challenges to Contemporary Dairy Analytical Techniques," 1984 Reading, England. These copies and a paper were presented to promote international cognizance of SMEDP.

Highlights of the 15th Edition

Adoption of the APHA classification of methods made it possible to combine all methods for a particular component into one chapter. Thus the appendices of recent editions are eliminated. Abbreviations and units are as specified by the CBE Style Manual Committee (CBE style manual: a guide for authors, editors, and publishers in the biological sciences. 5th ed. rev. and expanded. Bethesda, MD: Council of Biology Editors, Inc.; 1983). The use of L for liter instead of l is the most obvious change. Degrees Celcius (°C) is used throughout. Degrees Hortvet (°H) have been eliminated in compliance with AOAC recommendations. Generic descriptions have been used where possible. Proprietary or trade names are cited only in limited source situations, in which users might need ready access to supplier addresses. Mention of products or companies in the text does not constitute endorsement over those not mentioned.

One major change involves the adoption of a 25 to 250 colony count range for the Standard Plate Count (SPC) [Chapter 6].

Editorial changes have been made to improve clarity and readability. A new chapter, Chapter 2, incorporates essential principles of laboratory safety and quality assurance. The Alternative Count Methods, Chapter 7, now follows immediately the SPC Chapter 6. Chapter 9 is a new chapter that combines and adds tests for groups of organisms. Tests for all dairy food products were combined in Chapter 10. Both direct microscopic microorganism and somatic cell count methods have been combined into Chapter 11. The previous chapter on radionuclides, has been omitted and referred to in Chapter 18. To make this edition more available the APHA has provided a paperback edition for the first time, in addition to the traditional hardcover.

Acknowledgments

We are grateful to the 86 volunteers who helped with chapter committees and the consensus program to assure a quality up-to-date product. Technical Committee members worked diligently and generally met the deadlines agreed upon. Dr. Howard L. Bodily took care of most of the correspondence and consensus work and arranged committee meeting details. Dr. Ralston B. Read, Jr. provided grantor support and guidance throughout the project. Dr. Arnold E. Greenberg, Chairman, CLASP and Editorial Board, *Standard Methods for the Examination of Water and Wastewater* [5.14, #5] and Mrs. Seiko Brodbeck, Associate Executive Director for Programs, APHA, helped establish the consensus procedures for the Technical Committee.

We appreciate the work of the copy editor Ms. Lois M. Cox. Dr. Adrienne Ash, Director of Publications for the APHA provided helpful suggestions during the project. I am grateful to Dr. C. Anthon Ernstrom, Head of the Department of Nutrition and Food Sciences, Utah State University for providing word processing facilities, membership in USNAC, and for permitting me to devote the time required by this project.

I gratefully note the support of my wife, Fran, and my family for their encouragement and patience.

Logan, Utah
May 1984

Gary H. Richardson
Editor 15th Edition and Chairman
of the Technical Committee for
Standard Methods for the
Examination of Dairy Products

TABLE OF CONTENTS

Technical Committee v
Contributors vii
Consensus Reviewers xi
Preface xiii

1. STANDARD METHODS 1

    Introduction of *1*/ Function of Standard Methods *1*/ Classification of Procedures *2*/ Adopting New Methodology *4*/ Collaborative Studies *4*/ Uniformity of Methods *5*/ Split Sample Test Program *6*/ Relation of Farm and Plant Inspections with Laboratory Control *6*/ Accuracy of Methods for Measuring Sanitary Quality *6*/ References *8*

2. LABORATORY QUALITY ASSURANCE AND SAFETY 9

    Introduction *9*/ Laboratory Operational Management *9*/ Laboratory Facilities *15*/ Laboratory Equipment and Supplies *17*/ Control Tests and Record Keeping *25*/ Laboratory Safety *26*/ References *42*

3. PATHOGENS IN MILK AND MILK PRODUCTS 43

    Introduction *43*/ Raw Milk *44*/ Pasteurized Milk *53*/ Cheese *54*/ Cultured Milks *70*/ Butter *72*/ Nonfat Dried Milk *73*/ Frozen Desserts *78*/ References *79*

4. SAMPLING DAIRY AND RELATED PRODUCTS 89

    Fluid Milk and Cream Samples *89*/ Other Dairy Products *101*/ Sampling for Specific Laboratory Procedures *110*/ References *116*

5. MEDIA 119

    Introduction *119*/ Suitability of Water for Microbiological Applications *119*/ Preparation of Dilution Water *121*/ Water Toxicity Tests *122*/ Cleaning Glassware and Testing for Detergent Residues *122*/ Media *123*/ Basic Steps in Media Preparation *124*/ Adjustment of pH *125*/ Formulae and Directions for Preparation of Culture Media *125*/ Sterilization and Storage *129*/ Quality Control *130*/ References *131*

TABLE OF CONTENTS

6. MICROBIOLOGICAL COUNT METHODS.................... 133

   Introduction *133*/ Standard Plate Count *133*/ Preliminary Incubation for Raw Milk *146*/ Moseley Keeping Quality *147*/ Keeping Quality of Pasteurized Milk, Impedimetric Method *148*/ References *148*

7. ALTERNATIVE MICROBIOLOGICAL METHODS 151

   Introduction *151*/ Collection and Storage of Samples Before Testing *151*/ Statistical Protocol for an Analyst *152*/ Plate Loop Method *152*/ Spiral Plate Count Method *156*/ Oval Tube Method *164*/ Electrical Impedence Method *165*/ References *170*

8. COLIFORM BACTERIA 173

   Introduction *173*/ Definitions *174*/ General Interpretations *174*/ Sampling *175*/ Equipment, Supplies, and Media *175*/ General Procedure *175*/ Relative Values of Plate and Tube Methods *176*/ Coliform Test with a Solid Medium *177*/ Coliform Test with a Liquid Medium *179*/ Confirmed Test from a Solid Medium *179*/ Completed Test from a Liquid Medium *180*/ Reporting Results *180*/ Table of MPN Coliforms *182*/ Impedimetric Coliform Detection Method: Raw Milk *182*/ Hydrophobic Grid Membrane Filter Method *183*/ References *185*

9. TESTS FOR GROUPS OF MICROORGANISMS 189

   Thermoduric Bacteria *189*/ Thermophilic Bacteria *191*/ Proteolytic Microorganisms *191*/ Lipolytic Microorganisms *192*/ Enterococci *193*/ Aerobic Bacterial Spores *194*/ Psychotrophic Bacteria *194*/ Yeast and Mold Counts *197*/ References *198*

10. MICROBIOLOGICAL METHODS FOR DAIRY PRODUCTS........ 203

    Introduction *203*/ Evaporated, Concentrated and Sweetened Condensed Milks and Liquid Infant Formulas *203*/ Ultra-High Temperature Processed, Aseptically Packaged, Low-acid Products *207*/ Dry Dairy Products *208*/ Butter, Margarine and Related Products *210*/ Cheese and Other Cultured and Culture-added Products *211*/ Ice Cream and Related Food Products *214*/ References *216*

11. DIRECT MICROSCOPIC METHODS FOR BACTERIA
    AND SOMATIC CELLS..................................... 219

    Introduction *219*/ Application to Raw Milk *219*/ Application to Pasteurized Milk *220*/ Application to Dry Milk *220*/ Source of Error in the Microscopic Method *220*/ Procedure for Collecting Samples *221*/ Equipment and Supplies *221*/ Materials *226*/ Adjustment and Calibration of Microscope *226*/ Calculation of the Microscopic Factor (MF) *226*/ Using Transfer Instruments and Preparing Films *227*/ Handling of Slides During Staining *228*/ Preparation and Use of Stains *229*/ Storage of Slides *230*/ Examining Films for Bacteria or Somatic Cells *230*/ Field-wide Single Strip Method *231*/ Field Method *231*/ Strip Count Methods *233*/ Reporting Counts *236*/ References *236*

12. METHODS TO DETECT ABNORMAL MILK 239

   Introduction *239/* Screening Tests *239/* Confirmatory Tests *249/* References *256*

13. REDUCTION METHODS 259

   Introduction *259/* Standards *260/* Sampling *260/* Equipment and Supplies *260/* Preparing Dye Solution *261/* Testing Samples *261/* Methylene Blue Reduction Method *261/* Resazurin Reduction Method *262/* References *264*

14. DETECTION OF ANTIBIOTIC RESIDUES IN MILK
    AND DAIRY PRODUCTS 265

   Introduction *265/ Bacillus subtillis* Disc Assay *266/* Qualitative *Bacillus stearothermophilus* var. *calidolactis* Disc Assay *269/* Quantitative *Bacillus stearothermophilus* var. *calidolactis* Disc Assay *272/* Delvotest -P- Ampule and Multi Test Procedures *275/* The Charm Screening Assay for Beta-Lactam Residues *279/* Modified *Sarcina lutea* Cylinder Plate Method for Detection of Penicillin in Nonfat Dry Milk *281/* Specific Antimicrobial Cylinder Methods *285/* Detection of Sulfa Drugs and Antibiotics in Milk *285/* References *287*

15. MICROBIOLOGICAL TESTS FOR EQUIPMENT,
    CONTAINERS, WATER AND AIR 289

   Introduction *289/* Sample Collection *289/* Tests for Equipment Sanitation *289/* Standard Tests for Water Supplies *300/* Tests for Microbiological Quality of Air *300/* References *303*

16. SEDIMENT IN MILK 305

   Introduction *305/* General Methods *306/* Equipment and Supplies *306/* Preparation and Use of Standard Sediment Discs *308/* Procedure *308/* References *309*

17. PHOSPHATASE METHODS 311

   Introduction *311/* General Precautions for Methods Employing Disodium Phenyl Phosphate *312/* Controls Applicable to all Phosphatase Procedures *313/* Scharer Rapid Phosphatase Test *315/* Rapid Colorimetric Phosphatase Method *318/* Rutgers Phosphatase Test *320/* Phosphatase Reactivation in Dairy Products Treated by High-temperature, Short-time Methods *322/* References *324*

18. CHEMICAL AND PHYSICAL METHODS 327

Introduction *327*/ Acid Degree Value *327*/ Acidity *329*/ Ash and Alkalinity of Ash *336*/ Chloride *339*/ Chlorine, Available *344*/ Extraneous Matter: Cheese *346*/ Fat *347*/ Multi-component Methods *366*/ Moisture and Solids *373*/ Protein *382*/ Water: Added to Milk *394*/ Iodine: Selective Ion Procedure *398*/ Pesticide Residues in Milk *400*/ Radionuclides *401*/ References *402*

INDEX 405

CHAPTER 1
STANDARD METHODS

(G. H. Richardson, Tech. Comm.)

1.1 Introduction

Standard Methods for the Examination of Dairy Products (SMEDP) is a compilation of microbiological, chemical, and physical methods for analyzing dairy foods and dairy food substitutes that aid quality assurance programs. Qualities of dairy products that are discernible by the consumer will generally not be considered. Tests for specific nutrient constituents are not included. The major concern of *SMEDP* is with those tests for hidden attributes that can be done by laboratory procedures. Subjective tests for such variables as body, texture, and flavor generally are not adaptable to *SMEDP*. Some procedures described may be applied routinely for laboratory evaluation of products. Others are for special situations in which information of a type beyond that ordinarily obtained in routine control practices is needed.

The ultimate aim of control procedures should be to provide a product in which the original nutritive qualities, flavor, and appearance have been preserved and no harmful organisms or substances are present to adversely affect the consumer.

1.2 Function of Standard Methods

Dairy products commonly move through several steps from producer to consumer and may be transported beyond the regulatory jurisdiction within which they are produced. Thus, uniformity in procedures for examination and standards for acceptance becomes essential for orderly marketing.

SMEDP is a compendium of methodology and is not an attempt to establish legal standards for acceptability of products for either government or industry. Classification of methods in this edition [1.3] does, however, allow weighting of proven methods (i.e., Classes A or D) to help regulatory and industrial agencies select standardized methods for adoption. For example, the maximum permissible Standard Plate Count (SPC) on pasteurized milk, the maximum permissible amount of pesticide residue, the maximum limit of coliforms, and the minimum amount of milkfat in whipping cream are legal decisions beyond the scope of *SMEDP*. The Pasteurized Milk Ordinance[15] is an example of situations in which SPCs are suggested for adoption by regulatory agencies. *SMEDP* does deal with changes in methodology involving incubation temperature or time, composition of culture medium, or preparation of dilution blanks that may change significantly the count obtained on a sample. This in turn may determine whether the product will be acceptable within the current limits established by government or industry. A change in procedure may appreciably increase the stringency of a standard because more of the microorganisms are enumerated or a smaller amount of a chemical contaminant is detected than was possible by earlier methods. A recent example involved the increase in sensitivities of antibiotic methods when traditional test microorganisms were replaced [14.2 vs. 14.3].

Current levels of radionuclides in dairy foods are considered to be well below those that should cause concern.[4] Potential increases in anxiety about these and aflatoxins, however, have dictated inclusion in *SMEDP*.

Concern over small amounts of pesticide residues in dairy foods continues to be expressed as methodology becomes more sensitive. A committee of the National Academy of Sciences has stressed the futility of zero tolerance requirements. A standard method is included in Chapter 18 for some of these residues.

Chapters in this edition cover sampling of dairy foods, estimation of "total" microbial populations as determined by the SPC, detection of specific organisms, determination of several chemical constituents, and detection of foreign matter. The chapter arrangement and their contents represent the consensus of the Technical Committee, chapter contributors and consensus reviewers.

1.3 Classification of Procedures

The American Public Health Association's (APHA) Committee on Laboratory Standards and Practices (CLASP) has adopted "Categories of Methods Prepared for APHA Publications" as a reference. This unpublished report provides guidelines for establishing a classification for new methods and the phasing out of older technology.

The CLASP Committee has approved the addition of Class A2 and the use of Class D instead of C2 in *SMEDP*. This allows better "meshing" of *SMEDP* methodology with The Association of Official Analytical Chemists (AOAC),[7] which uses "Official First Action" (APHA Class A2) and "Official Final Action" (APHA Class A1). The APHA classification also separates a method being phased out (APHA Class D) from one being introduced (APHA Class C).

The *SMEDP* classifications are defined as follows:

1.3 Classification of Procedures

Class O
A method or procedure that has been subjected to a thorough evaluation, has been widely used, has thereby demonstrated its value by extensive application, but may not have been formally, collaboratively tested. This classification will include methods that are referred to as standard methods in current APHA publications; essentially it is a "grandfather clause".

Class A1
A method or procedure that has been subjected to a thorough evaluation, has demonstrated its application for a specific purpose on the basis of extensive use, and has been successfully, collaboratively tested.

Class A2
A method or procedure that has been subjected to a thorough evaluation and has been successfully studied collaboratively.

Class B
A method that has been used successfully in research or other situations, has been devised or modified explicitly for routine examination of samples, has had limited evaluation, and has not been tested collaboratively.

Class C
An unproved or suggested method not previously used but one that has been proposed by recognized laboratory workers as useful and gives promise of being suitable.

Class D
A method that previously has been placed in Classes O, A, or B but which, through technological advances or significant change in numerical level of acceptable exposure or other circumstances, is being superseded by a method of a higher classification. A Class D method may be removed from subsequent editions of *SMEDP*.

Classes O, A1, and A2 are standard methods.

This scheme allows for a progression from Class C to A1, thereby permitting an unproven method or procedure to be made available pending further evaluation. Consensus reviewers were encouraged to react to the classifications assigned, and the Technical Committee then gave final approval or disapproval to the assignments. The following were assigned Class D status and removed from this edition: the Filter DNA and the OSCC II methods for somatic cell determinations, and the interval plating technique for water toxicity.

Valid applications of all classes of methods can be found throughout the dairy industry. Sampling and analytical costs might dictate using a lower class method than would be considered "official" for certain quality assurance purposes. But the use of Class A or O methods would certainly be best whenever court cases might develop or government regulatory agencies are involved that have established acceptance of standardized and tested methodology. This classification scheme allowed elimination of the appendices or "yellow pages" used in the last two editions and incorporation of all methods relevant to a given subject into a single chapter.

1.4 Adopting New Methodology

Methods in *SMEDP* often have quasi-legal status even though procedures to designate a method as "Standard" are far removed from due legal process. Regulatory ordinances and codes frequently contain stipulations that *SMEDP* shall be used to determine whether a product satisfies certain aspects of the regulations. Some jurisdictions have sought protection from capricious changes in *SMEDP* by specifying a particular edition as a reference.

Decisions to change *SMEDP* have been made only after interested parties and all consensus reviewers had an opportunity to study and react to the proposal(s). The Technical Committee has generally required publication of the method and of a collaborative study before accepting a new method or modifications of old ones and has sought to establish criteria for retaining previously accepted methods.

Since the 13th edition of *SMEDP,* some chemical and physical procedures have been included. In some instances, procedures that have not been considered for official action by AOAC are incorporated because of their particular applicability to a situation or product.

Considerable changes have been made in the recommendations for and regulations of antibiotics and similar compounds that may be encountered in dairy foods. Improvements in analytical procedures have led to greater sensitivity and a considerable reduction in the amounts of residue that are detectable. In this state of flux, a procedure accepted today may be outdated tomorrow. A precedent was therefore established to publish such changes in *SMEDP,* in the *Journal of Food Protection,* between *SMEDP* editions.[8,12] Requests for reactions from the field may be similarly published.[9] The editor should be contacted for assistance in seeking approval for a new method.

1.5 Collaborative Studies

A published method should be collaboratively studied before being accepted as a standard method. Youden and Steiner[16] have given details for planning a study and analyzing the results. Their format is not rigid, but the following are essential ingredients of an acceptable collaborative study:

A. A minimum of five collaborators should participate.

B. A minimum of three samples in duplicate should be analyzed. (It is suggested that the number of collaborators multiplied by the number of determinations per collaborator equal 60.) Samples should cover the concentration range expected of the method. Duplicate portions of the sample should be unknown to the collaborator.

C. Estimates of experimental error should be computed from the duplicate observations.

D. Estimates of variation are computed among analysts. Variance components among analysts, between replicates and for each analyst times sample interaction are evaluated for method reproducibility.[16]

Values for variance of reproducibility can be compared with previously reported results. For example, though the study of Donnelly et al.[1] was not a collaborative study, variance components among analysts were calculated to be 0.007 and between replicates to be 0.005 for the SPC as computed from the $\log_{10}$ SPC. A newly developed Spiral Plate alternative method[6] [Chapter 7] was collaboratively studied for milk[10] and foods.[5] The variance values were similar to those of the Donnelly et al. study (i.e., 0.012). An F test can be used to examine the total variance in both studies.

Chapter 7 includes a procedure whereby an analyst can test his or her performance of an alternative procedure.

1.6 Uniformity of Methods

Emphasis must be placed on uniformity of procedures employed in various laboratories.[3] If products are to be moved from one plant to another, or from one regulatory jurisdiction to another, uniformity of testing procedures is essential so that a reasonable degree of agreement can be achieved among different laboratories. Minor variations in technique can cause significant changes in analytical results. The Laboratory Quality Assurance Branch of the Food and Drug Administration, in cooperation with several states, has been an important factor in standardizing laboratory procedures in recent years.

In 1905, the APHA appointed a committee to standardize methods for bacteriological examination of milk.[11] Studies by this and subsequent committees have led to successive editions of *SMEDP* that have brought standardization to laboratory methods. There is continued need to emphasize adherence to standard methods and improve existing methods.

Where milk laboratories are accredited by some central authority, operations are more uniform. Accreditation is based upon qualifications and experience of personnel, on adequate facilities, and on operational compliance with *SMEDP*. Currently, industrial, independent, commercial, state, and local government labs are accredited to officially grade milk and milk products by the procedures of the National Conference of Interstate Milk Shippers.[13,14] General acceptance of determinations made in such laboratories enables milk control agencies and dealers to accept and move milk and milk products more freely.

The need for manufacturers to standardize culture media between lots is apparent [5.9G]. Standardization is necessary for SPC and all selective media.

Failure to monitor media changes can produce uncontrolled increases or decreases in microbial numbers and adversely affect official legal limits.

1.7 Split-Sample Test Program

The National Conference of Interstate Milk Shippers approved a provision in 1950 that individual states may accept results of local laboratories that comply with *SMEDP* and that check closely with results on split samples a minimum of once a year.[2,13,14]

The split-sample program allows administrators to determine whether work is being done satisfactorily by analysts under their supervision. Systematic errors can easily be identified and eliminated. Random errors can be minimized.

The program[14] consists of annually sending split samples to be analyzed by all laboratory methods for which the analysts of accredited laboratories are certified. The samples are representative of the major types of milk and milk products. They include replicate samples yielding a normal range of results and replicate samples yielding results above and below the normal range. Coordination of the split-sample program is a function of the FDA under agreements with the National Conference on Interstate Milk Shippers.

1.8 Relation of Farm and Plant Inspections to Laboratory Control

Farm inspections are essential because the quality of milk is determined initially where it is produced.[2] Similarly, processing plant inspections are important. Laboratory data, however, are needed to guide inspectors to improve milk quality and correct sanitation failures. Farm and plant inspection programs should be closely allied with laboratory tests.

1.9 Accuracy of Methods for Measuring Sanitary Quality

Determinations on individual samples by any one method should be interpreted *solely on the merits of that method*. Regardless of the method by which the limits for acceptability are officially established, enforcement officers as well as dealers or producers, when shown tests results, can come to any of several conclusions. They can tell whether a product: a) has good margins of safety in terms of sanitary codes, b) falls just within "acceptable" limits, c) marginally violates codes, or d) is grossly contaminated. The cardinal objective is to classify samples into one of these groups.

Dairy foods are often held refrigerated for extended periods before reaching the consumer and this practice has increased the importance of measuring bacteria (psychrotrophs) that can cause undesirable changes in the foods during such storage. Some of these microorganisms produce colonies poorly, if at all, when incubated at 32°C, and may reduce indicator dyes very slowly. Under some conditions, these organisms may stain less satisfactorily for the direct microscopic count than do many of the organisms that predominate at higher holding temperatures. While some psychrotrophic microorganisms

1.9 Accuracy of Methods

cause rancidity and bitterness after multiplication to large numbers, most pathogenic bacteria grow slowly or not at all at these temperatures.

Laboratory procedures to evaluate dairy products cover materials of widely different physical and chemical composition, and products in which certain consitituents of milk are present in amounts that represent concentration in some instances and dilution in others. Details of procedures for sample preparation applicable to one product may therefore be quite unsuitable for another. Various constituents that may be added in the formulation of certain products can complicate examination procedures. Not only the final product, but also the individual materials used to make the final product, must therefore be examined. For some products these examination procedures are mostly chemical.

Microbiological methodology differs somewhat among products because of variations in processing, holding conditions, and other factors. The presence of yeasts or molds, the "total" bacterial population, or the numbers of thermoduric, thermophilic, or psychrotrophic bacteria in a product may be of greater significance in some circumstances than in others. In this edition specialized presentations are made for different products, with cross-references to basic procedures.

Methods are updated with each edition of *SMEDP*. Considering the foregoing discussion, it is imperative to encourage adherence to *SMEDP* and to be very careful when interpreting results. Decisions as to which procedure to use for control of dairy food quality, and values to be considered satisfactory for compliance must be made by the regulatory agency, whether industrial or governmental. For example, the *Grade A Pasteurized Milk Ordinance*[15] recommends maximum permissible numbers of bacteria, maximum permissible amounts of certain undesirable chemicals, and minimum amounts for some desirable constituents, as determined by procedures outlined in *SMEDP*.

Routine determinations for the presence in dairy foods of pathogenic bacteria, rickettsiae, protozoa, and viruses are impractical with available techniques, except perhaps for lactic bacteriophage. Therefore, procedures that detect conditions under which disease agents might be present are used widely. Since pasteurization is almost universally employed to kill pathogens, phosphatase tests are frequently employed to determine the effectiveness of pasteurization. Adequacy of protection against post-pasteurization contamination commonly is evaluated by testing for coliform bacteria, which are destroyed by proper pasteurization. Presence of non-sporeforming psychrotrophic bacteria also may be interpreted as the result of post-pasteurization contamination because most of these organisms are unable to survive this heat treatment. Psychrotrophic bacteria (including some strains of coliform organisms), however, do grow in refrigerated dairy foods. High numbers of these organisms found in dairy foods so held may not be indicative of the amount present immediately after pasteurization.

The assumption that care in excluding microorganisms might, in general, result in exclusion of potentially harmful types appears to be justified. Thus the determination of the "total" microbial population of raw milk by the SPC, microscopic count, or dye-reduction method probably still has some public health significance.

1.10 References

1. DONNELLY, C.B.; HARRIS, E.K.; BLACK, L.A.; LEWIS, K.H. Statistical analysis of standard plate counts of milk samples split with state laboratories. J. Milk Food Technol. **23**:315–319; 1960.
2. DONNELLY, C.B.; PEELER, J.T.; BLACK, L.A. Evaluation of state central milk laboratories by statistical analyses of standard plate counts. J. Milk Food Technol. **29**:19–24; 1966.
3. FAULKNER, J.D.; TAYLOR, D.W.; SCHLAFMAN, I.H. The USPHS method of rating milk supplies and its use in the interstate milk shipper program. J. Milk Food Technol. **26**:277–281; 1963.
4. FRIES, G.F.; BURMANN, F.J.; COLE, C.L.; SIMS, J.A.; STODDARD, G.E. Radioiodine in milk of cows under various feeding and management systems. J. Dairy Sci. **49**:24–27; 1966.
5. GILCHRIST, J.E.; DONNELLY, C.B.; PELLER, J.T.; CAMPBELL, J.E. Collaborative study comparing the spiral plate and aerobic plate count methods. J. Assoc. Offic. Anal. Chem. **60**:807–812; 1977.
6. GILCHRIST, J.E.; CAMPBELL, J.E.; DONNELLY, C.B.; PEELER, J.T.; DELANEY, J.M. Spiral plate method for bacterial determination. Appl. Microbiol. **25**:244–252; 1973.
7. HORWITZ, W., ED. Official methods of analysis, 13th ed. Washington, DC: Assoc. Offic. Anal. Chem.; 1980.
8. MARTH, E.H.; REINBOLD, G.W.; MARSHALL, R.T.; MATHER, D.W.; CLARK, W.S., JR.; DIZIKES, J.L.; HAUSLER, W.J. JR.; RICHARDSON, G.H.; ULLMANN, W.W. An addition to the thirteenth edition of Standard Methods for the Examination of Dairy Products: Single-service plastic vials for samples of raw milk and cream. J. Milk Food Technol. **37**:159; 1974.
9. MESSER, J.W.; CLAYPOOL, L.L.; HOUGHTBY, G.A.; MIKOLAJCIK, E.M.; SING, E.L.; MARTH, E.H. Request for comments on method to detect penicillin in raw milk as proposed for inclusion in Standard Methods for the Examination of Dairy Products. J. Food Prot. **40**:270–271; 1977.
10. PEELER, J.T.; GILCHRIST, J.E.; DONNELLY, C.B.; CAMPBELL, J.E. A collaborative study of the spiral plate method for examining milk samples. J. Food Prot. **40**:462–464; 1977.
11. PRESCOTT, S.C. The need for uniform methods in the sanitary examination of milk. p. 246. Public Health Papers and Reports of the APHA 31 (part 2); 1905.
12. RICHARDSON, G.H.; MARTH, E.H.; MARSHALL, R.T.; MESSER, J.W.; GINN, R.E.; WEHR, H.M.; CASE, R.; BRUHN, J.C. Update of the fourteenth edition of Standard Methods for the Examination of Dairy Products. J. Food Prot. **43**:324–328; 1980.
13. RUPPERT, E.L.; TAYLOR, D.W.; SCHLAFMAN, I.H. The interstate milk shipper certification program—a cooperative accomplishment. J. Milk Food Technol. **28**:379–382; 1965.
14. U.S. DEPT. OF HEALTH AND HUMAN SERVICES. Evaluation of milk laboratories. No. 017-001-00412-0. Washington, DC: U.S. Gov. Printing Office; 1978.
15. U.S. DEPT. OF HEALTH AND HUMAN SERVICES. Grade A pasteurized milk ordinance. No. 017-001-00419-7. Washington, DC: U.S. Gov. Printing Office; 1980.
16. YOUDEN, W.J.; STEINER, E.H. Statistical manual of the AOAC. Washington, DC: Assoc. Offic. Anal. Cheml; 1975.

CHAPTER 2

LABORATORY QUALITY ASSURANCE AND SAFETY

R. D. Fischbeck, and H. M. Wehr
(H. M. Wehr, Tech. Comm.)

2.1 Introduction

Quality assurance and safety are vital components of laboratory operation. Without proper attention to these areas, the validity of analytical results can be questioned and the safety of laboratory staff can be jeopardized.

Quality assurance is a systems approach to achieve quality results. Quality control encompasses the individual components (e.g., control tests, record keeping) of the quality assurance system. Quality assurance thus is the sum of all activities designed to ensure that information generated by the laboratory is correct.

This chapter is not an exhaustive review of laboratory quality assurance and safety; excellent reviews are available for this purpose.[7,13,14] Rather, an overview emphasizing the needs of dairy products analysis will be presented covering aspects of laboratory operational management, facility recommendations, basic equipment/supply requirements, control tests and record keeping, and laboratory safety. While much of the information presented is applicable to all laboratories, portions of the material may be applicable only to those doing diversified microbiological and chemical analyses. The topics discussed in this chapter, when combined with close attention to standard methods of analysis, form the basis of good laboratory practice.[4]

2.2 Laboratory Operational Management

A. Laboratory organization:

The organization of the laboratory is important to its successful operation. Of the many possible organizational structures for a laboratory, the most common center around either projects or functions.

Project organizational structure concentrates on the type of work being done for the laboratory's clients irrespective of the technology being employed. Examples of this type of structure include methods development groups, quality control laboratories, or special project groups, and can involve both chemical and microbiological analyses.

A function organization is developed around similarities of techniques, procedures, and instrumentation. A project will be divided so that each function (e.g., sample receiving, preparation, microbiological analysis, and gas chromatographic analysis) is done by a separate section of the laboratory.

Each approach offers advantages and disadvantages. Project organizational structure maximizes project control by concentrating all aspects of the project within a single laboratory operating unit but requires various project groups to either share or duplicate equipment. Organization by function concentrates services into specialized groups whose equipment and personnel are dedicated to a specific technology, but makes management and control of individual projects more difficult because responsibility is usually spread over several laboratory sections.

Laboratory organizational structure is really a function of the philosophy of the managers and of the nature or type of work being done. Organizational structures cannot be static; as the nature of the work changes over time, managers must reevaluate their organizational structure.

B. *Laboratory staffing, personnel selection, and training:*

Laboratory staffing is a function of the organizational structure of the laboratory and the diversity of operations being performed. It typically includes technical staff, management, and support personnel. Chemists, microbiologists, laboratory technicians, and assistants form the basis of the technical staff, while section leaders or project managers, and laboratory administrators form the management team. Support functions can include personnel, accounting, data processing, purchasing, and maintenance staff. Staffing is principally a function of the laboratories technical requirements. As the size of a laboratory increases, more administrative support functions become necessary.

Personnel selection can be done formally or informally, depending upon the complexity of the laboratory operation. In small organizations, the informal approach (almost a trial/error process depending largely on personal interactions) may be adequate. The formal approach, however, depends upon a written job or position description for each individual employed within the organization. The position description allows both the applicant and the laboratory to understand what is required by the job and whether the applicant is qualified. The position description should detail the position title, reporting relationships, position responsibilities, and qualifications desired (education and experience, special needs of the position).

Staff selection is undertaken based on the position description. Care must be taken in the screening and interviewing process to ensure that candidates meet the requirements of the position as closely as possible. It is undesirable to staff a position with a person who is either underqualified or overqualified. Both types of staffing errors will lead to disappointments, frustrations, and higher than normal turnover of the laboratory staff.

Once a position is filled, the management must train the individual to

2.2 Laboratory Operational Management

perform his/her specific responsibilities. Formal training within the laboratory is essential and can be accomplished in one of two ways.

The new individual may be trained by working with an experienced individual or a staff trainer who is familiar with the responsibilities of the position. Alternatively, training may be through use of written descriptions of methods of analyses, instrument operating manuals and procedures, policy and procedural texts, slide/movie technique programs, etc.

In practice, good training employs both an individual and a written approach. The new employee should not be allowed to work without close supervision until the training process is complete and the responsibilities have been comprehended and mastered.

Management must ensure that the newly hired individual has demonstrated an ability to perform the job before discharging its responsibilities to him/her. On-job activities of the new employee should then be monitored until it is obvious that the quality of his/her work is satisfactory. If a problem is encountered, the employee needs to be informed and shown again how properly to perform that specific responsibility.

Initial training, as well as periodic retraining, should be well documented in the employee's personnel records. Such a requirement should be an essential aspect of the laboratory's quality assurance program. Not only does such a requirement ensure that management is performing its training responsibility, but it also allows supervisors to review the qualifications and specific experience levels of personnel in unfamiliar departments.

The section/group leaders should make training activities, and their documentation, part of each employee's overall performance evaluation. The laboratory manager should similarly review the section/group leaders' performances, in part noting their attentiveness to, and effectiveness in, training activities of their respective employees.

C. Standard operating procedures:
1. General: A number of different types of procedures are necessary to describe and document the multitude of functions covered by an analytical laboratory. Procedures routinely fall into the following general divisions: policy, administrative, technical operations, equipment operation, and analytical methods.

    Procedural documentation—from dishwashing to balance calculation and maintenance, to analytical assay procedures—is essential to overall laboratory quality control. Written and approved procedures should be readily available to laboratory personnel, and should be reviewed and updated regularly. Inaccurate or outdated procedures within the laboratory can cause severe quality problems.
2. Methods: Laboratory analytical methods are of two types: 1) standard or official methods such as Class A or O methods in this book, and 2) methods developed or modified by the laboratory or taken from the literature. The need for the latter category normally arises

because the method compendium does not have a method for an analysis that the laboratory has been requested to perform, or the laboratory cannot use the compendium method due to circumstances (e.g., equipment availability).

Whether official or other methods are employed, the laboratory in either instance should be able to demonstrate with historical validation data that it can perform the analytical determination described by the method. The validation data should be evaluated for precision, both within a trial and between trials. In addition, recovery or standard-addition studies should be performed whenever appropriate.

When a modified method is employed, the responsibilities of the laboratory go beyond the validation requirement. The laboratory in this instance must show that the modified and the reference methods are equivalent. Equivalence is determined by showing, with statistically evaluated data, that the modified method is capable of generating data with precision and accuracy that are equal to or exceed those generated by the reference method.

All validation data, equivalence data, and written procedures should be reviewed periodically by management and quality assurance personnel. Procedures should not only be reviewed for technical content, but also for format and readability. This process of review should be repeated on a regular basis to ensure that the current written procedures still accurately reflect the laboratory's current practices.

3. Records and reports: The laboratory should have a secure records archive in which its original, raw data are maintained.

    A system of cataloging records by department, by sample number or data, or by assay, should be developed so that records may be easily located when the need arises and so that review and ultimate disposal may be accomplished in an orderly fashion. Raw data should be stored in clearly labeled, record-retention boxes and housed in a vermin-free area.

    Raw data include not only the final report, but also recorded data such as microbial colony counts, optical densities, peak heights, strip charts, calculations, etc. All data that are necessary to reconstruct the reported result can be considered raw data. The laboratory should establish a records-retention policy that clearly defines the period of time for which records will be kept. This period is a function not only of the laboratory's needs, but also of state and federal regulations.

    The end product of laboratory analysis is the report of results. The format of the report should assure that data are clearly presented. At a minimum, the analysis report should contain the following information: Laboratory name and address, laboratory sample num-

2.2 Laboratory Operational Management

ber, time and date of sample collection, person collecting sample, date of sample receipt, date of final report, sample description, assay name(s), assay result(s)/unit(s), and analyst signature. Incoming sample logs, section logs, laboratory worksheets, the final analytical report, and other pertinent forms, should provide sufficient information to document the chain of sample custody from sample collection through receipt, analysis, storage, and disposition.

4. Analytical quality control: Analytical quality control is needed to ensure that laboratory personnel are performing the analyses in such a manner that the accuracy and precision of the method are as good as the method is capable of providing. Additionally, the laboratory must ensure that, as analysts change over time, the precision and accuracy of results do not change greatly. Variation in technique in such procedures as pipetting, use of volumetric glassware, weighing, colony counting, peak measurement, etc., can exist among analysts, thereby leading to differences in final results. These variations must be minimized.

The laboratory should establish a daily, routine, quality control program that monitors assay/analyst quality by generating data on instrument performance and analysis results, based on known control samples, sterility controls, blanks, known additions, etc. The data should be sufficient so that acceptance/rejection of laboratory results on unknown samples can be made based on results obtained from the quality control program. Quality control data should be preserved historically.

Laboratory proficiency testing should also be employed on a periodic basis. Proficiency samples that give acceptable limits of performance can, in some instances, be obtained externally. In other instances, no such sample is obtainable. The laboratory then must prepare its own proficiency samples, either by pooling fluid samples or preparing samples of known quantities of test compounds by standard additions. In the case of self-prepared samples, the laboratory must generate sufficient data to be statistically significant, reflecting precision across days and between analyses.

A special type of proficiency testing commonly employed is the split sample program [1.7].[12] Samples are split by the laboratory, and results are obtained by different analysts within the same laboratory, or between laboratories. These data are analyzed statistically to determine which analysts and/or laboratories are yielding data that are significantly different. When evaluating data between laboratories, it is essential that all analysts clearly understand that the procedure specified must be followed uniformly. Although this caution appears obvious, it is probably the single most serious obstacle to obtaining comparable data between laboratories in split sample programs.

D. Sample control and preparation:

It is axiomatic that the laboratory result is only as good as the sample obtained for analysis [Chapter 4]. Proper sample control, involving sample indentification, preparation, storage, and disposal is important to maintaining the integrity of the sample and, thus, to the ultimate generation of quality analytical results. Written procedures for sample receipt, identification, subsampling, preparation, storage, distribution, retention, and ultimate disposal should be standard. All the many approaches to sample control that exist in various laboratories have several factors in common.

The process of sample control begins with collection. Key factors that affect sample quality at this stage include proper shipment of samples, cross contamination during shipment due to improper packaging, spillage or breakage during shipment, dirty or improper sample containers, or vague or nonexistent sample identification. Standards for sample identification and supporting instructions should be established. Examples of information that may be required include: sample description, approximate sample size (at shipment), shipping conditions required (ambient temperature, refrigerate, frozen), sample collector, time of collection, safety precautions (describe), stability information (heat, light, air, etc.), analysis requested, specific methods to be employed (optional), client contact and phone, date results needed, and where results are to be sent.

The sample receipt function is primarily one of inspection to ensure that all the required information is present. Special attention should be directed to the sample size at shipment, shipping conditions required, safety precautions, and stability information. Time of receipt at the laboratory may also be important. If anything is unusual in any of these areas, the client and/or the laboratory should be notified immediately.

If the sample has been properly received with sufficient information to perform the analyses requested, the laboratory normally must translate the request into a format that is acceptable to the laboratory. If the laboratory analysis request form is properly designed, it may be sufficient to pass on the request form or copies thereof along with the sample for analyses. In other instances, various forms of laboratory worksheets are prepared that describe the sample and specify the analysis to be performed. The system employed must be developed with the needs of the analytical laboratory in mind.

Whether or not samples are subdivided by the laboratory, the laboratory may place incoming samples into containers of standardized size to facilitate their orderly flow and storage within the appropriate work areas. Since these sample containers are usually sized to provide only the minimum amount of sample required for analysis, sample labels cannot usually contain all of the information present on the anlysis request form. Many laboratories identify samples by a unique sample number, which is assigned to that sample at receipt. In addition to sample number, some laboratories also list sample description, storage conditions, and abbreviated analysis requests on the label. Since the label will not normally provide space for sufficiently detailed information, the laboratory must have a system to match label information (e.g., sample number) with the analysis request form or worksheet system.

If samples are required to be subdivided or prepared for various groups within the laboratory, it is normally most efficient and most appropriate from a sample quality standpoint, to do so in a centralized, sample-preparation area. If more than one preparation technique is employed on the same sample because of differing analyses, the subsamples must be clearly identified as to the analysis intended. Laboratory workers must be aware of the differences between subsamples and honor the label designations.

The laboratory analyst must be charged with the responsibility of ensuring that samples are maintained within the laboratory under the conditions specified. A sample that requires refrigeration and is found at room temperature after being left out on the balance table overnight should not simply be placed back into the refrigerator. If a sample must be kept frozen before analysis and multiple analyses are to be performed on the sample, the laboratory must either request individually frozen subsamples or ensure that all analyses are started concurrently with the initial thawing of the sample. Multiple thawing and refreezing of samples can adversely effect sample stability and homogeneity.

The above examples point to the need for the laboratory to design a stringent sample control procedure, which must periodically be evaluated and modified to meet the needs of the laboratory and its clients. Since mistakes will happen, the laboratory should also maintain a reserve portion of the sample (preferably unaltered from its original state) for re-analysis, if that is required due to a problem encountered by the laboratory. This reserve sample can also serve as the laboratory retention sample.

The sample tracking system within the laboratory is of critical importance to efficient laboratory operation. Significant amounts of time can be wasted if samples are stored haphazardly. If samples are coded upon receipt with a chronological number, tracking is facilitated. Samples can thus be stored in refrigerators in numerical order. If the laboratory analysts place samples into their appropriate positions after use, the system can be self-sustaining. Responsibility for monitoring the orderliness of the sample storage area should be assigned to someone in the laboratory. This individual should also be responsible for removing samples from the storage area for disposal as directed.

2.3 Laboratory Facilities

Laboratory work areas should meet the following criteria. For microbiological testing purposes, items 2.3C, 2.3D, and 2.3F must be adhered to.

A. Work space:

Although specific space requirements are difficult to define because of differing functions among laboratories, it is essential that adequate area be provided for analytical functions, support equipment, and related activities. The recommended bench space per analyst is six linear feet exclusive of space devoted to support equipment or activities; specialized analyses may require additional space per analyst.

B. *Laboratory use only:*

The area designated for laboratory use should be used only for analytical and related functions. The laboratory should not be a portion of a larger area used for processing, warehousing, or other non-laboratory activities. The location of a laboratory within a larger building should minimize personnel flow. Ideally, different analytical functions should be separated. Preferably, microbiological and chemical testing should be located in separate laboratory areas.

C. *Ventilation and air quality:*

Laboratories should be well ventilated and free of dust and drafts. Temperatures should normally be maintained between 16 and 27°C; extremes should be avoided.

In chemical laboratories, hazardous or flammable solvent use should be restricted to hoods. Chemical storage areas should be well ventilated.

In microbiological laboratories, when plating, the microbic density of air should be determined. The number of microorganisms (bacteria, yeasts, and molds) in air in plating areas should not exceed 15 colonies per plate during a 15-min period [15.5C]. When plating throughout the day, prepare one plate for samples plated in the morning and one for samples plated in the afternoon. Maintain records of all air quality tests. Steps to correct excessive air contamination include: 1) improve laboratory cleanliness, especially dust removal; 2) close windows and reduce drafts; and 3) replace air filters or otherwise clean the air-handling system.

D. *Lighting intensity:*

Laboratory lighting should be even, screened to reduce glare, and should provide at least 50, and preferably 100, footcandles intensity at working surfaces.

E. *Workbenches:*

Work areas shall be a table, bench, or other rigid support surface that is level with ample surface. Bench tops should be set at a height of 91 to 96 cm with a depth of 71 to 76 cm. Desk tops or sit-down benches can be set at a height of 65 to 79 cm as needed to accommodate microscopy, plate counting, computer usage, or paperwork activities. Bench tops should be of stainless steel, epoxy plastic, or other smooth, impervious material that is inert and corrosion resistant. There should be minimal seams.

F. *Storage space:*

Adequate storage space should be provided for laboratory supplies. Storage areas should be kept neat, clean, and orderly, and should be adequate for the protection of glassware or other laboratory supplies or apparatus. Flammable solvent areas, if provided, should be continuously well ventilated and temperature controlled. Additionally, such areas should be provided with an appropriate fire retardant system and be posted to indicate storage of flammable materials. Electrical grounding connections should be available for use where flammable solvents are transferred. Acids and bases should be stored separately in dry, well ventilated areas on lower shelves. Safety showers, eye wash fountains, and other protective equipment (face masks, rubber aprons, etc.) should be available and operable. Radioactive materials, if used, should be stored in conformity to license requirements.

G. Utilities:

Adequate utilities should be available for all laboratory operations. Grounded electrical connections of proper voltages, and hot and cold water are required. Gas, air, vacuum, and steam may be needed. Sufficent electrical outlets should be provided to prevent circuit overloading. Adequate utility outlets should be provided to prevent draping of cords or tubing. As a minimum, utilities shall meet all local and state codes.

H. Cleanliness and maintenance:

High standards of cleanliness are essential in laboratory operations. Disorder, dust, and the buildup of unneeded materials can interfere with the quality of analytical testing. Laboratory benches, shelves, floors, windows, and storage areas should be cleaned and inspected on a regular basis. Doors and windows should be screened, insects and rodents must be kept out. Clutter should be eliminated by making an effort to clean up work areas immediately after each use and by conducting a weekly clean-up of the entire laboratory area.

Laboratory equipment should be properly maintained through regular inspection and servicing. Regular maintenance of laboratory facilities should be provided to prevent unnecessary disruption of, or damage to, air handling equipment, and electrical and other utilities. Painted or finished surfaces should be maintained to prevent peeling or cracking.

2.4 Laboratory Equipment and Supplies

As applicable, equipment and supplies used by laboratories should meet the following criteria. For microbiological testing purposes, the specifications noted in this section must be followed.

A. Autoclaves:

Autoclaves shall be of sufficient size to prevent crowding of the interior and be constructed to provide uniform temperatures within the chamber up to and including the sterilizing temperature of 121°C [5.10A]. Autoclaves must be equipped with a pressure gauge and a properly adjusted safety valve. Additionally, they must be equipped with an accurate mercury-filled thermometer, bimetallic helix dial thermometer, or temperature recorder-controller properly located to register the minimum chamber temperature (usually on exhaust). Unless automatic recording devices are employed, start-up time, shut-down time, and temperatures must be recorded for each usage and records maintained. If automatic recorders are used, charts recording usage must be maintained. The accuracy of the chamber thermometer should be checked at least semi-annually using a maximum registering thermometer, with the results recorded and maintained. Sterilization should be checked weekly using spore strips or suspensions, with the results recorded and maintained. To insure safe and adequate operation, autoclaves should be periodically inspected and serviced by qualified service representatives.

B. *Analytical balances:*

Analytical balances with a sensitivity of 0.1 mg or less at a 10-g load must be used when weighing 2 g or less. For larger quantities, a balance with a sensitivity of 100 mg (0.1 g) at a 200-g load must be used. Balances should be checked monthly with a set of certified weights (S or S-1 weights traceable to NBS standard weights) and records of checks maintained. This recommendation is especially important for electronic balances where component failure can lead to unnoticed measurement inaccuracies. Balances and weights should be cleaned after each use. Weights should be protected from the laboratory atmosphere, manual contact, and corrosion. Balances should be serviced and checked for accuracy at least annually by a qualified maintenance representative.

C. *Colony counters and tallies:*

A standard colony counter, Quebec model, is preferred,[10] or one providing equivalent magnification and visibility. Grid plates of Quebec-type darkfield or equivalent model colony counters should not be excessively scratched or etched. Internal mirror placement and cleanliness should be checked periodically to insure optimal illumination. Hand or probe tallies should be used and checked periodically to insure accuracy. If instrumental colony counters are used, they must be evaluated periodically to provide results equivalent to manual counts [6.2A2, 7.3].

D. *Dilution bottles:*

Dilution bottles (a capacity of about 150 mL) shall be of borosilicate glass, polysulfone, polycarbonate, polypropylene, or other autoclaveable, non-toxic plastic material, and closed with Escher-type stoppers or screw caps. Use friction-fit liners in screw caps, as required, to make the closure leakproof. Be sure that each batch of dilution blanks is properly filled [5.3B]. Dilution bottles should be marked indelibly at 99 ± 1 mL graduation level. Plastic caps for bottles or tubes and plastic closures for sample containers should be treated, when new, to remove toxic residues. This may be accomplished by autoclaving them twice while they are submerged in water, or exposing them to two successive washings in 82°C water containing a detergent. Toxic residues can be evaluated as described in 5.4D.

The volume of fill and pH of dilution blanks must be checked and records

2.4 Laboratory Equipment and Supplies

maintained for each batch of blanks prepared. Blanks filled must be discarded if < 97 mL or > 101 mL. Generally, the volume of one of every 25 blanks should be checked; for very large batches, this frequency may be reduced if test results warrant.

E. Dispensing apparatus:

The accuracy of a dispensing apparatus should be checked with the use of a National Bureau of Standards (NBS), Class A, graduate cylinder at the start of each volume change and periodically thereafter. Moving parts should be lubricated according to the manufacturer's recommendations. Leaks, lose connections, or malfunctions should be immediately corrected.

F. Hoods and laminar flow cabinets:

Hoods and laminar flow cabinets should be located away from room exits and traffic corridors as a fire safety measure and because traffic movement can disrupt hood face velocity. Fume hood face velocity should be at least 30.5 m per min for hoods and 23 m per min for Class A cabinets and should be checked periodically with an accurate airflow measurement device. Fume hoods and exhaust systems should be fabricated of corrosion resistant material and, except for balancing dampers, should possess no air flow restrictive devices. Air discharge points should be located to prevent reentry of discharged air/fumes through building air intakes. Filters, if so equipped, should be checked monthly for plugging or dirt accumulation, and cleaned or replaced as needed. Sterilization lamps, if provided, should be checked quarterly with a short-wavelength, ultraviolet meter and replaced if output falls below 80% of rated value.

G. Glassware:

Laboratory glassware, including pipettes, dilution bottles, and petri dishes should be free of scratches, chips, cracks, bubbles, or other defects. Damaged glassware should be discarded. Graduations on pipettes, dilution bottles, and other measurement items should be clearly, distinctly, and indelibly marked with a contrasting color. Each batch of cleaned glassware should be checked to assure absence of detergent or other toxic residues [5.5].

H. Incubators:

Incubators or incubator rooms must be properly constructed and controlled, and must be of sufficient size to prevent crowding of the interior. Incubator rooms, if used, must be well insulated, equipped with properly distributed heating units, and have forced air circulation.

Load incubators so that there will be at least 2.54 cm of space between adjacent stacks of plates and between walls and stacks.

Keep incubators in rooms where temperatures are within the range of 16 to 27°C. The temperature within the incubator must not vary more than ± 1°C[1]. Chamber temperatures must be measured by no fewer than two thermometers, one on an upper and one on a lower shelf. Thermometer bulbs must be submerged in water (use partial immersion type thermometers) using stoppered 20- x 150-mm test tubes. Test points should be increased proportionately in walk-in incubators. Temperatures must be recorded twice daily (early mornings after incubators have been closed all night, and again just before the end of the work day). Optionally, use automatic devices of predetermined accuracy for controlling and recording termperatures in incubator and/or incubator rooms. Check accuracy of these devices weekly by comparing readings with those from accurate thermometers and record the results. When standard thermometers are used, minimum- and maximum-registering thermometers may be placed in each incubator to indicate undetected gross temperature deviations. Do not depend on readings from these special thermometers for daily records of temperatures. As necessary, adjust thermoregulators to maintain proper temperatures. All temperature records must be maintained.

Humidity levels of incubators and incubator rooms must be maintained such that Standard Plate Count (SPC) agar plates do not lose more than 15% of agar weight after 48 h incubation at $32 \pm 1°C$. Agar weight loss should be determined quarterly and results recorded and maintained.

I. Petri plates:
Petri plates with bottoms of at least 80-mm inside diameter, 12-mm deep, with interior and exterior surfaces of the bottoms being free from bubbles, scratches, or other defects must be used. Petri dishes may be made of glass or non-toxic, sterilized plastic.[2,6]

J. Petri dish containers:
Containers of stainless steel or aluminum, with covers, are preferred. Char-resistant sacks of high-quality sulfite pulp kraft paper may also be used. Single-service petri dishes should be stored in their original sterilized plastic sleeves.

K. Pipette containers:
Stainless steel or aluminum pipette containers are preferred. Copper containers shall not be used. Char-resistant, high-quality sulfite pulp kraft paper may be used. Sterile plastic or paper pipette covers are satisfactory for single-service plastic pipettes.

L. pH meters:
pH meters should have division increments readable to at least 0.1 unit for both analog and digital type meters. pH measurement should be made at room temperature if possible; otherwise, compensate for temperature [5.8]. The instrument should be standardized whenever used or at least daily if multiple uses occur. Standardize at least two buffer solutions of known pH at the temperature of meter use; one standard should be at pH 7.0, the second at least 2 pH units away, towards the pH of the product or solution to be tested to properly set the intercept and slope of the instrument. The meter must always be calibrated first with the pH 7.0 buffer, then with the second buffer. Standard buffer solutions should be replaced frequently. Meters should be checked at least annually by a qualified service representative.

2.4 Laboratory Equipment and Supplies

M. Pipettes:

Pipettes[3,8,9] should be nontoxic, with straight walls, and ground or fire polished tips (glass), conform to prescribed specifications, be clearly marked, and undamaged [2.4G].

The major types of laboratory pipettes are classified based on the method of draining.

1. To contain (TC) pipettes: Pipettes calibrated to hold or contain the amount specified by the calibration. The pipette *must* be completely emptied to provide the stated volume.
2. To deliver (TD) pipettes: Pipettes designed to release the calibrated amount when the pipette tip is held against the receiving vessel wall until draining stops. Bacteriological, Mohr, seriological, and volumetric pipettes are of this type.
3. Blow-out pipettes: These pipettes are intended for rapid use in serology and are emptied by forceful blow-out. The pipettes do not deliver the calibrated amount until they are completely emptied. Pipettes of this type are not used in the microbiological examination of milk.

    To contain and *to deliver* are indicated by the letters TC or TD marked, respectively, on the stem with other calibration information. *Blow-out* pipettes are marked with a double-band etching or fired-in marking on the neck.
4. Dual purpose pipettes: A pipette combining three calibrations. The pipette has an upper graduation mark that is the *to-deliver* and *blow-out* line, a lower graduation that is the *to-contain* line, and carries the double band for *blow-out*. Pipettes of this type are not used in the microbiological examination of milk.
5. Pipette tolerance values: Several tolerance specifications exist to describe pipette accuracy. When accuracy of measurements is critical, only Class A pipettes with the following tolerance limits should be used (Ref. NBS specifications for Mohr measuring pipettes, Class A Volume Tolerances Circular 602 and Federal Specification NNN-P-350).

Capacity/Grad (mL)	Tolerance (mL)
1/10 in 1/100	±0.0025
2/10 in 1/100	±0.004
1 in 1/10	±0.01
1 in 1/10	±0.01
2 in 1/10	±0.01
5 in 1/10	±0.02
10 in 1/10	±0.03
17.6 in 1/10	±0.05
25 in 1/10	±0.05

For microbiological examination of milk, use only transfer pipettes calibrated for bacteriological use and conforming to APHA specifications (Fig. 2.1). Use only pipettes with unbroken tips and having graduations distinctly marked to contrast sharply with the color of milk or diluted milk, Discard pipettes with damaged tips.

6. *Mechanical pipettes:* Numerous semiautomated repeating pipettes exist that can be used in analyses. Volumes can be adjusted on some. Sterile disposable tips exist which can make them adaptable to microbiological techniques.

N. Refrigerators:

A refrigerator to cool and maintain the temperature of samples between 0 and 4.4°C until tested is required. The refrigerator may also be used to store prepared media. Sufficient refrigeration and/or freezer storage capacity should be provided to prevent crowding. Temperatures should be measured using no less than two thermometers, one on an upper and one on a lower shelf [2.4H]. Temperatures should be recorded daily and records maintained. (Freezer temperatures, determined by placement of a thermometer directly on a shelf of the unit, should be similarly maintained). Refrigerated storage of materials should be such that contamination of stored items does not occur (e.g., liquid samples for analysis should not be stored above prepared media). Flammable solvents should be stored in explosion-proof or flammable-storage refrigerators.

O. Thermometers:

Thermometers of appropriate range may be mercury-filled (or have a distinctively colored fluid with a freezing point lower than $-1°C$), or of an adjustable type, with a graduation interval not to exceed 1°C. Unless otherwise specified, accuracy should be checked at least every two years with a thermometer certified by the NBS or one of equivalent accuracy. Dial and electronic thermometers should be checked more frequently, at least once every 6 mo. Records of thermometer checks must be maintained. Where a record is desired of temperature in refrigerators, autoclaves, hot-air ovens, or incubators, automatic temperature-recording instruments may be used. If used, such devices should be compared weekly against an accurate thermometer and the results of such comparisons recorded and filed.

Two general types of mercury thermometers are in use in laboratories: those calibrated for total immersion and those designed to be partially immersed. Partial-immersion thermometers have a line completely around the stem of the thermometer at the point to which they should be immersed. If this line is absent, the thermometer is designed for total immersion. As an example, a partial-immersion thermometer should be used in an incubator or refrigerator because only part of the thermometer is immersed in a control vial of water, the temperature of which is being measured. Conversely, a thermometer placed flat on the bottom or shelf of a water bath is totally

2.4 Laboratory Equipment and Supplies

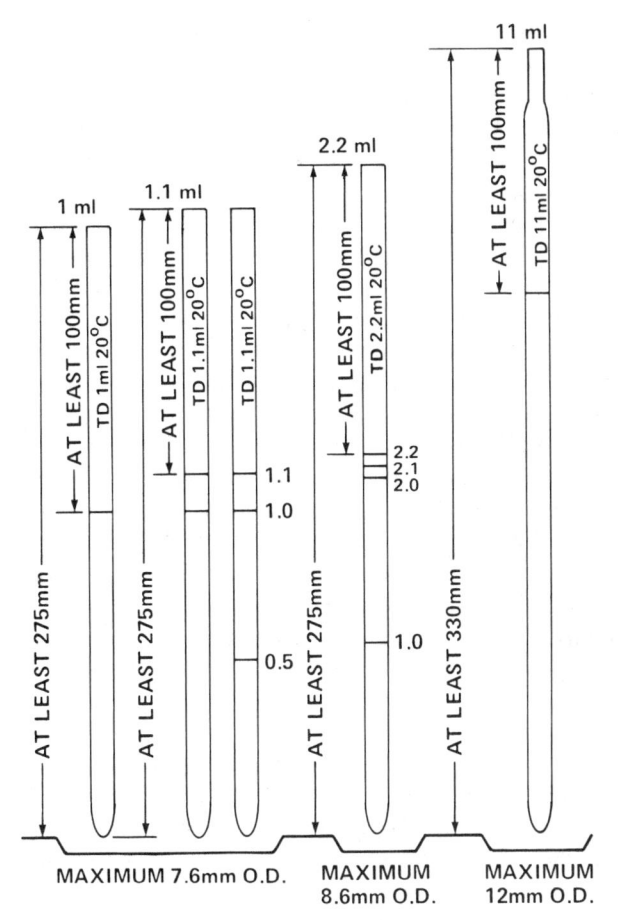

ALL PIPETS ARE CALIBRATED TO DELIVER INDICATED AMOUNT OF MILK AT 20°

Pipet size	Accuracy	
1.0 and 1.1 ml	± 0.025 ml	From any graduation to tip
2.2 ml	± 0.040 ml	and between successive
11 ml	± 0.20 ml	graduations

Figure 2.1*. Bacteriological transfer pipettes. Delivery 1 mL; delivery 1.1 mL in 0.1- and 1-mL amounts; delivery 1.1 mL in 0.1-, 0.5 and 0.5-mL amounts; delivery 2.2 mL successively in 0.1-, 0.1-,1- and 1-mL amounts; and delivery 11 mL.

---

*Volume delivered in 4-sec maximum, last drop of undiluted milk blown out, or last drop of diluted milk touched out, l-mL tolerance ± 0.025 mL. To allow for residual milk and milk dilutions on walls and in tip of the glass pipette under rapid transfer techniques described in section [6.2G] such pipettes shall be calibrated to contain 1.075 mL of water at 20°C. If the pipette is of styrene plastic, it should be calibrated to contain 1.055 mL of water at 20°C.

immersed in the warm (or cold) environment and should be of the total-immersion type.

The easiest way to check the calibration of laboratory thermometers is to put them in a water bath, either partially or totally immersed in the water, according to the way they will be used in the laboratory. Also place in the water bath a thermometer certified by the NBS. Most, but not all, of these thermometers are calibrated for total immersion. Vigorous stirring of the water in the bath is essential to insure uniform temperature during thermometer calibration.

Check the calibration by comparing the temperature reading on the certified thermometer with that of the laboratory thermometer at or very near the temperature the thermometer will be used to measure (e.g., an incubator thermometer should be checked at 32°C because this is the temperature of interest[1]). If the thermometer is to be used for several purposes, it should be checked at the different temperatures of use. If there is a difference in the temperature reading between laboratory and certified thermometers after the reading of the certified thermometer has been corrected as indicated by the certificate, attach a tag to the laboratory thermometer to show the amount of correction that should be applied to obtain an accurate measurement of temperature.

P. *Sterilizing ovens:*

Sterilizing ovens shall be of sufficient size to prevent crowding and shall be constructed to give uniform and adequate sterilizing temperatures throughout the chamber, with vents suitably located to insure prompt and uniform heat penetration [5.10B]. Ovens should be equipped with an accurate thermometer of an adequate range (to 220°C), suitably located to register minimal temperatures. Start-up, shut-down, and maximum temperature reached must be recorded and the records maintained. Oven sterilization performance must be tested periodically (preferably quarterly) using spore strips or suspensions, and results recorded.

Q. *Water baths:*

Water baths should be of an appropriate size for the required workload with a suitable water level maintained. Baths must be thermostatically controlled at the appropriate temperatures. Recirculators or agitators may be necessary to prevent temperature stratification. For holding melted media, the bath must be maintained at 44 to 46°C. Temperatures should be monitored by maintaining an accurate thermometer immersed in the water bath. Only racks constructed of stainless steel, rubber, plastic or plastic coated substances, or other corrosion-proof material should be used.

R. *Water stills, reverse osmosis units, and deionizers:*

Water stills, reverse osmosis units, and deionizers should be cleaned, and/or have their cartridges replaced at intervals recommended by the manufacturer or as indicated by control test results. As appropriate, monitor water for conductance and residual chlorine levels at least monthly and/or whenever reagent or media preparation is done. Water deionization systems should be tested at least monthly for excessive[5] microbial counts. Test water for the presence of toxic or bacterial growth enhancement substances at least once every 6 mo [5.4]. Records of water quality test results must be maintained.

S. Specialized instrumentation:
Allow only trained personnel to utilize equipment. Follow manufacturer's recommended procedures for operation and maintenance. Utilize control tests as appropriate to monitor instrument operation.

2.5 Control Tests and Record Keeping

Control tests and record keeping are a vital part of laboratory quality assurance. Use of control tests and associated records assists in avoiding problems, and in resolving difficulties should they arise. For microbiological testing purposes, all control tests and record keeping must be adhered to.

Control tests and record keeping for the following areas are described in the chapter sections indicated.

Air bacterial quality; 2.3 C
Autoclave time/temperature; 2.4 A
Autoclave sterilization adequacy; 2.4 A
Balance sensitivity; 2.4 B
Dilution bottle fill; 2.4 D, 5.3 B
Incubator temperature; 2.4 H
Refrigerator, freezer temperature; 2.4 N
Thermometer calibration 2.4 O
Sterilizing oven time/temperature; 2.4 P
Water quality; 2.4 R

The following additional control tests and records are recommended.

A. Sterility controls:
Sterility controls should be prepared for each batch of media, dilution blanks, pipettes, blenders, or other items used in microbiological analysis. Records should be kept.

B. Sample receiving/laboratory analytical section logs:
An incoming log for samples should be maintained. The log should show the type of sample, sample and/or report number, assigned laboratory number, date of receipt, incoming sample temperature (if appropriate), from whom/where received, whether sample was received in a sealed container and, if appropriate, the section of the laboratory to which it was assigned. Additionally, individual laboratory section logs are recommended, such logs may indicate much of the same information shown in the incoming sample log, but additionally should show the date(s) analysis was initiated and completed, and the tests required, and may show test results.

C. Analytical worksheets:

Worksheets for sample analysis should be maintained [2.2C3]. Worksheets should show raw data and calculations needed to determine, at some later time, precisely how the analytical results were obtained.

D. Residual acid/alkali:

The presence of residual acid or alkali on each batch of cleaned glassware should be determined using bromthymol blue indicator [5.5C]. Results of all tests must be recorded and maintained.

E. Media pH:

The pH of each prepared batch of media or other appropriate reagents should be measured [5.8]. The results must be recorded and records maintained.

F. Inventory control:

Inventory control records should be maintained for microbiological media, reagents and other chemicals. Each container (and its packing box if it is so stored) should be marked with the date of receipt. Open containers should be marked with the date of opening. Inventory control assists in the usage of media before expiration dates, prevents over or under ordering, and prevents excessive degradation of sensitive chemicals. Inventory control for other supply items including disposable and reusable glassware and plasticware is also recommended.

2.6 Laboratory Safety

A. Introduction:

Laboratory safety is an essential element of laboratory operation. A vigorously enforced safety program is essential for the protection of the personnel and the facility.

A laboratory safety program should be established by management and made available as a manual to everyone in the laboratory. This action program should be reviewed and amended as required to meet changing needs, and supplemented by specific procedures and policies for individual departments and projects.

Some of the general areas of laboratory safety that should be considered when preparing a written safety program are described in Section 2.6. This is not an all-encompassing treatment of the subject. Texts on laboratory safety should be consulted for additional information.[11,15,16]

Creating and maintaining a successful program requires a daily commitment from every individual to four basic aspects of safety: 1) prevent accidents from occurring, 2) deal quickly and efficiently with accidents that do occur, 3) provide employees with safety information and education, and 4) ensure that safety procedures are closely followed.

Regular evaluation of personal safety practices and housekeeping procedures in the work area are essential to assure laboratory safety.

2.6 Laboratory Safety

B. Policy:

Overall policy regarding laboratory safety should also be established and incorporated into the program. Because safety results as much from an attitude of mind as it does from physical or engineering factors, results depend upon the efforts of each individual to eliminate unsafe acts and conditions. Continuing safety requires a genuine interest in safety from all employees at all levels of responsibility.

The laboratory safety policy should charge employees with following safe work methods and with being concerned about the safety not only of themselves, but also of their fellow workers.

C. First aid:

The American Red Cross offers excellent first aid training in nearly every major city. It is strongly recommended that selected personnel within the laboratory be provided with this training, which will lead to their certification by the Red Cross.

When an accident occurs, however, professional help should be summoned immediately from the local fire department, rescue squad, or hospital. Unless absolutely necessary to prevent further injury, an accident victim should not be moved by untrained personnel. Keep the victim on his/her back, quiet, comfortable, and warm until professional help arrives.

First aid procedures should only be administered by those who are adquately qualified by professional training. There are, however, a few general actions that can and must be taken until trained help arrives:

1. Asphyxia or stoppage of breathing: Remove objects such as false teeth, gum, or food from the victim's mouth. Start artificial respiration or cardiopulmonary resuscitation (CPR) at once and continue until the victim breathes on his/her own or until professional help arrives. All laboratory personnel should recognize when a fellow employee is choking and know what to do when it occurs.
2. *Severe bleeding:* Lay the victim down, Keep a bleeding limb elevated. Press firmly and directly on the wound with a clean cloth. Keep pressure on until the bleeding stops or professional help arrives. Do not apply tourniquets.
3. *Eye injuries:* Acid, alkali, or petroleum products should be flushed from the eye with copious amounts of water for at least 15 min (preferably from a standard eyewash fountain).

    Foreign objects embedded in the eye should *not* be removed. The eye should be covered loosely and professional help sought.
4. *Burns:* For chemical burns, flush the area immediately with cool water. Remove or cut away clothing. For thermal burns, immerse the affected area in ice water. Get professional help. If clothing adheres to the skin in a burned area, do not remove it. Do not put any creams, etc. on the burns.

5. *Swallowed poisons:* Determine what has been ingested. Seek professional assistance by using the telephone number of the local poison control center. This number is usually listed inside the front cover of the telephone directory.

The 24-h poison control center emergency number in Atlanta, GA, is (404) 588-4400.

D. *Reporting of accidents:*

All accidents should be reported to management, whether or not they result in injury. "Close calls" must be noted so that preventive measures can be taken in the future.

E. *Evacuation plan and fire emergencies:*

An evacuation plan is essential to facilitate a rapid and orderly exit from the laboratory building in case of an emergency due to a fire, gas, or similar threatening condition.

To prevent the loss of life or disabling injury due to uncontrolled exit procedures: 1) *Everyone* must learn all possible exits from the building. 2) *Everyone* must be aware of the location of all fire extinguishers and fire alarms in the work area. 3) *Everyone* should remain calm during an emergency and leave the building in an orderly fashion by a predetermined route.

Regularly scheduled (e.g., every 6 mo) fire and evacuation drills are essential to rehearse the evacuation plan. No one should make use of an elevator during an evacuation. The fire alarm signals the need for the building evacuation. The evacuation plan should be written and contain floor plans which clearly show the proper routes of evacuation. The plan, in addition to being an integral part of the safety manual of the facility, also must be posted in the work area.

If evacuation is necessary, predetermined members of management, fire marshals, etc., should check to ensure that everyone is out of the work area. If any problems arise, they should be reported immediately to fire department personnel upon their arrival.

When there is any doubt that a fire can be brought under control locally with available personnel and equipment, the following actions should be taken: 1) Notify the fire department and activate the alarm system. 2) Confine the emergency (close doors and fire doors between laboratories). 3) Assist any injured personnel during evacuation. 4) Evacuate the building.

Fire extinguishers should be used to extinguish small fires whenever possible. In case of burning clothing, use an emergency safety shower or wrap the individual in a fire blanket.

Small fires can be extinguished by use of the proper extinguisher. The discharge of the extinguisher should be directed at the base of the flame. To ensure that the proper type of extinguisher is used, compare its use label with the descriptions below:

Class A fires—ordinary combustible solids such as paper, wood, coal, rubber and textiles.

2.6 Laboratory Safety

Class B fires—petroleum hydrocarbons (diesel fuel, motor oil, and grease) and volatile flammable solvents.

Class C fires—electrical equipment.

Class D fires—combustible or reactive metals (such as sodium and potassium), metal hydrides, or organometallics (such as alkylaluminum).

It is recommended that all-purpose fire extinguishers (ABC type) be available to handle any type of fire that might be encountered.

F. Chemical spills:

The laboratory should have a written spill control procedure as part of its safety program manual. Knowledgeable chemists and microbiologists should be assigned to handle spill control, evacuation, and cleanup. A spill-control policy should include consideration of the following points: 1) Prevention—storage, operating procedures, monitoring, inspection, and personnel training. 2) Containment—engineering control on storage facilities and equipment. 3) Cleanup—measures and training of designated personnel to help reduce the impact of a chemical spill. 4) Reporting—provisions for reporting spills both internally (to identify controllable hazards) and externally (for example, to state and federal regulatory agencies).

G. Personnel protective equipment:

1. Eye protection policy: Eye protection for laboratory workers and visitors is frequently essential. Exceptions may be granted for areas within the laboratory where potentially dangerous chemicals and/or procedures are not normally employed (conference rooms, libraries, office areas, microscope rooms, etc.).

   Safety glasses should comply with existing standards set by the Occupation and Education Eye and Face Protection Standard (287.1) established by the American National Standards Institute.

   Safety glasses should be made available to visitors upon entering the laboratory. Inexpensive plastic safety glasses should be obtained for this purpose.

   Side shields can be used to afford some protection from objects that approach from the side. Side shields do not, however, afford adequate protection from splashes. When working in an area where a splash can occur, full-face shields should be required.

   Contact lenses should not be worn in the laboratory. Chemical vapors and gasses can be concentrated under such lenses causing permanent eye damage. Furthermore, in the event of a splash to the eye, it is extremely difficult to remove a contact lens. Persons attempting to wash the eyes of an unconscious person often will be unaware that the person is wearing contact lenses. Persons who are required to wear contact lenses should make management aware of the situation so that alternative safety precautions can be devised.

2. Hand protection: Three types of gloves are recommended for laboratory use to prevent injury to the hand:

a. Insulated gloves should be used when handling hot flasks, crucibles and beakers. These gloves are relatively stiff and should not be used to pick up small objects. Tongs should be used for small objects.
b. Surgical gloves of natural rubber or synthetic rubber are useful when dexterity is needed. These should be used to keep corrosive, toxic, and carcinogenic materials off of the hands. The gloves are disposable and should be discarded after use.
c. Heavy rubber gloves are recommended for handling concentrated acids, bases or other corrosives or when handling broken glassware.
3. Clothing and footwear: Adequate laboratory clothing should be worn in the laboratory. Avoid wearing loose or ragged clothing. Skimpy clothing such as shorts, halter tops, etc., also should be prohibited since they do not afford adequate body protection. Loose or torn clothing and unrestrained long hair can easily catch fire, dip into chemicals, or become ensnared in apparatus and moving machinery. Laboratory clothing that has the potential of becoming contaminated should not be worn outside the working area. Finger rings should be avoided around moving equipment. Perforated shoes or open sandals should not be worn. Laboratory coats, rubber aprons, etc., should be used as the situation dictates.
4. Smoking policy and food consumption: Smoking should be strictly prohibited in laboratory areas. Smoking can cause fires. This is particularly true when organic flammable solvents are present. Consumption of food and drink should be discouraged or prohibited. Food and drink consumption within the laboratory increases the risk of accidental ingestion of potentially dangerous laboratory chemicals or microorganisms. Food and sample storage should not be mixed.

H. Safety rules:

The laboratory's Safety Manual should include rules to be noted by all laboratory personnel. Management should be responsible that these rules are followed and enforced. It is impossible to establish a set of rules that covers every possible laboratory situation. Sections 1 to 17 below, however, reiterate some general guidelines from Prudent Practices For Handling Hazardous Chemicals in Laboratories[15] (Reprinted by permission) that should be kept in mind:

1. General principles: Every laboratory worker should observe the following rules:
    a. Know the safety rules and procedures that apply to the work that is being done. Determine the potential hazards (e.g., physical, chemical, biological) and appropriate safety precautions before beginning any new operation.

2.6 Laboratory Safety

    b. Know the location of and how to use the emergency equipment in your area, know how to obtain additional help in an emergency, and be familiar with emergency procedures.

    c. Know the types of protective equipment available and use the proper type for each job.

    d. Be alert to unsafe conditions and actions and call attention to them so that corrections can be made as soon as possible. Someone else's accident can be as dangerous to you as any you might have.

    e. Avoid consuming food or beverages or smoking in areas where chemicals are being used or stored.

    f. Avoid hazards to the environment by following accepted waste disposal procedures. Chemical reactions may require traps or scrubbing devices to prevent the escape of toxic substances.

    g. Be certain all chemicals are correctly and clearly labeled. Post warning signs when unusual hazards, such as radiation, laser operations, flammable materials, biological hazards, or other special problems exist.

    h. Remain out of the area of a fire or personal injury unless it is your responsibility to help meet the emergency. Curious bystanders interfere with rescue and emergency personnel and endanger themselves.

    i. Avoid distracting or startling any other worker. Practical jokes or horseplay cannot be tolerated at any time.

    j. Use equipment only for its designed purpose.

    k. Position and clamp reaction apparatus thoughtfully in order to permit manipulation without the need to move the apparatus until the entire reaction is completed. Combine reagents in appropriate order, and avoid adding solids to hot liquids.

    l. Think, act, and encourage safety until it becomes a habit.

2. Health and hygiene: Laboratory workers should observe the following health practices:

    a. Wear appropriate eye protection at all times.

    b. Use protective apparel, including face shields, gloves, and other special clothing or footwear as needed.

    c. Confine long hair and loose clothing when in the laboratory.

    d. Do not use mouth suction to pipette chemicals or to start a siphon; a pipette bulb or an aspirator should be used to provide vacuum.

    e. Avoid exposure to gases, vapors, and aerosols. Use appropriate safety equipment whenever such exposure is likely.

    f. Wash well before leaving the laboratory area. However, avoid the use of solvents for washing the skin; they remove the natural protective oils from the skin and can cause irritation and inflam-

mation. In some instances, washing with a solvent might facilitate absorption of a toxic chemical.
3. Food handling: Contamination of food, drink and smoking materials is a potential route for exposure to toxic substances. Food should be stored, handled and consumed in an area free of hazardous substances.
   a. Storage and consumption of food and beverages should be restricted to well-defined areas. No food should be stored or consumed outside of those areas.
   b. Areas where food is permitted should be prominently marked and a warning sign (e.g., EATING AREA—NO CHEMICALS) posted. No chemicals or chemical equipment should be allowed in such areas.
   c. Neither consumption of food or beverages nor smoking should be permitted in areas where laboratory operations are being carried out.
   d. Glassware or utensils that have been used for laboratory operations should never be used to prepare or consume food or beverages. Laboratory refrigerators, ice chests, cold rooms, and such should not be used for food storage; separate equipment should be dedicated to that use and prominently labeled.
4. Housekeeping: There is a definite relationship between safety performance and orderliness in the laboratory. When housekeeping standards decline, safety performance inevitably deteriorates. The work area should be kept clean, and chemicals and equipment should be properly labeled and stored.
   a. Work areas should be kept clean and free from obstructions. Cleanup should follow the completion of any operation or at the end of each day.
   b. Wastes should be deposited in appropriate receptacles.
   c. Spilled chemicals should be cleaned up immediately and disposed of properly. Disposal procedures should be established and all laboratory personnel should be informed of them. The effects of other laboratory accidents should also be cleaned up promptly.
   d. Unlabeled containers and chemical wastes should be disposed of promptly according to appropriate procedures. Such materials, as well as chemicals that are no longer needed, should not be allowed to accumulate in laboratory.
   e. Floors should be cleaned regularly; accumulations of dust, chromatography adsorbents, and other assorted chemicals pose respiratory hazards.
   f. Stairways and hallways should not be used as storage areas.
   g. Access to exits, emergency equipment, and controls, should never be blocked.

2.6 Laboratory Safety

    h. Equipment and chemicals should be stored properly; clutter should be minimized.
5. Equipment maintenance: Good equipment maintenance is important for safe, efficient operations. Equipment should be inspected and maintained regularly. Servicing schedules will depend on both the possibilities and the consequence of failure. Maintenance plans should include a procedure to insure that a device that is out of service cannot be restarted.
6. Guarding for safety: All mechanical equipment should be adequately furnished with guards that prevent access to electrical connections or moving parts (such as the belts and pulleys to a vacuum pump). Each laboratory worker should inspect equipment before using it to ensure that the guards are in place and functioning.

    Careful design of guards is vital. An ineffective guard can be worse than none at all because it can give a false sense of security. Emergency shutoff devices may be needed in addition to electrical and mechanical guarding.
7. Shielding for safety: Safety shielding should be used for any operation that has a potential for explosion such as: (a) whenever a reaction is attempted for the first time (small quantities of reactants should be used to minimize hazards), (b) whenever a familiar reaction is carried out on a larger than usual scale (e.g., 5 to 10 times more material), and (c) whenever operations are carried out under non-ambient conditions. Shields must be placed so that all personnel in the area are protected from hazard.
8. Glassware: Accidents involving glassware are a leading cause of laboratory injuries!
    a. Careful handling and storage procedures should be used to avoid damaging glassware. Damaged items should be discarded or repaired.
    b. Hands should be protected adequately when inserting glass tubing into rubber stoppers or corks or when placing rubber tubing on glass hose connections. Tubing should be fire-polished or rounded and lubricated, and hands should be held close together to limit movement of glass should fracture occur. The use of plastic or metal connectors should be considered.
    c. Glass-blowing operations should not be attempted unless proper annealing facilities are available.
    d. Vacuum-jacketed glass apparatus should be handled with extreme care to prevent explosions. Equipment such as Dewar flasks should be taped or shielded. Only glassware designed for vacuum work should be used for that purpose.
    e. Heavy gloves should be used when picking up broken glass. (Small pieces should be picked up with a vacuum cleaner.)

    f. Proper instruction should be provided in the use of glass equipment designed for specialized tasks, which can represent unusual risks for the first-time user. (For example, separatory funnels containing volatile solvents can develop considerable pressure during use.)
9. Flammability hazards: Because flammable materials are widely used in laboratory operations, the following rules should be observed:
    a. Do not use an open flame to heat a flammable liquid or to carry out a distillation under reduced pressure.
    b. Use an open flame only when necessary and extinguish it when it is no longer actually needed.
    c. Before lighting a flame, remove all flammable substances from the immediate area. Check all containers of flammable materials in the area to ensure that they are tightly closed.
    d. Notify other occupants of the laboratory prior to lighting a flame.
    e. Store flammable materials properly.
    f. When volatile flammable materials may be present, use only nonsparking electrical equipment.
10. Cold traps and cryogenic hazards: The primary hazard of cryogenic materials is their extreme coldness. They, and the surfaces they cool, can cause severe burns if allowed to contact the skin. Gloves and a face shield may be needed when preparing or using some cold baths.

    Neither liquid nitrogen nor liquid air should be used to cool a flammable mixture in the presence of air because oxygen can condense from the air, which leads to an explosion hazard. Appropriate dry gloves should be used when handling dry ice, which should be added slowly to the liquid portion of the cooling bath to avoid foaming over. No worker should lower his/her head into a dry ice chest; carbon dioxide is heavier than air, and suffocation can result.
11. Systems under pressure: Reactions should never be carried out in, nor heat applied to, an apparatus that is a closed system unless it is designed and tested to withstand pressure. Pressurized apparatus should have an appropriate relief device. If the reaction cannot be opened directly to the air, an inert gas purge and bubbler system should be used to avoid pressure buildup.
12. Waste disposal procedures: Laboratory management has the responsibility for establishing waste disposal procedures for routine and emergency situaitons and communicating these to laboratory workers. Workers should follow these procedures with care to avoid any safety hazards or damage to the environment.
13. Warning signs and labels: Laboratory areas that have special or unusual hazards should be posted with warning signs. Standard signs and symbols have been established for a number of special situations.

2.6 Laboratory Safety

These include radioactivity hazards, biological hazards, fire hazards, and laser operations. Other signs should be posted to show the locations of safety showers, eye wash stations, exits, and fire extinguishers. Extinguishers should be labeled to show the type of fire for which they are intended. Waste containers should be labeled for the type of waste that can be safely deposited.

The safety- and hazard-sign systems in the laboratory should enable a person unfamiliar with the usual routine of the laboratory to escape in an emergency (or help combat it, if appropriate).

When possible, labels on containers of chemicals should contain information on the hazards associated with use of the chemical. Unlabeled bottles of chemicals should not be opened; such materials should be disposed of promptly and require special handling procedures.

14. Unattended operations: Frequently, laboratory operations are carried out continuously or overnight. It is essential to plan for interruptions in utility services such as electricity, water, and inert gas. Operations should be designed to be safe under such circumstances, and plans should be made to avoid hazards in case of failure. Wherever possible, arrangements for routine inspection of the operation should be made and, in all instances, the laboratory lights should be left on and an appropriate sign should be placed on the door.

One hazard frequently encountered is failure of cooling water supplies. A variety of commercial or homemade devices can be used that: a) automatically regulate water pressure to avoid surges that might rupture the water lines or b) monitor the water flow so that its failure will automatically turn off electrical connections and water supply valves.

15. Working alone: Generally, it is prudent to avoid working in a laboratory building alone. Under normal working conditions, arrangements should be made between individuals working in separate laboratories outside of working hours to crosscheck periodically. Alternatively, security guards may be asked to check on the laboratory worker. Experiments known to be hazardous should not be undertaken by a worker who is alone in a laboratory.

Under unusual conditions, special rules may be necessary. The supervisor of the laboratory is responsible for determining whether the work requires special safety precautions such as having two persons in the same room during a particular operation.

16. Accident reporting: Emergency telephone numbers to be called in the event of fire, accident, flood or hazardous chemical spill should be posted prominently in each laboratory. In addition, the numbers of the laboratory workers and their supervisors should be posted.

These persons should be notified immediately in the event of an accident or emergency.

Every laboratory should have an internal accident reporting system to help discover and correct unexpected hazards. This system should include provisions for investigating the causes of injury and any potentially serious incident that does not result in injury. The goal of such investigations should be to make recommendations to improve safety, not to assign blame for an accident. Relevant federal, state and local regulations may require particular reporting procedures for accidents or injuries.

17. Everyday hazards: Finally, laboratory workers should remember that injuries can and do occur outside the laboratory or other work areas. Safety should therefore be a goal in offices, stairways, corridors and other places, where safety is largely a matter of common sense. Constant awareness of everyday hazards can prevent costly accidents.

I. Corrosive materials:

Substances that can damage body tissue or are corrosive to metal by direct chemical action are classified as corrosive chemicals. Corrosive injury may be minor (skin irritation) or more severe (burns, disruption of body tissue).

Corrosive chemicals act on body tissue through inhalation, ingestion or direct contact with the skin or eyes.

Corrosive liquids are the most common type of compounds causing corrosive injury in the laboratory. Most of these injuries result from direct contact with the skin or eyes.

Typical corrosive liquids include:
1. Mineral acids such as hydrochloric acid (HCl), hydrofluoric acid (HF), nitric acid (HNO_3), sulfuric acid (H_2SO_4) and acetic acid (CH_3COOH).
2. Strong bases such as sodium hydroxide (NaOH), potassium hydroxide (KOH), ammonium hydroxide (NH_4OH), quaternary ammonium hydroxides; certain chlorinated hydrocarbons, terpenes, liquified phenol, bromine and thymol chloride.

Bases often cause more injury than acids because they burn deeper and cause less pain. The pain often subsides, while the base continues to react with the tissue.

Proper precautionary measures must be taken to prevent corrosive injury. Such measures include wearing adequate protective equipment (e.g., gloves, goggles, face shields) and using suitable respiratory protection when there is danger of inhalation of corrosive fumes. The label of the container should always be checked before the contents are used, to familiarize the user with the potential risks involved.

In the event of skin or eye contact with corrosive liquids, the exposed area should be flushed with copious amounts of water for a minimum of 15 min.

Corrosive solids cause injury based on their respective solubility in skin moisture and the mode of contact. Typical corrosive solids include alkaline hydroxides (NaOH, KOH), alkaline carbonates (Na_2CO_3), alkaline sulfides (Na_2S), alkaline earth hydroxides ($BaOH_2$, $CaOH_2$) and alkaline earth carbonates ($CaCO_3$).

Corrosive action of solids can be minimized by rapid removal upon contact. They are dissolved rapidly, however, by moisture present in the respiratory and alimentary systems.

In addition to injuries caused by direct contact, corrosive solids pose other dangers because:
1. Solutions of corrosive solids are readily absorbed through the skin.
2. Many solid corrosives cause delayed injury. Caustic alkalines may not produce immediate pain.
3. Corrosive solids often produce dust; therefore, the risk of inhalation is present.
4. Molten corrosive solids (phenols) greatly increase the potential risk since they can act as liquid corrosive materials.

Protective measures include adequate protective equipment (including respiratory), adequate ventilation to control corrosive solid dust, and full eye protection (goggles, face shields).

In the event of exposure, the affected area should be immediately washed with copious amounts of water for a minimum of 15 min. The affected area should then be washed with soap and water before a final rinse with water. Immediate medical attention should be sought.

Corrosive gases pose the most serious hazard associated with corrosive chemicals. This is because the gas may be both directly inhaled and absorbed into the body through the skin after dissolution by skin moisture.

Adequate respiratory protection as well as adequate ventilation and protection for the skin and eyes are essential when handling corrosive gases.

Corrosive chemicals are also highly reactive. This reactivity can cause surrounding materials to give off corrosive, toxic, and flammable vapors upon contact with the corrosive chemical. Extreme care must be exercised in handling these chemicals.

J. Flammable chemicals:

Flammable solvents are used in a variety of analytical procedures and also in the cleaning of apparatus that has contained materials insoluble in water. A fire hazard exists whenever a flammable liquid has sufficient volatility so that the vapors will mix with air in ignitable proportions.

Three elements are necessary for a fire: fuel, oxidizer, and ignition source. Fire thus can be prevented or extinguished by:

1. Removing the oxidizer by working in an inert atmosphere. This is often difficult and/or impractical. This is the principle employed when using a fire extinguisher.
2. Removing the fuel, which is an easier approach. Control of solvent vapors can be achieved through adequate ventilation. Solvents should be stored in isolated areas away from reactive substances such as oxidizers (e.g., nitric acid). When working with flammable solvents, fume hoods should be employed. Safety containers are proper for storage of solvents in solvent cabinets; however, working-size containers of 1 L or less should be used in laboratory work areas.

    The following safety guidelines should be adhered to when using flammable solvents:
    a. Safety cans should be used to store flammables in laboratory areas.
    b. Cradle 18.9-L safety cans in a tilt frame so that the user can operate the self-closing faucet with one hand.
    c. Laboratory safety-disposal cans should be used for waste disposal.
    d. Approved solvent-storage cabinets should be used for flammable storage.
    e. Reactions and distillations involving flammables should be set up in a metal pan (to contain spills) and preferably in a hood.
    f. Glass bottles containing flammables should be protected by using a metal bucket or rubber acid carrier.
3. Removing the ignition source, which is often the easiest measure to initiate. Solvents should be stored in a cool area to prevent vaporization. The following guidelines should be observed:
    a. Keep ignition sources remote from all equipment containing flammable liquids, without precluding the proper heating of reactants or distillation devices.
    b. Prohibit smoking or open flames in areas containing solvents.
    c. Remove any sparking equipment from the laboratory bench where work is being done and from the floor area of the room (many solvents are heavier than air and therefore sink to the floor).
    d. Ground drums and transfer containers to prevent sparking (static buildup).
    e. Keep essential sparking equipment (switches, motors, etc.) explosion-proof or enclosed so that flammable vapors cannot come in contact with the spark.
    f. Avoid overheating of samples containing flammable vapors during oven drying.

K. Toxic chemicals:

Toxic substances can cause damage by direct chemical action with body systems. Almost any substance can be considered toxic when it is ingested

2.6 Laboratory Safety

in concentrations above "tolerance" limits. By interfering with cell function, toxic substances disrupt overall body functions.

The toxicity of a substance is related to the duration of exposure of the body to the substance. Two substances may have the same toxicity rating for a single exposure; however, their effects following prolonged or repetitive exposures may be quite different. Acute toxicity refers to toxic action over a short-term exposure to a substance on single contact, ingestion, or inhalation. Chronic toxicity refers to the toxic action of chemicals upon the biological system due to multiple or prolonged exposures. Chronic toxicity is often not recognized until its effects reach an advanced stage, when permanent damage has been done to the body.

Chemicals whose toxic effects are known, are commonly described in terms of LD_{50}, LC_{50} or TLV. LD_{50} (lethal dose, 50% kill) is the dose which, when administered to test animals, will cause 50% of the test population to die. LD_{50} is expressed in milligrams of toxic substance per kilogram of body weight. The route of administration (oral, intravenous, etc.) is normally specified. LC_{50} (lethal concentration, 50% kill) is the concentration of toxic material in the air or water that will cause fatality to 50% of the test population when administered to test animals for a specified period of time (exposure). TLV (threshold limit value) is a measure of the average safe level of exposure to a toxic substance over a prolonged period of time (usually 8 h per day on a daily basis). The units of a TLV are normally expressed in parts per million (ppm) and/or milligrams per cubic meter (mg/m^3). TLVs are normally used in reference to airborne concentrations of toxic substances (gases, solvents, mists, dusts, etc.)

Publications are available that list the known toxic properties of laboratory chemicals[15]. Workers using hazardous chemicals should be aware of the potential dangers. Substances of unknown toxicity should be treated as if they were toxic.

The body may be protected from toxic substances by employing one or more of the following safety measures, as appropriate:
1. Cover exposed areas of skin (gloves, aprons, caps, etc.).
2. Do not work with toxic chemicals if you have an open sore or wound.
3. Wear safety goggles or safety glasses to prevent absorption through the eyes.
4. Use an adequate respirator when appropriate.
5. Exercise care when handling contaminated laboratory apparatus.
6. Change protective garments as required.
7. Use safety cabinetry.
8. Keep food, drink and cigarettes out of laboratory work areas.

9. Do not use laboratory glassware for eating or drinking.

L. Compressed gases:

Compressed-gas cylinders are commonly used in most analytical laboratories. These cylinders contain gas under very high pressure. If the tank is suddenly ruptured, the release of energy can cause extreme damage. Care must be taken to protect the valve system and flow regulator when a compressed-gas cylinder is being used.

The following safety measures should be routinely used in conjunction with compressed-gas cylinders.

When storing such cylinders:
1. Adequately post the areas with the names of gases being stored.
2. Be sure temperatures do not exceel 38°C.
3. Floors should be level, smooth and free from dampness.
4. Store cylinders in an upright position, chained to a support to prevent falling.
5. Do not mix flammable and oxidizing gases in the same storage area.
6. Limit storage in laboratory areas to cylinders in use.

When handling and using such cylinders:
1. Do not handle by the cap or valve.
2. Use a cylinder transport cart, and secure the cylinder to the cart with a belt or chain.
3. Keep the protective cap firmly in place during transport.
4. Do not allow the cylinder to be dropped or to strike objects during transport.
5. Cylinders should be chained in place or otherwise secured at all times.
6. Use cylinders only in an upright position.
7. Be sure all cylinder valves are closed when not in use.
8. Use only the regulator specified for the particular gas. Do not use adaptors, or otherwise modify regulators to fit particular situations.
9. Electrically ground all cylinders containing flammable gasses.

M. Radioactive materials:

The use of radioactive materials in the laboratory is regulated by the Nuclear Regulatory Commission (NRC). The possession and/or use of radioactive material is allowed only when covered by a state and/or federal license. Such a license specifies the type and quantity of radioactive material, the conditions of use, required test programs, and the conditions of storage and disposal. Possession, use, or disposal by any means other than those specified in the license is a violation of federal law and is punishable by fine and/or imprisonment.

A requirement for obtaining the license is to have appointed a Radiation Safety Officer for the laboratory. This person is to be responsible for all laboratory use of radioactive materials and sources of radiation. The specific

2.6 Laboratory Safety

radioactive safety program for the laboratory should be designed and monitored by the Radiation Safety Officer.

Laboratories conducting pesticide residue, antibiotic, or other analyses in which radioactive materials may be employed, either in detector devices or radioactive chemicals, should be familiar with the safety and operating requirements for their usage.

N. Personnel practices in microbiological laboratories:

The following specific guidelines may serve as a basis for developing a microbiological laboratory safety program:

1. Working surfaces should be disinfected immediately before and after use.
2. Laboratory clothing should be worn at all times in the work area.
3. Hands and exposed areas of the arms of the analyst should be washed immediately before and after handling potentially dangerous microbiologic materials.
4. Spills of contaminated materials should be flooded immediately with disinfectant solutions and cleaned up promptly.
5. All labware containing incubated culture material should be heated to 85°C prior to being washed. Containers of primary dilutions made from food and drug samples need not be sterilized unless there is reason to suspect the presence of pathogens.
6. Materials should not be placed in autoclaves overnight anticipating autoclaving the next day.
7. Contaminated glass from broken laboratory ware should be picked up with forceps, autoclaved, and then disposed of properly.
8. Centrifuging should be done only in suitable closed tubes.
9. Contaminated pipettes should be placed, immediately after use, in a tray or pipette jar containing a suitable disinfectant.
10. Loops and needles should be allowed to cool before introduction into inoculated media.
11. All filter flasks should be fitted with a trap to prevent escape of contaminated material into the vacuum system.
12. All broth, agar or suspensions, whenever inoculated from suspect material, should be considered hazardous.
13. Contaminated materials and toxin-producing cultures should not be removed from the laboratory except under controlled conditions.
14. Laboratory clothing that is contaminated with potentially dangerous microorganisms should be sterilized before laundering.
15. Surgical gloves should be worn when working with pathogens if the analyst has noticeable skin cuts, lesions or abrasions.
16. Sniffing unknown cultures may result in infection. Do not sniff inoculated media.

2.7 References

1. BABEL, F.J.; COLLINS, E.B.; OLSON, J.C.; PETERS, I.I.; WATROUS, G.H.; SPECK, M.L. The standard plate count of milk as affected by the temperature of incubation. J. Dairy Sci. 38:499–503; 1955.
2. BOLAFFI, A.; LITSKY, W. Studies on the use of plastic petri dishes for the cultivation of bacteria. J. Milk Food Technol. 22:67–70; 1959.
3. FELLAND, H.S.; NADIG, J.W. A comparative study of the volumetric deliveries from glass and plastic pipettes. J. Milk Food Technol. 28:124–126; 1965.
4. FISCHBECK, R.D. Good laboratory practice regulations and their impact on laboratory organization and operation. Amer. Lab. 12(9):125–130; 1980.
5. GREENBERG, A.E.; CONNORS, J.J.; JENKINS, D., EDS. Standard methods for the examination of water and wastewater, 15th ed. Washington, DC: Amer. Public Health Assoc.; 1980.
6. HUHTANEN, C.N.; BRAZIS, A.R.; ARLEDGE, W.L.; COOK, E.W.; DONNELLY, C.B.; FERRELL, S.E.; GINN, R.E.; LEMBKE, L.; PUSCH, D.J.; BREDVOLD, E.; RANDOLPH, H.E.; SING, E.L.; THOMPSON, D.I. A comparison of horizontal versus vertical mixing procedures and plastic versus glass petri dishes for enumerating bacteria in raw milk. J. Milk Food Technol. 33:400–404; 1970.
7. INHORN, S.L., ED. Quality assurance practices for health laboratories. Washington, DC: Amer. Public Health Assoc.; 1978.
8. MARTH, E.H.; PROCTOR, J.F.; HUSSONG, R.V. Effect of pipette type and sample size on plate count results. J. Milk Food Technol. 28:314–319; 1965.
9. PRICKETT, P.S.; MILLER, N.J. A modified A.P.H.A. milk dilution pipette. Amer. J. Public Health and the Nation's Health 21:1374–1375; 1931.
10. RICHARDS, O.W.; HEIJN, P.C. An improved darkfield Quebec colony counter. J. Milk Food Technol. 8:253–256; 1945.
11. SAX, N.I. Dangerous properties of industrial materials, 4th ed. New York, NY: Van Nostrand Reinhold Company; 1975.
12. U.S. DEPT. OF HEALTH AND HUMAN SERVICES. Milk laboratory evaluation forms. Cincinnati, OH: Food and Drug Admin.; 1980.
13. U.S. ENVIRON. PROT. AGENCY. Microbiological methods for monitoring the environment, water and wastes. Cincinnati, OH: EPA-600/8-78-017.; 1978.
14. U.S. ENVIRON. PROT. AGENCY. Handbook for analytical quality control in water and wastewater laboratories. Cincinnati, OH: EPA-600/4-79-019; 1979.
15. U.S. NAT. ACAD. SCI. Prudent practices for handling hazardous chemicals in laboratories. Washington, DC: Nat. Res. Council.; 1981.
16. WINDHOLZ, M., ED. The Merck index, 9th ed. Rahway, NJ: Merck and Co.Inc.; 1976.

CHAPTER 3
PATHOGENS IN MILK AND MILK PRODUCTS

E.H. Marth
(Tech Comm. E.H. Marth)

3.1 Introduction

In 1983 in the U.S., approximately 11 million cows produced about 140 billion pounds of milk.[7] Nearly 150,000 people worked in several thousand factories to convert this milk into an array of milk products. Production, processing and distribution, including retail sale of milk products, was overseen by inspectors from more than 15,000 state, county, local and municipal health and sanitation jurisdictions.[7] Despite such diversity in production, processing, distribution and regulatory control, milk and milk products are among the safest of foods available to the consumer. Not only are they nearly always safe, the products generally also are of excellent quality when such attributes as nutritional value, flavor, odor, appearance and keeping quality are considered.

Milk and milk products did not always have this excellent record; in fact they were once major vehicles for dissemination of pathogens. The diseases once commonly spread by milk, beside the common forms of food poisoning and food infection, include typhoid fever, scarlet fever, septic sore throat diptheria, tuberculosis, and undulant fever (brucellosis).[29]

The single development that has eliminated virtually all disease problems of microbial origin is the widespread use of pasteurization in the processing of milk. Development and use of standardized laboratory procedures and their periodic publication in *Standard Methods for the Examination of Dairy Products* enabled (and continues to do so) processors to exert suitable control over adequacy of pasteurization and quality of raw and finished products. Other factors which have and continue to be important in production of safe and high-quality dairy products are: (a) improvements in sanitation on farms and in dairy factories, (b) virtual elimination of some diseases from dairy cattle, (c) development, periodic revision and adoption

---

This chapter is a contribution from the College of Agriculture and Life Sciences, University of Wisconsin-Madison, Madison, Wisconsin.

by states of the Pasteurized Milk Ordinance, (d) development of the Conference on Interstate Milk Shipments and (e) educational programs for producers, processors and consumers which have been prepared and disseminated by universities, professional organizations, trade associations, and various publications. This was recently described in some detail by Marth.[105]

The price of safe milk and milk products is eternal vigilance. However, this vigilance can only be exercised when appropriate information is available. This chapter has been prepared with that view in mind. The information is arranged according to type of product and the discussion of some public health problems is more extensive than that of others. This is related to recent experience on the relative importance of the problems and also to the availability of information. When possible, causes of problems have been identified so appropriate action can be taken to prevent the problems in the future.

3.2 Raw Milk

Milk is sterile as it is secreted by the specialized secretory cells in the mammary gland. As the milk moves through the mammary gland, it becomes contaminated with bacteria that reside within the udder. Most often these bacteria are non-pathogenic microccoci, streptococci and bacilli. Sometimes, however, pathogens may be present.

The health of the cow is the first, and, perhaps, principle factor which is related to the absence or presence of pathogens in raw milk. Mastitis, an inflammatory disease of the udder, is likely to be accompanied by development of millions of pathogens in the infected quarter and the resultant discharge of large numbers of these pathogens in the milk.[29] Pathogens that can cause mastitis and thus appear in milk include *Staphylococcus aureus, Streptococcus agalactiae,* other streptococci, *Escherichia coli, Pseudomonas aeruginosa, Listeria monocytogenes* and others.[29] Cows infected with such pathogens as *Brucella* spp., *Coxiella burnetti* or *Mycobacterium bovis* can shed them in their milk.[29]

The environment of the dairy cow is a second source of microorganisms in milk, including some pathogens. Specific items in the environment that can contaminate the surfaces of teats, the udder and adjacent skin and hair and thus ultimately the milk include soil, litter, feed, water, and feces.[29] *Bacillus cereus* and *Clostridium perfringens* in milk can come from soil, litter, feed or feces. Salmonellae may have their origin in feed or feces, but sometimes they are shed from the udder. *Yersinia* and *Campylobacter* may enter milk from water or feces of infected animals. Other pathogens that may enter milk from feces include *E. coli* and enterococci.[29] *L. monocytogenes* has been associated with feed, particulary silage, and thus may enter milk from this source.[118] Other potential sources of pathogens in milk include improperly cleaned and sanitized milk-handling equipment, and workers who milk the cows, handle milk or handle equipment associated with the milking process. Contamination from such workers has lead to milkborne outbreaks of typhoid fever, diphtheria, septic sore throat, scarlet fever, shigellosis, staphylococcal enterotoxicosis, hepatitis A and poliomyelitis.[29]

The molds, *Aspergillus flavus* and *Aspergillus parasiticus* produce a group of secondary metabolites that are toxic and carcinogenic.[17] Collectively, these metabolites are called "aflatoxin" and the most common forms are B_1, B_2, G_1 and G_2, with aflatoxin B_1 being most toxic and carcinogenic. Growth of *A. flavus* or *A. parasiticus* on a dairy product can result in the presence of aflatoxin in that product. Growth of mold on feed can result in the presence of aflatoxin B_1 in that feed. When the cow consumes the feed with the aflatoxin, mixed function oxidases in the liver of the cow convert some of the aflatoxin B_1 to aflatoxin M_1, which is about as toxic as B_1, but is somewhat less carcinogenic. Aflatoxin M_1 is excreted in milk, but the cow is a good "filter" and only about 1 to 3% of the amount of aflatoxin B_1 that is ingested appears in milk as aflatoxin M_1. Aflatoxin M_1 can appear in milk in as little as 4 h after the cow consumed aflatoxin B_1.[18] It is evident from the foregoing discussion that the potential public health problems associated with raw milk continue to be numerous. From this it is readily apparent that the drinking of raw milk remains a distinct hazard to the health of the consumer. Not all of the problems described thus far are of equal importance and so only a few will be discussed in greater detail.

A. Salmonellae:

Occurrence of salmonellae in raw milk has been verified repeatedly by investigators who studied certain outbreaks of salmonellosis in humans. Some of these incidents will be described. A comparison of two surveys on foodborne illnesses in England and Wales showed that the incidence of salmonellosis attributable to raw milk has not changed significantly during the interval between 1951 and 1965. Cockburn and Simpson[35] found that 4 of 235 and 4 of 274 food-poisoning incidents in 1951 and 1952 were caused by salmonellae. In 1965, according to Vernon,[179] the number of outbreaks had increased to 4,091 and of these 6 resulted from consumption of raw milk contaminated with *Salmonella typhimurium*.

Information on the concentration of salmonellae in raw milk supplies seems to be lacking in the literature and only a few surveys have been made on the kind of salmonellae present. Varela and Olarte[178] examined Mexican certified milks and recovered 11 different salmonellae from 25 of 520 samples. Organisms identified included: *Salmonella essen* (occurred 6 times); *S. paratyphi* (occurred 5 times); *S. derby* (occurred 3 times); *S. typhi, S. bredeny* and *S. newport* (each occurred twice), and *S. abony, S. muenchen, S. bovismordificans,* and *S. senftenberg,* (each occurred once). Ritchie and Clayton[150] examined 762 milks from England, Scotland and Ireland for *Salmonella dublin* and recovered the organism from one sample. Another survey to determine the incidence of salmonellae in Northern Ireland milk supplies was conducted by Murray.[123] He was unable to recover these organisms from bulk-collected samples, but did find *S. dublin* in milk from two individual producers.

Raw milk is most frequently, although not exclusively, contaminated by cows. Other sources include human carriers, water supplies, and equipment. Some outbreaks of salmonellosis attributable to raw milk contaminated from the various sources will be described below.

1. Infected cows and calves as sources of contamination: Salmonellae infections in dairy cattle have been observed and reported by numerous investigators. Some of the reports have resulted from surveys conducted on cattle, whereas others had their origin in outbreaks of human illness attributable to consumption of raw milk. Reports on surveys will be considered first and details of foodborne illnesses associated with raw milk contaminated by the cow will be given later.

    a. Surveys: Field[51] conducted a survey in Wales and found evidence of salmonellosis on 70 farms. Calves were affected on three of the farms and cattle from 1 to 14 y of age on the other 67 farms. The infective organism on 64 farms was *S. typhimurium*. Eight outbreaks of bovine salmonellosis were investigated by McFarlane and Rennie.[109] Three of the outbreaks occurred in adult cattle and five in calves. Organisms isolated from the diseased cattle include: *S. typhimurium, S. orientals, S. newington, S. anatum, S. muenchen,* and *S. london*.

    Murdock and Gordon[122] examined fecal samples from apparently healthy cows in an attempt to learn about the incidence of *S. dublin*. Results of tests on 272 fecal samples revealed the presence of *S. dublin* in 7.4% of the specimens. A similar study on 1,250 cows was carried out by Smith and Buxton[164], who found *S. dublin* in fecal samples from 6 cows. Post-mortem examinations on 17 cows culled because of chronic mastitis were conducted by Zagaevskii,[192] who recovered *S. dublin* from the udders of five of the animals. Nyström et al.,[128] working in Norway, noted *S. dublin* infections in 69 dairy herds that were tested.

    A rather extensive survey on dairy cattle in Germany was carried out by Rasch.[148] He examined 1,427 dairy herds and found that 325 cows in 277 herds were carriers of salmonellae. Guinee et al.,[68] working in the Netherlands, conducted a survey on the incidence of salmonellae in certain tissues obtained from cows and calves. Salmonellae were recovered from the mesenteric lymph nodes of two of 600 normal cows and 11 of 265 normal calves. A somewhat more detailed study was made on 1,504

3.2 Raw Milk

normal calves at the time of slaughter. Tests on the mesenteric and portal lymph nodes, gall bladder, and feces revealed the presence of salmonellae in one or more samples taken from each of 216 calves. When an even more intensive study was made on 416 calves (samples were taken from musculature, diaphragm, spleen, liver, gall bladder, portal lymph nodes, feces, and all mesenteric nodes), salmonellae were found in one or more samples from each of 63 calves. Salmonellae isolated from the calves belonged to 32 different serotypes, with *S. typhimurium* appearing in 61.4% and *S. dublin* in 8.5% of the infected animals. The musculature and organs of infected calves rarely contained salmonellae, but the lymph nodes, gall bladder, and intestinal contents were frequently infected. According to Rothenbacher,[153] *S. typhimurium* infections were responsible for many deaths among dairy calves.

b. Raw milkborne illnesses: The literature reports numerous outbreaks of human illness attributed to raw milk contaminated with salmonellae by the cow. Numerous serotypes of *Salmonella* have been noted in these outbreaks. Included are *S. dublin*,[6,33,39,67,71,99,108,165,170,176] *S. thompson*,[189] *S. heidelberg*,[40,93] *S. newport*,[23] *S. enteritidis*,[20] and *S. typhimurium*.[5,127,149] The reports suggest that *S. dublin* was recovered with greater frequency than were other salmonellae when raw milk contaminated with these organisms was associated with human illness. Goats also have been reported as able to contract *S. dublin* infections.[62]

Many authors report that cows were shedding salmonellae in their milk and some suggest these organisms may cause symptomless cases of mastitis. There is not complete agreement, however, and some authors feel that contamination with fecal material from infected cows may be an important means by which the organisms enter milk. Scott and Minett[159] conducted some experiments in an attempt to better understand the behavior of salmonellae in cows. They drenched a cow with *S. typhi* at concentrations far greater than would be ingested if the animal consumed contaminated water. No illness resulted from this treatment nor could the organism be recovered from feces or milk. Two calves received a similar treatment; one contracted the disease, whereas the other calf shed the organism in its feces for 48 h. The authors also infused *S. typhi* cells directly into the udder and found that: (a) ten quarters ceased to excrete *S. typhi* in ten days, (b) two quarters of one cow excreted the organism for 27 d, and (c) two quarters of one cow excreted the organism for 85 d.

Cows may become infected (or contaminated) with salmonellae

from a variety of sources, including feed and water. Hutchinson,[77] described an outbreak of human illness which may have had its origin in a feed material. Examination of feces from 20 persons who consumed raw milk from a common source revealed *S. heidelberg*. Upon further investigation, the organism was recovered from milk samples and also from the feces of several cows suffering from enteritis. The cows were on the farm that furnished the raw milk in question. Farm personnel also excreted *S. heidelberg*. The original source of infection is believed to have been barley, which was fed to cows and which contained the pathogen. The barley probably became infected from rats, since their fecal material was evident in the grain.

Calves on a dairy farm can serve as sources of salmonellae which contaminate milk. Such incidents have been described in the literature. Dobrier,[42] reported on an outbreak of human illness caused by *S. enteritidis* present in raw milk. Inspection of the farm revealed that calves were suffering from an illness caused by this organism. Since hygienic practices on the farm were substandard, it was believed that the organisms from the calves entered the milk, and, hence, caused illness in consumers.

An outbreak of gastroenteritis in which diseased calves may have been important occurred in the United States in January and February of 1966 and was described by Holterman and Mather.[76] Eleven persons in five families were affected and all of them had consumed milk produced on a single farm. The farmer responsible for the milk was not in the business of selling raw milk but gave his surplus to relatives and friends. Although the exact route of contamination remained unknown, several possibilities were apparent to the authors. They found that the farmer was busy throughout the winter making trips to a beef-cattle feeding lot several miles from his farm. Apparently, 16 calves were in the feeding lot and three died early in January from diarrheal disease. Fecal samples were taken from 6 of the calves 5 weeks after recovering from the disease and *S. typhimurium* was recovered from one sample. The farmer admitted that he often worked late in the evening with the calves while they were ill and, after returning home, had neglected to wash his hands before milking. Milk was collected in open buckets and then poured into large widemouthed jars. In addition to possible contamination from the hands, manure and other barnyard materials could have easily fallen into the bucket during the milking process. The milk was cooled slowly after collection; more than 2 h were required to bring the temperature to 4.4°C, thus allowing considerable time for bacterial growth. Additional information on salmonellae in milk and milk products can be found in a review by Marth.[106]

B. *Staphylococcus aureus:*

Staphylococcus aureus, a gram-positive, coccus-shaped bacterium, can produce an enterotoxin (a protein with a molecular weight of approximately 28,000 to 34,000—there are several enterotoxins). This enterotoxin, when ingested, can cause such symptoms as nausea, vomiting, retching and often diarrhea.[116] Recovery often follows in 24 h but several days may be required. The extent to which these symptoms may appear and the severity of illness are determined chiefly by the amount of toxin that was ingested and the susceptibility of the individual who becomes a victim of the disease.[116] Outbreaks of this disease are infrequently associated with raw milk, but have been caused by toxic cheese.[3,73,194,195]

1. Occurrence of staphylococci in milk: Raw milk may become contaminated with staphylococci from several sources but the principal one is likely to be the mastitic bovine udder. Mastitis is often caused by *S. aureus* and continues to be a major problem in dairy cattle. For example, in Wisconsin at least 16% of all dairy cows suffered from mastitis;[15] to be sure, not all of these cows were infected with *S. aureus*. Mastitis also seems to appear more regularly in large (91% of Wisconsin herds) than in small (38%) dairy herds. Since the trend is toward maintenance of ever larger herds on dairy farms, it is almost certain that mastitis will continue as a problem in animal health and also will result in the presence of *S. aureus* in some lots of raw milk.

   The occurrence of staphylococci in raw milk has been verified by results of numerous surveys. Among the first to do so was the survey by Williams.[184] He studied 10 herds and found that more than 50% of the cows in these herds were shedding staphylococci in their milk. Staphylococcal counts in excess of 1000/mL were not uncommon and isolates obtained were generally coagulase-positive staphylococci. Williams[184] did his work before antibiotics were used regularly to treat mastitis; hence the problem of antibiotic resistance in staphylococci of bovine origin and the increase in frequency of staphylococcal mastitis had not yet occurred. Twenty years after the work of Williams, samples of raw milk were tested by Clark and Nelson.[34] These investigators found that their samples contained 25 to 3300 coagulase-positive staphylococci/mL.

   Occurrence of staphylococci in raw milk would be of less concern if enterotoxigenic strains were never present. Unfortunately, this is not true; data obtained by several researchers have demonstrated that some staphylococci from the bovine udder are enterotoxigenic. Bell and Veliz[21] found that enterotoxin was produced by 25 of 35 cultures of staphylococci isolated from the udder. Enterotoxigenicity of staphylococci obtained from raw milk was determined by Casman,[31] who found that 4.2% of 190 staphylococcal cultures from raw milk were able to produce enterotoxin A on B. About 75% of the toxigenic strains produced enterotoxin A, whereas the remainder formed enterotoxin B. Olson et al.[129] tested 157 cultures of staphylococci from mastitic udders. Eleven of the cultures produced enterotoxin C, 11 others yielded enterotoxin D, 1 formed both C and D, and none produced enterotoxins A and B.

Data cited here demonstrate that: (a) staphylococci are likely to be present in raw milk, and (b) some of the staphylococci in raw milk are likely to be enterotoxigenic.

2. Growth of staphylococci in milk: The foregoing considerations would be of less consequence if staphylococci could not grow in raw or processed milk since growth is needed for synthesis of enterotoxins. In general, heated milks are excellent substrates, but raw milk is less favorable for staphylococcal growth.

Clark and Nelson[34] held raw milk for 7 days at 4° and 10°C. Staphylococci failed to grow at 4°C but grew, although generally slowly, at 10°C. The initial number of staphylococci was not indicative of whether or not growth would occur during the holding period. Although the aerobic plate count of all test milks was low initially, the count was uniformly high after 4 d at 10°C. In one instance, marked growth of staphylococci was evident, which suggests that the kind of bacteria present and growing in milk may affect proliferation of staphylococci.

Both growth and enterotoxin production by staphylococci in milk were investigated by Donnelly et al.[45] They inoculated low- and high-count raw milk with *S. aureus* and then held the milks at different temperatures. From their data it is again evident that behavior of staphylococci can differ in milks with similar initial numbers of bacteria. Although the indigenous flora grew in both samples of milk, staphylococcal growth was more pronounced in one low-count milk than in the other such milk. This more extensive growth by the staphylococcus resulted in enterotoxin production in one but not in the other milk. Similar results were obtained when incubation was 25° and 35°C.

C. Campylobacteriosis:

The frequency with which *Campylobacter* has been isolated, in recent years, from stools of patients suggests that this organism is responsible for as many cases of gastroenteritis as is *Samonella*.[46] *Campylobacter jejuni* appears to cause most cases of human campylobacteriosis.

Campylobacter is widespread in the environment and outbreaks of illness caused by the organism have been traced to contaminated water, contami-

3.2 Raw Milk

nated foods (including raw milk), and contact with infected animals or persons.[46,151,163]

Milkborne outbreaks of campylobacteriosis have been reported in the United States, England, Wales Scotland and Canada. The vehicle in all outbreaks was either raw milk or certified raw milk.[29]

D. *Streptococcus agalactiae*:

Streptococcus agalactiae, a Lancefield group B streptococcus, is a major cause of bovine mastitis. Many persons have viewed this organism as unimportant in human health. This view may not be correct.

Recently Ross[152] reviewed the literature dealing with this organism. From his review it is evident that adult males and females often are carriers of this organism. Such carriage occurs primarily in the genital tract but also in the respiratory tract. Adults with some underlying disease or deficiency (compromised host) are most susceptible to infection by *S. agalactiae*. Such infections have been associated with pyelonephritis of bacterial origin, pneumonia, endometritis and to a lesser extent meningitis and septic arthritis.

If a woman harbors the organisms in her genital tract, and in particular in the vagina, and gives birth to an infant, it is possible for the infant to become contaminated with *S. agalactiae* during the birth process. Infection of infants has been reported to in the range of about 1 to 3 per 1000 live births. It appears as if the frequency of infection of infants has grown in the last decade. Disease in infants can be of the "early onset" or "late onset" type. The "early onset" type occurs within the first 5 days after birth, is characterized by a fulminating septicemia and possibly meningitis, and has a mortality rate of 75% within 24 to 36 h after onset of the disease. The "late onset" form of the disease usually appears after the 10th day of life as a purulent meningitis. It can appear as late as 12 weeks after birth and this form of the disease has a mortality rate of 15-20%.

The question of whether all, some or none of the strains of *S. agalactiae* that cause mastitis also cause human illness is not fully resolved. It has been observed that infection with *S. agalactiae* is more common in humans that drink raw milk than in those consuming pasteurized milk[185]. Hence, it has been suggested that in passage through the human body, the strains become modified and are changed from the bovine to the human type. Evidence that this happens is lacking. However, it is known that human types can establish themselves in the bovine mammary gland and cause mastitis which is indistinguishable from that produced by the bovine strains. Although more work needs to be done for absolute certainty, it appears prudent to deal with *S. agalactiae* as if it were a milkborne pathogen.

E. *Aflatoxin M_1*:

The first residue of aflatoxin observed in a tissue of dairy cows and the one most extensively studied is aflatoxin M_1 (AFM_1) in milk. Milk is probably the only human food of mammalian origin known to be contaminated naturally with AFM_1. AFM_1, the hydroxylated metabolite of aflatoxin B_1

(AFB$_1$) resides mainly with the protein fraction of the milk.[1] Allcroft and Carnaghan[1] reported that certain milk extracts from cows fed aflatoxin-contaminated groundnut (peanut) meal induced liver lesions when administered to ducklings, and the lesions were identical with those caused by AFB$_1$. De Iongh et al.[41] showed that the toxicity of this milk resided in a blue-violet fluorescent compound, termed the "milk toxin." These workers demonstrated conversion of AFB$_1$ and AFM$_1$ in the lactating rat.

Several studies have been undertaken to determine the relationship between ingestion of aflatoxin and excretion of AFM$_1$ in milk. Allcroft and Roberts[2] measured the amount of AFM$_1$ in milk from cows given diets containing various levels of aflatoxin. The daily intake of AFB$_1$ ranged from 0.875 to 24.5 mg. Results indicated that the amount of AFM$_1$ excreted in milk was in direct proportion to the intake of AFB$_1$. Results of a similar study by Masri and co-workers[107] supported the findings of Allcroft and Roberts[2], for the former researchers also found a linear relationship to exist between aflatoxin intake and the amount of AFM$_1$ in milk. Excretion of AFM$_1$ into milk, as a percentage of aflatoxin consumed, varies from less than 1 to 3%.

AFM$_1$ appears in cows' milk within 2 days of ingestion of feed contaminated with AFB$_1$,[218] or within 4-5 h of an orally administered dose of purified AFB$_1$.[2,18] In most instances, assuming significant quantities of aflatoxin are consumed, AFM$_1$ will appear in milk in at least 12 h post-feeding—a time corresponding to the normal milking following feeding.[83] One reason for the rapid appearance of AFM$_1$ in milk may be the rapid transport of AFM$_1$ from the rumen. Polan and co-workers[137], working with [^{3}H]AFB$_1$, found absorption of tritium, based on direct counting of blood serum, to increase from 20 cpm/mL at 2 h post-administration to 40 cpm/mL at 4 h. However, further in-vivo radioactive tracing is necessary before such a hypothesis is proven.

Polan et al.[137] undertook a study to determine the minimum intake of AFB$_1$ which would produce a detectable level of AFM$_1$ in cows' milk. Cows fed a concentrate containing 50 ppb of AFB$_1$ had only traces (ca. 0.01 µg/L) of AFM$_1$ in their milk. No AFM$_1$ was detected in milk from cows receiving 10 ppb of AFB$_1$.

Using regression analysis to express the relationship between intake of AFB$_1$ and concentration of AFM$_1$ in milk, Polan et al.[137] showed that AFM$_1$ can be expected to appear in milk when the content of AFB$_1$ in concentrate feed exceeds 46 ppb. Extrapolated to the total ration intake, this would represent 15 ppb or 230 µg of AFB$_1$ per day. Their results, in general, agree with those of Allcroft and Roberts[2], who suggested that the content of AFB$_1$ should not exceed 50 ppb in the concentrate feed to ensure presence of only negligible amounts of aflatoxin in milk.

Excretion of AFM$_1$ by the cow occurs relatively quickly.[98] AFM$_1$ will disappear from milk in 3 to 4 d after elimination of the aflatoxin-contaminated diet.[101] Polan et al.[137], who found the concentration of AFM_1 in milk usually plateaued by day 4 of a feeding trial, observed no AFM_1 in milk 2 d after treatment ceased. Keyl and Booth[82] reported a drop in amount of AFM_1 in milk after treatment was discontinued and this happened at a time similar to that reported by Polan et al.[137] Allcroft and Roberts[2] detected no aflatoxin in milk 5 d post-treatment. Other studies,[107] in which daily intakes of AFB_1 ranged between 20 and 30 mg, showed very low levels of toxin to be present in the milk at 7 d post-aflatoxin ingestion. A conclusion of practical significance from these data is that milk produced by cows that have consumed aflatoxin-contaminated feed would be expected to be essentially free of AFM_1, within a short time after contaminated feed was no longer consumed.

Factors influencing the appearance of AFM_1 in milk and reasons for differences among the studies are only partially understood. The concentration of AFM_1 in milk seems to depend more on intake of AFB_1 than on milk yield.[177] However, the toxin content of milk appears to increase rapidly when milk yield is reduced as a result of high toxin intake.[107] Rate of metabolism by the liver and rate of excretion by other routes (urine and feces) are also likely to influence the toxin level in milk.[140] Factors that may account for differences in the results of the several studies cited are most likely associated with individual variation among the cows tested, differences in the quantity and purity of aflatoxin administered and differences in analytical methods. Additional information on aflatoxin in milk and milk products appears in a recent review by Applebaum et al.[17]

3.3 Pasteurized Milk

Use of pasteurization insures that nonsporeforming pathogens likely to be present in raw milk are killed and that the milk is safe for consumption.[29] To remain safe, pasteurized milk must remain free of pathogens, e.g. must not be recontaminated, before it is consumed.

Two recent outbreaks of yersiniosis (caused by *Yersinia enterocolitica*) were associated with pasteurized milk.[29] Why the milk contained yersiniae is not completely clear, but recontamination of the milk after pasteurization seems likely. A recent survey of 100 samples of pasteurized milk in the Madison, Wisconsin area indicated that the samples were free of virulent strains of *Y. enterocolitica*.[120] However, one sample yielded an avirulent strain of the bacterium.

In the summer of 1983, an outbreak of listeriosis occurred in Massachusetts. Forty-nine patients were hospitalized with septicemia or meningitis caused by *L. monocytogenes*. Forty-two of the patients were adults who had an underlying disease or were taking medications that caused immunosuppression, and seven were newborns. Fourteen of the patients died, a mortality rate of 29%. Epidemiologic studies revealed that development of disease was strongly associated with consumption of a particular brand of whole or 2% pasteurized milk. An inspection of the facility where the milk was processed revealed no evidence of improper handling or pasteurization. The dairy factory received its milk from specific farms, on several of which veterinarians had diagnosed listeriosis in dairy cows during the outbreak. However, the route by which *L. monocytogenes* entered the pasteurized milk has not been clarified.

3.4 Cheese

Natural (e.g. ripened) cheese is commonly made from whole or partially skimmed milk which is raw, pasteurized or given a sub-pasteurization heat treatment adequate to inactivate *S. aureus*. Other ingredients used to make cheese commonly include lactic acid bacteria, a milk clotting agent which may be of animal or microbial origin, and salt. If cheese is made from raw milk or milk that received a sub-pasteurization heat treatment, the cheese must be held at least 60 d at or above 1.7°C. The purpose of this is to allow for inactivation of pathogens during the ripening process. Unfortunately the specified conditions do not always cause the demise of pathogens if they are in the cheese.

Although rare, when one considers the large amount of cheese that is produced, outbreaks of foodborne illness have been associated with cheese made from both raw and pasteurized milk. Staphylococcal food poisoning and salmonellosis are the most common problems. Control of foodborne pathogens in cheesemaking involves (a) use of pathogen-free raw, heat-treated or pasteurized milk, (b) preventing contamination of cheese milk with a pathogen, (c) use of an active starter culture so growth of a pathogen, if present, is prevented, (d) ripening the finished cheese under conditions that will insure inactivation of pathogens, and (e) handling the finished cheese in ways so that it is not contaminated with pathogens before it reaches the consumer.

A. Staphylococcal food poisoning:

Although outbreaks of staphylococcal food poisoning have been associated with rennet-type cheese, this problem can be largely avoided if: (a) cheese is made from milk that was stored to preclude staphylococcal growth and then was given a heat treatment, such as pasteurization, that is adequate to inactivate staphylococci; (b) heated milk is not contaminated with staphylococci; and (c) an active starter culture that produces sufficient acid during cheesemaking is used. Since illness has been associated with cheese, several investigators have studied the behavior of *S. aureus* during the manufacture of different kinds of cheese.

    1. Heat inactivation of staphylococci in cheese milk: Since a heat treatment of some sort is commonly given to milk before cheese is made, it is appropriate to briefly consider use of heat to inactivate

staphylocci in milk. According to data by Thomas et al.,[171] the D value (time in minutes at a given temperature to reduce the population by 90%) for *S. aureus* in skim milk at 60°C is 3.44 and at 65.6°C it is 0.28. The z value (degrees Fahrenheit needed for the thermal death cureve to traverse one log cycle), according to Thomas et al.[171] was 9.17. Heinemann[74] tested the same strain of *S. aureus* that was used by Thomas et al.[171]. When the bacterium was heated in raw milk, it was completely inactivated after 80 min at 57.2°C, 24 min at 60°C, 6.8 min at 62.8°C, 1.9 min at 65.6°C, and 0.14 min at 71.7°C. Heinemann[74] determined the z value to be 9.2, a figure that agrees well with that of Thomas et al.[171].

Zottola et al.[196,197] isolated staphylococci from milk and cheese and then tested them for heat resistance in raw milk. When the test population exceeded 10^6/mL, 116 of 236 strains survived heating at 63.9°C for 21 sec; whereas when populations were below 10^6/mL, 108 of the strains survived exposure to 65.6°C for 21 sec. These heat treatments were used to simulate conditions which sometimes exist in cheesemaking when milk receives a sub-pasteurization treatment. This is done because many cheesemakers believe ripened cheese such as Cheddar is more flavorful when it is made from milk that has received a sub-pasteurization rather than a full pasteurization heat treatment.

2. Starter cultures and staphylococci: It was mentioned earlier in this section that use of an active lactic starter culture contributes to production of safe cheese. Why this is so will be evident from the discussion that follows.

Reiter et al.[145] worked with a lactic starter culture that was to be used for cheesemaking. This culture inhibited growth of *S. aureus* in raw, pasteurized, and steamed milk. When lactic acid developed by the starter culture was neutralized as it was formed, inhibition was found to be a function of more than just pH. The quantity of lactic acid bacteria added to milk can influence growth, enterotoxin production, and even survival of *S. aureus*. For example, Richardson and Divatia[148] noted that adding 0.001 and 0.01% of a milk culture of lactic streptococci to milk containing approximately 10^6 *S. aureus*/ml served to appreciable retard growth of the staphylococcus. Lack of growth by the staphylococcus during the first 8 h of incubation and subsequent inactivation were observed when the inoculum was 0.1, 0.5, or 1.0%. Enterotoxin appeared when the inoculum was 0.001 and 0.01% but not when the three larger inocula were used. When the number of staphylococci in milk was reduced, a smaller inoculum of lactic acid bacteria was needed to inhibit or inactivate *S. aureus*.

According to a report by Jezeski et al.[79] an actively growing culture of *Streptococcus lactis* inhibited and sometimes inactivated

S. aureus when both organisms were together in sterile or steamed skim milk. Enterotoxin was produced by *S. aureus* when it grew alone or in the presence of the streptococcus and a homologous bacteriophage. Absence of the bacteriophage enabled the lactic acid bacterium to prevent enterotoxin production by the staphylococcus. Haines and Harmon[69] inoculated ATP (All Purpose Tween) broth with *S. aureus* and *S. lactis* to determine when growth and enterotoxin production by *S. aureus* would be inhibited. They learned that: (a) no enterotoxin was produced and growth of *S. aureus* was retarded similarly when the initial pH values were 6.0, 6.5, and 7.0; (b) no enterotoxin was produced at either 25° or 30°C, but growth of *S. aureus* was inhibited somewhat more at 25° than 30°C; (c) enterotoxin was produced at 30°C when the number of *S. lactis* to *S. aureus* was 10:90, but not when it was 50:50 or 90:10; (d) different strains of *S. lactis* were equally inhibitory to *S. aureus* and prevented toxin production; and (e) different strains of *S. aureus* were inhibited in a similar way by *S. lactis*.

Further evidence for the inhibitory properties of lactic streptococci comes from a report by Gilliland and Speck[63]. They tested six different lactic streptococcus cultures for their ability to inhibit *S. aureus*. Inhibition of *S. aureus* was almost complete regardless of the starter culture used or the time, within limits, required for acid formation. These authors suggested that repression of staphylococci by lactic streptococci may involve production of antibiotics, hydrogen peroxide, and volatile fatty acids in addition to lactic acid. Of these, undoubtedly acid production is the single most important factor which causes the staphylococci to be inhibited.

Minor and Marth[115] did a series of studies to more clearly define the role of acid in controlling growth of staphylococci in milk. They gradually added acid over a 4, 8 or 12 h period to milk which was inoculated with *S. aureus*. When acid was gradually added over a 12 h period, 90% reduction in growth of *S. aureus* was achieved if the final pH values were 5.2 for acetic, 4.9 for lactic, 4.7 for phosphoric and citric and 4.6 for hydrochloric acid. To achieve a 99% reduction during a 12 h period, the final pH value had to be 5.0 for acetic, 4.6 for lactic, 4.5 for citric, 4.1 for phosphoric, and 4.0 for hydrochloric acid. A final pH value of 3.3 was required for 99.9% reduction in growth with hydrochloric acid, whereas the same result was obtained at a final pH value of 4.9 with acetic acid. Correspondingly lower final pH values were required to inhibit growth within 8 and 4 h periods.

3. Behavior of staphylococci in cheese: Data obtained by Tuckey et al.[175] indicate that some growth of staphylococci, if present, is possible even when Cheddar cheese is made by a normal process.

3.4 Cheese

Hence it is necessary to inactive staphylococci before cheesemaking begins. The initial increase (after 2 h) in staphylococcal population is largely attributable to the concentration effect when curd is formed. Further growth continued through the manufacturing process until salt was added to the curd. The inhibitory effect of salt was limited, as might be expected, and growth of staphylococci again was evident when cheese was removed from the press. Takahashi and Johns[166] and Walker et al.[182] made similar observations when they studied Cheddar and Colby cheese, respectively.

Additional work was done by Tatini et al.[168]. Their data indicate the importance of both the starter culture and the initial concentration of staphylococci in determining whether or not a cheese becomes toxic. These authors concluded that Colby cheese made normally would be toxic if it contained at least 15 million staphylococci/g. The value for Cheddar cheese was 28 million/g. If the starter culture failed to perform properly, then toxic cheese could result if 3 to 5 million staphylococci/g were present. Finally, it should be mentioned that the number of staphylococci decreases as the cheese ripens. However, if they are present initially, it is not usual for them to persist in well-ripened cheese.

Aged Cheddar cheese is sometimes added to another food product to impart a desired flavor. Procedures to accelerate ripening of the cheese have been sought so that relatively fresh and less expensive, cheese could be used for this purpose instead of the more costly aged product made by conventional processes. One method to prepare a liquid cheese product with a Cheddar-like flavor in fewer than 7 d has been described by Kristoffersen et at.[97]. Basically, the process involves making a slurry from 2 parts of 24 h-old salted, unpressed Cheddar curd and 1 part of a solution of 5.2% sodium chloride. The slurry then is held at 30°C. Addition of 10–100 ppm of reduced glutathione enhances development of the Cheddar flavor. Gandhi and Richardson[59] inoculated similar slurries with *S. aureus* and found that the bacterium could grow and produce enterotoxin in 24 to 48 h at 32°C when the slurries contained 45 to 60% moisture. Treatment of the inoculated slurry with 0.5% hydrogen peroxide did not eliminate the staphylococci, but addition of 0.2 or 0.3% sodium sorbate inhibited their growth and thus prevented production of enterotoxin. The work of Gandhi and Richardson[59] emphasizes the need to evaluate new processes for potential health hazards.

Tatini et al.[169] also studied the potential for enterotoxin production by added staphylococci in the manufacture of brick, Swiss, Mozzarella and blue cheese. No enterotoxin appeared in their experimental lots of Mozzarella or blue cheese. Environmental conditions in these two types of cheese were such that the staphylococci, even though some growth occurred, were unable to produce enterotoxin. In contrast to these observations, the investigators noted that staphylococcal growth and enterotoxin production did occur during the manufacture of brick and Swiss cheese. The kinds of starter culture used, the initial population of staphylococci, and the population achieved by 7 h after cheese was made appear to have determined whether or not the cheese was toxic. These authors evaluated several toxigenic strains of *S. aureus* when they experimented with Swiss cheese. As might be expected, when some strains were used, the cheese became toxic, whereas others failed to produce detectable enterotoxin even though growth was evident during the cheesemaking process. The authors concluded that staphylococcal populations of 3–8 x 10^6/g could result in toxic Cheddar, Colby, brick, or Swiss cheese, depending on the kind and activity of the starter culture that was used and the strain of toxigenic *S. aureus* that was present.

B. *Salmonellosis:*

Outbreaks of salmonellosis have been associated with consumption of different cheeses. The earlier literature is primarily concerned with typhoid fever (a form of salmonellosis), whereas more recent investigations indicate that salmonellae other than *Salmonella typhi* may be associated with cheese-induced illness.

1. Cheddar cheese: Gauthier and Foley[60] described an epidemic of typhoid fever in Canada. Forty cases were involved and six deaths resulted. The only food common to all the patients was Cheddar cheese, made locally from raw milk and consumed when it was 10 d old. Although the factory where the cheese was made lacked adequate sanitation, the source of the epidemic was a known typhoid carrier who, against orders from public health authorities, milked cows whose milk was used by the factory to produce Cheddar cheese.

   Another outbreak of typhoid fever attributable to Cheddar cheese occurred in Quebec, and was described by Foley and Poisson[52]. The original source of infection was never traced, although it was found that the cheesemaker's wife had an active case. Nevertheless, the authors believe she was not responsible for infecting the cheese. Foley and Poisson[52] also recommended use of a 3-mo ripening period when Cheddar cheese is made from raw milk.

   Of 507 cases of typhoid fever in Alberta between 1936 and 1944, 111 were caused by Cheddar cheese, according to Menzies.[111] Samples of cheese from the last three outbreaks in 1944 were recovered and tested for *S. typhi*. The organism was found in 30-day-old cheese, but could not be recovered from 48- and 63-d-old cheese. As a consequence of this outbreak, Alberta halted sale of

3.4 Cheese

cheese made from raw milk unless the cheese was ripened for at least 3 mo.

Survival of *S. typhi* in Cheddar cheese was studied by Ranta and Dolman[142] and Campbell and Gibbard.[30] The former authors mixed *S. typhi* with Cheddar cheese and found that the organism survived for 1 month at 20°C. Inoculation of *S.typhi* onto the surface of cheese was accompanied by similar survival at room temperature, longer survival at refrigerator temperature, and penetration of the organism to a depth of 4 to 5 cm into the cheese after 17 days.

Campbell and Gibbard[30] inoculated milk with *S. typhi* and used it to make Cheddar cheese. All cheeses were ripened for 2 wks at 14.4° to 15.6°C, after which 1 cheese from each duplicate set was transferred to storage at 4.4° to 5.6°C. At the lower temperature, 7 of 10 cheeses contained viable *S. typhi* cells for more than 10 mo, whereas at the higher temperature the organism generally disappeared after 3 mo of ripening. Size of inoculum and acidity of the cheese did not appear to affect the longevity of *S. typhi*.

Goepfert et al.[66] investigated the behavior of *S. typhimurium* during manufacture and ripening of Cheddar cheese. Pasteurized milk was inoculated with *S. typhimurium* when the lactic starter culture was added. A slight increase in number of salmonellae occurred during the time between inoculation and cutting of the curd, followed by a rapid increase during the interval between cutting the curd and draining the whey. After accounting for concentration of cells through coagulation, an average of 3.5 generations of salmonellae developed during this period. Salting of the curd was associated with a reduction in the growth rate, and ripening of the cheese was accompanied by a decrease in the salmonellae population. Survival of *S. typhimurium* exceeded 12 weeks at a ripening temperature of 12.8°C and 16 weeks at 7.2°C. Limited tests demonstrated that acetate accumulating in ripening cheese may contribute to the demise of salmonellae.

Park et al.[135] made Cheddar cheese using a slow acid-producing strain of *S. lactis* to determine what might happen if salmonellae were present in cheese produced with an abnormal fermentation. Salmonellae grew rapidly during manufacture of cheese and limited additional growth occurred in cheese during the first week of ripening at 13°C, after which there was a gradual decrease in population of *S. typhimurium*. *S. typhimurium* survived during ripening of the low-acid cheese for up to 7 mo at 13°C and 10 mo at 7°C.

Park et al.[133] also made cold-pack cheese food that was contaminated with *S. typhimurium* and stored it at 4.4° and 12.8°C. Rapid decrease in number of salmonellae occurred during the first week of storage regardless of temperature or composition of the product.

Viable salmonellae could not be recovered, after 3 wks at 12.8°C or 5 wks at 4.4°C, from cheese food adjusted to pH 5.0 with lactic acid and containing 0.24% potassium sorbate. Substituting sodium propionate for sorbate resulted in 14 and 16 wks of survival by salmonellae when cheese food was held at 12.8° and 4.4°C, respectively. Partial and complete replacement of lactic acid by acetic acid was accompanied by somewhat longer survival of *S. typhimurium* than when only lactic acid was used. Elimination of added acid from the cheese food resulted in survival of *S. typhimurium* for 6 to 7 weeks when potassium sorbate was present, for 16 and 19 wks when sodium propionate was used, and in excess of 27 wks without any preservative. The ability of potassium sorbate to inactivate *S. typhimurium* was confirmed in another study by Park and Marth.[131]

Effects of lactic acid bacteria on *S. typhimurium* were determined by Park and Marth.[130] They observed that *S. cremoris, S. lactis*, and mixtures of the two, when added to milk in the amount of 0.25%, repressed growth but did not inactivate *S. typhimurium* during 18 h of incubation at 21° or 30°C. Increasing the inoculum of the mixed lactic cultures to 1% resulted in inactivation of *S. typhimurium* at 30°C. When added at the level of 1%, *Streptococcus thermophilus* was more detrimental to *S. typhimurium* at 42°C than was *Lactobacillus bulgaricus*. Mixtures of *L. bulgaricus* and *S. thermophilus*, when added to contaminated milks at levels of 1 and 5%, caused virtually complete inactivation of *S. typhimurium* during the interval between 8 and 18 h of incubation of 42°C.

2. Colby cheese: A typhoid fever epidemic started in January 1944 in the northern part of Indiana and covered 18 to 20 counties.[147] Approximately 250 cases and 13 deaths were recorded in this outbreak. Thomasson[172] noted that the carrier was never traced but illness was associated with consumption of Colby cheese. The cheese, made by a single dairy in the area, was produced from raw milk preheated to 32.2° to 37.8°C and was not allowed to ripen before sale.

Tucker et al.[174] recorded an incident in which 384 cases of illness in Kentucky were caused by consumption of 12- to 14-day-old Colby cheese infected by *S. typhimurium*. Investigation revealed that a dead mouse had been removed from a 1000-gal. vat of milk used to produce the cheese. Tests on infected cheese demonstrated that *S. typhimurium* survived for 302 days during storage at 6.1° to 8.9°C.

3. Mold-ripened cheese: Mocquot el al.[119] made blue cheese from milk inoculated with 10^4 and 10^6 *Salmonella* sp. cells per ml and observed the behavior of the organisms during manufacture and ripening of the cheese. The death rate of *Salmonella* was related to the pH value of 24-h-old cheese and increased with a decrease in the pH.

3.4 Cheese

Survival and death of the organism were similar in the inner and outer portions of the cheese. The percentage survival of salmonellae in 6-day-old cheese was less than 0.01%.

Camembert cheese was responsible for an extensive outbreak of illness in Germany with more than 6000 cases involved. According to Bonitz[22] the cheese was infected with *S. bareilly*, and rennet was thought to be the original source of the contaminant. After conducting additional experiments, Bonitz[22] changed his mind and postulated that infection probably entered the cheese via the glue used to fasten labels to the individual cheeses.

4. Other cheese: Occurrence of *Salmonella* in a variety of other ripened cheeses has been reported. Many of these cheeses are not common in the United States, but the information may be useful. In some instances, investigators simply stated that salmonellae were recovered from cheese, without specifying the type of cheese being studied.

Bruhn et al.[28] studied survival of salmonellae in Samsoe cheese. This cheese has a pH value of 5.15 to 5.20 after 24 h and contains 44 to 46% moisture. Samsoe cheese was ripened at 16° to 20°C for 5 to 6 wks, after which it was held at 10° to 12°C for an additional 7 to 10 wks. The authors noted that a 60-day period was necessary to achieve a 10,000-fold reduction in number of viable salmonellae. The death rate was less rapid at 10° to 12°C storage than at 16° to 20°C.

In May of 1944, an unusually high number of typhoid fever cases were reported in four counties of California. Investigation, according to Halverson,[70] showed that the sources of infection were Romano Dolee, Teleme, and high-moisture Jack cheeses, all made from unpasteurized milk. Over 90% of the affected persons had consumed one or several of the cheeses just mentioned. This outbreak provided the impetus for the state of California to pass laws controlling the manufacture and sale of cheese in that state.

Wahby and Roushdy[181] studied the survival of *Salmonella enteritidis, S. typhi,* and *Salmonella paratyphi* B in Domiatti cheese, an Egyptian dairy product. When the cheese was held at 20° to 25°C, *S. enteritidis* survived for 17 d, *S. typhi* for 12 d, and *S. paratyphi* B for 27 d.

Forty samples of Kareish cheese (another Egyptian dairy product) were examined chemically and bacteriologically by Moutsy and Nasr.[121] They observed that the cheese contained from 2.33 to 11.38% salt, 0.75 to 2.7% titratable acidity, 68 million to 6.3 billion bacteria per g, and 1000 to 100,000,000 coliforms per g. One sample yielded *S. typhimurium*.

The behavior of *S. typhimurium* and *S. enteritidis* in Kachkaval cheese (a hard cheese) was studied by Todorov.[173] He added 5 million to 350 million salmonellae per ml of milk, which was then made into Kachkaval cheese. Heating of the curd in a water bath at 71° to 72°C for 70 to 90 sec did not destroy the salmonellae. Their survival in the cheese ranged from 4 to 20 d, depending on the initial level of contamination.

Vizir[180] studied behavior of *Salmonella breslau*, a contaminant of Brynza cheese, in different media fortified with salt. At 12° to 16°C, the organism remained alive in milk with 5, 10 and 15% salt and in broth with 5 to 10% salt during the 45 d of the experiment. Lactic acid at a concentration of 0.9% or higher destroyed the organisms in 5 to 10 days, regardless of temperature. Zagaevskii[193] noted that *S. typhimurium* and *S. dublin* remained viable in Brynza cheese for up to 22 mo.

An outbreak of typhoid fever attributable to a cheese made in a Norwegian home was reported by Hemmes.[75] A man ill with typhoid fever was a part of the household when the cheese in question was made. Studies on the aged cheese revealed that viable *S. typhi* was present after 40 but not 55 d.

C. *Enteropathogenic Escherichia coli*:

Enteropathogenic *Escherichia coli* can be defined as any strain of *E. coli* having the potential to cause diarrheal disease. Enteropathogenic *E. coli* strains (EEC) have been divided into 2 groups, based on the type of disease produced. Those causing a disease with cholera-like symptoms (watery diarrhea leading to dehydration and shock) also produce enterotoxins, and thus are called toxigenic EEC. These strains have been implicated as the cause of "infantile diarrhea" and "traveler's diarrhea."[47,157] Those strains causing a shigella-like illness (diarrhea with stools containing blood and mucus) are called invasive EEC because of their ability to penetrate the epithelial cells of the colonic mucosa. These strains do not produce an enterotoxin. Invasive EEC are associated with dysentery-like disease in people of all ages.

1. Incidence of *E. coli* and coliforms in cheese: Presence of coliforms in cheese has been studied for over 90 years. Early investigations were concerned with prevention of gassy defects in curd and cheese caused by coliform bacteria.[72,104,156] Since coliforms must reach numbers close to 10^7/g to cause gassiness in Cheddar cheese, cheese of normal appearance can still have substantial number of *E. coli* present.[191] Even with pasteurization, post-pasteurization contamination of milk with coliforms can be great enough to cause cheese to become gassy.[49] Yale and Marquardt[191] found that Cheddar cheese of high quality could contain up to 57,000 coliforms/g in the curd.

A survey of coliform bacteria in Canadian pasteurized dairy products by Jones et al.[80] showed that 18.7% of the coliforms isolated

3.4 Cheese

from these products were of intestinal origin. Three serotypes or 2% of the *E. coli* isolates were enteropathogenic. Lightbody[100a] found that 97% of Queensland Cheddar cheese contained coliforms after 2 to 3 weeks of aging. *E. coli* biotype 1 was found in 70% of the samples. Cheese samples with more than 10^6 coliforms/g were of poor quality. Some high grade cheese had more than 1000 coliforms/g. After further studies of Queensland Cheddar cheese, Dommet[44] reported that improper pasteurization of milk, unsanitary equipment, and contaminated starter cultures were all responsible for coliform contamination of the cheese. In recent surveys of Canadian cheese varieties, Elliott and Millard[48] noted that 15% of retail cheese samples contained over 1500 coliforms/g, and Collins-Thompson et al.[37] found 18.1% of soft cheeses and 13.6% of semisoft cheeses exceeded 1600 total coli-forms/g.

Recently Frank and Marth[56] examined 106 samples of commerical cheese for presence of fecal coliforms and EEC. Included in their survey were samples of Camembert, Brie, brick, Muenster, and Colby cheese. Of the samples tested, 58% contained fewer than 100 fecal coliforms/g, but 17% contained more than 10,000/g. No EEC serotypes were found in any of the cheese. A similar survey by Glatz and Brudvig[64] also demonstrated the absence of EEC from commercial cheese.

2. Inhibition of *E. coli* by lactic acid bacteria: Frank and Marth[54,55] examined the effects of lactic acid bacteria on *E. coli* when both organisms were in skim milk. With no lactic acid bacteria present, the generation times of pathogenic and nonpathogenic strains of *E. coli* ranged from 28 to 35 min when incubation was at 32°C and from 66 to 109 min at 21°C. Addition of 0.25 or 2.0% of a commerical starter together with *E. coli* served to completely inhibit growth of *E. coli* in 6–9 h of incubation at 32°C. At 21°C, *E. coli* often had difficulty initiating growth in the presence of the lactic acid bacteria. *S. cremoris* and *S. lactis* were equally inhibitory to *E. coli* at 32°C. At 21°C, *S. cremoris* was more inhibitory to *E. coli* than was *S. lactis*, but a commercial mixed-strain lactic starter culture was more inhibitory than was either of the pure cultures.

3. Behavior of EEC in cheese: The first well-documented outbreak of foodborne illness caused by *E. coli* in the United States occurred between October 30th and December 10th, of 1971, and involved at least 387 persons in 107 episodes of gastroenteritis.[103] The food implicated in these episodes was Camembert cheese imported from a single manufacturer in France and distributed in the United States between September 26th and November 14th, 1971. Episodes were scattered across the country. The attack rate in these cases was 94% for exposed persons. *E. coli* 0124 was isolated from 14 of 49 fecal

specimens obtained from patients involved in the outbreak. No other enteric pathogen was identified in the stools. Seventy-one cheese samples from 13 production days bracketing the implicated days of manufacture were tested and 25 yielded *E. coli* 0124 and two *E. coli* 0125. All reported cases of the illness were traced to cheese manufactured on two production days. The 45 samples obtained for these two production days yielded 23 *E. coli* 0124 isolates and one *E. coli* 0125 isolate. Several of these samples contained *Enterobacter hafniae* strains which cross-reacted with *E. coli* 0112:B11 antisera. Counts revealed 10^5 to 10^7 *E. coli* 0124 per gram of cheese. Ostensibly, a malfunctioning filtration system for filtering river water used in cleaning was to blame. However, bacteriological analysis of the river water and stools from workers obtained at the time of this problem were examined and no known enteric pathogens were isolated, although some workers left the plant and could not be found for stool samples to be taken.

The starter cultures, salt, equipment and in-line dairy products were studied at each level of processing. *E. coli* 0124 was found in a curdling tank and in the cheese. Counts on cheese in the ripening process showed 10^2 and 10^3 coliforms per gram. *E. coli* 0124 isolated from patients and cheese was invasive but not enterotoxigenic.

Since cheese containing 10^5 EEC/g has been implicated in causing foodborne illness,[103] some studies have been conducted to determine their survival in the manufacture and storage of different cheeses.[57,58,126,134] Park et al.[134] studied the fate of six strains of enteropathogenic *E. coli* in the manufacture and ripening of Camembert cheese. *E. coli* was inoculated at about 100/mL of pasteurized milk. Growth was slow until after the curd was cut and hooped. Populations in excess of 10^4/g occurred in some cheeses 5 h after cheesemaking began. *E. coli* populations began to decrease after overnight growth. The pH also decreased at this time to about 5.0 or less. (The pH of the cheese averaged 4.65 at this point.) Salting of the cheese and 1 day of ripening at 15.6°C resulted in a further decrease in viable *E. coli*. The decline persisted during the rest of the week (15.6°C) and during storage at 10°C. Between 0 and 9 weeks were required at 10°C to reduce *E. coli* populations to a nondetectable level, depending on the strain used. The strain which persisted longest was *E. coli* 0128:B12, which survived more than 7 wks but less than 9 wks at 10°C. When *S. cremoris* C$_1$ was substituted for a commercial lactic starter culture, *E. coli* populations exceeded 10^4/g at 9 wks of storage at 10°C. When cheese was made from milk with penicillin, *E. coli* populations reached 10^9/g in 24 h and decreased to 10^7/g by 9 wks of storage at 10°C.

Frank et al.[57] also studied survival of different EEC strains during

Camembert manufacture. In these studies, one pathogenic strain (B$_2$C) and two nonpathogenic strains (H-52 and B) survived past 4 wks at levels ranging between 1/g and 1000/g. Hence, these organisms could have been present in the cheese at the time of consumption. Growth and survival curves were nearly parallel when inocula of 10^2, 10^3 and 10^4 *E. coli* B$_2$C/mL were used. When less starter culture was used (0.25% v/v instead of 2.0% v/v), *E. coli* strain B$_2$C increased four log cycles in number compared to two log cycles during normal manufacture. During ripening, numbers of *E. coli* B$_2$C decreased less when less starter was used. Under these conditions the pH was only lowered to 6.4 instead of 5.0 after 6 h of manufacture. However, the cheese had a normal pH thereafter. Because ripening proceeds from the outside to the inside of the cheese wheel, the authors determined survival of *E. coli* at different places in the cheese wheel during ripening. From the bactericidal unripened core outward conditions became more favorable for survival, and growth of *E. coli* was observed at the outer surface. Survival of several enteropathogenic *E. coli* strains was related also to pH. The higher the pH, the longer the strains survived in the cheese.

Recently Rash and Kosikowski[144] examined the behavior of EEC in Camembert cheese made from milk after it was concentrated by ultrafiltration. This work also demonstrated the importance of pH in controlling growth of EEC during cheesemaking and in determining survival of EEC during ripening of the cheese.

Frank et al.[58] also studied survival of three enteropathogenic *E. coli* strains in the manufacture and ripening of brick cheese. Growth of two of these strains was 10 times greater during the initial hours of manufacture than in Camembert manufacture, probably as a result of a higher temperature and less rapid decline in pH during brick cheese manufacture. Inhibition of enteropathogenic *E. coli* occurred at pH 5.2–5.5, as opposed to 5.0–5.2, for the same strains in Camembert cheese manufacture. The strains studied were inactivated faster in Camembert than in brick cheese. However, the pH of unripened Camembert cheese was lower. Little difference in survival was noted when the pH of the brick cheese ranged from 5.15 to 5.3. The pH of most of the brick cheese samples was 5.3 after 2 wks of ripening. After 7 wks (2 weeks of ripening at 15.5°C; 5 weeks at 7°C), *E. coli* populations ranged between 20,000 and 700/g. In this cheese, as well as in Camembert cheese, *E. coli* B$_2$C survived best. (In this instance, highest numbers appeared at 7 weeks.) Growth of *E. coli* on the surface of brick cheese was much less than on the surface of Camembert cheese. Several factors may have caused this observation. A major difference between the surface environment of brick cheese and that of Camembert cheese is the absence of

Penicillium camemberti on brick cheese. Yeasts and micrococci predominate on the surface of brick cheese during the early ripening stages and cause the pH to rise to 5.4–65.5. On Camembert cheese surfaces the pH is much higher. The microflora on brick cheese surfaces may inhibit growth of *E. coli* by competing for nutrients or by producing of metabolites toxic to them.

Recently a study was completed by Kornacki and Marth[94] on survival of EEC and nonpathogenic *E. coli* in Colby-like cheese made from pasteurized whole milk artifically contaminated with 100 to 1000 *E. coli*/mL. Numbers of *E. coli* increased 100- to 1000-fold, depending on the strain, to about 1×10^6/g of curd, in most instances, by the end of the cook (3.5–3.9 h). After this point numbers of *E. coli* declined over a period of several weeks. The survival of EEC in Colby-like cheese differed among the strains tested with enterotoxigenic strain B_2C surviving better than enterotoxigenic strain H 10407, which survived better than enteroinvasive strain 4608. After 11 wks *E. coli* B_2C was <10/g in one cheese stored at $3 \pm 1°C$ and about 3×10^3/g in another. In one instance, a low level (<350/g) of enterotoxigenic *E. coli* H10407 survived 13 wks at $3 \pm 1°C$. However, this organism was usually inactivated between 4 and 6 wks at $3 \pm 1°C$. In all lots, enteroinvasive *E. coli* 4608 was inactivated between 2 and 4 wks. When one compares these results to those of Frank et al.,[57] it appears that Camembert cheese was more inhibitory to EEC than was Colby or brick. For instance, when *E. coli* B_2C was added to milk at about 100/mL, it survived better in Colby than in Camembert. Furthermore, numbers of *E. coli* B_2C in Colby-like cheese after 7 weeks were comparable to those in brick cheese at about the same time (20,000/g). Frank et al.[58] suggested that the lower pH of unripened Camembert cheese accounted for the poorer survival of EEC in Camembert than in brick cheese. This is also likely to be true for Colby-like cheese.

Storage of Colby-like cheese at $3 \pm 1°C$ or $10 \pm 1°C$ had little effect on survival of EEC in this study. The pH of the cheese appeared to have the most important role in survival of *E. coli* in Colby-like cheese. Washing the curds more than 20 min resulted in cheese with high pH values and high coliform counts (1×10^8/g) which persisted at high levels for many weeks. The results of this study emphasized the importance of pH regulation, proper sanitation, and proper manufacturing procedures in the manufacture of Colby-like cheeses, and suggest that intermediate moisture cheeses like Colby could be potential vehicles for the transmission of EEC under certain conditions.

It is evident that a variety of factors may affect survival of coliforms in cheese. These include temperature of storage,[162] initial

3.4 Cheese

levels of coliforms in milk and cheese,[44,50,57,191] length of storage,[50,57,58,94,134] stage of manufacture,[94,134,191] amount of and type of starter culture,[57,134,191] *E. coli* strain differences,[57,58,94,134] pH,[57,58,94] salting,[134] washing,[94] presence of penicillin in the milk supply,[134] location in the cheese[57] and type of cheese.[58,64,94] It is therefore necessary to use care in the manufacture of cheese to promote conditions which diminish the possibility of having large numbers of potentially pathogenic *E. coli* present in the cheese. Additional information on EEC can be found in a recent review by Kornacki and Marth.[95]

D. Botulism:

This serious and often fatal disease results after ingestion of a toxin produced by the anaerobic spore-forming *Clostridium botulinum*. In the United States, there have not been any known cases of botulism attributable to consumption of natural or process cheese.[96] One outbreak of botulism in the United States in 1951 and another in Argentina in 1974 were associated with consumption of process cheese spread. This product also has an excellent safety record since there have been no known problems caused by it in the US in more than 30 y. Process cheese spread can be meet legal standards if it has as much as 60% water and has a pH value of up to 5.2. At these limits, the product has less inhibitory potential than if the amount of water present and the pH were reduced.

Under the Good Manufacturing Practices for canned foods in the United States,[14] process cheese spreads would be classified as a low-acid food requiring a heat process sufficient to destroy spores of *C. botulinum* unless it can be demonstrated that the spores cannot grow in the cheese spread. Since the heat treatment currently used is insufficient to destroy spores of *C. botulinum*, the situation just described has prompted renewed interest in the behavior of *C. botulinum* in cheese spreads.

Kautter et al.[81] inoculated jars of several varieties of process cheese spread (pH 5.05–6.32, water activity 0.930–0.953) with 24,000 spores of *C. botulinum* per jar. One variety, cheese-with-bacon, also received 460 spores per jar. When stored at 35°C, 46 of 50 jars of Limburger and 48 of 50 jars of cheese-with-bacon spread became toxic after 83 and 50 d, respectively. One jar of cheese-with-bacon spread that received the smaller inoculum became toxic during 6 mo of storage at 35°C. No jars of three other varieties of cheese spread (Cheez Whiz, Old English, and Roka Blue) became toxic when subjected to the same treatment.

Another series of experiments was done by Tanaka et al.[167] in which spores of *C. botulinum* (1000/g) were added to process cheese spread when it was manufactured. Moisture content and pH of various batches ranged from 49.7 to 59.2% and 5.80 to 6.28, respectively. Cheeses were incubated at 30°C for up to 48 wks. No samples became toxic under these conditions when the product contained 52 or 54% moisture and was made with sodium phosphate as the emulsifier. Use of sodium citrate as the emulsifier resulted in a product that remained toxin-free at 52% moisture and one that became toxic at 54% moisture. Products with 58% moisture became gassy and toxic regardless of the emulsifier that was used.

E. *Aflatoxin:*
   1. Aflatoxin M_1 in natural and process cheese: Occurence of AFM_1 in milk and other products suggests that the toxin would also be found in cheeses. Although this is true, there seem to be fewer incidents reported than with milk. From September 1972 until December 1974, 80 samples of fresh cheese, 77 of hard cheese, 65 of Camembert and 134 of process cheese were tested in West Germany by Polzhofer[138] for the presence of AFM_1. Hard cheese had the most AFM_1 (1.30 μg/kg of cheese), whereas all types of cheese which were positive for AFM_1 had at least 0.1 μg/kg.

   A similar survey was done in West Germany in the summer of 1976.[92] In this survey, 197 commercial cheese samples were analyzed for AFM_1. Of the samples tested, 136 contained AFM_1 and the average concentration was 0.09 μg per kg of cheese. Results of this survey also showed that soft cheeses had appreciably less AFM_1, both qualitatively and quantitatively, than did other types of cheese.

   In addition to knowing that AFM_1 occurs in cheese, it is also valuable to know how the concentration of the toxin differs from that of the milk from which the cheese was made. There is also the question of what effects processing of the cheese might have on AFM_1 concentration.

   The possibility that AFM_1 somehow binds or has an affinity for casein[24] suggests that there is likely to be an enrichment of toxin in the final curd over that found in milk. Kiermeier and Buchner[87] demonstrated this when naturally contaminated milk was used to make uncured, Tilsit and Camembert cheeses. The AFM_1 content was 3.2-fold higher in the uncured, 3.7-fold higher in the Tilsit and 3.7-fold higher in Camembert cheese, as compared to the milk from which the cheeses were made. Similar observations were made for other varieties of cheese by Brackett and Marth.[25,26,27]

   AFM_1 is quite stable for at least 3 mo in cheese made from naturally contaminated milk.[87] Kiermeier and Buchner[87] noted that AFM_1 concentrations rose 24 to 28% in uncured cheese, 45 to 54% in Camembert and 23% in Tilsit cheese during the first 2 wks after manufacture, and then decreased 4 to 12%, 25% and 25%, respectively.

   The AFM_1 concentration appeared to be greater in the rind of Camembert cheese than in the center, but was actually about the same when expressed on the basis of dry weight of cheese. In addition, washing the curd decreased the AFM_1 content by 22%.

3.4 Cheese

Distribution of AFM_1 between whey and the final cheese varies, depending on the type of cheese being made.[87] Binding of AFM_1 to milk proteins is suggested as a reason for more toxin going into the curd than into the whey. However, the observation may also be related to analytical techniques that were used. McKinney and Cavanagh[110] showed that 75% of AFM_1 occurred in the curd and 25% occurred in whey when one extraction method was used for analysis. In contrast, 39.6% of the toxin was found in the same curd and 60.4% of the toxin was found in the same whey when a different extraction method was used. Thus one should not be surprised to see variability in results reported in the literature. Obviously, more work on binding of AFM_1 to milk proteins and use of optimal analytical techniques needs to be done before one can be sure of what is happening in a product.

There is little information in the literature regarding the effect of temperature of AFM_1 in cheese or cheese products. However, one laboratory reported that heat treatment of uncured cheese (5 min at 80°C) did not reduce its AFM_1 content.[91] Polzhofer[139] obtained similar results when processed cheese was produced from artificially contaminated Emmentaler (Swiss) cheese. Heating times of 3 or 30 min at 90°C did not destroy AFM_1. These results are in contrast to those of Purchase et al.[141] in which pasteurization temperatures (62°C/30 min, 72°C/45 s, 80°C/45 s) decreased the AFM_1 content from 36 to 68%. Kiermeier and Buchner[87] found a loss of AFM_1 of about 12% at 28 and 35°C. Various milk clotting agents apparently have little effect on the AFM_1 content or on its distribution in cheese.[87] Neither increasing amounts of rennet nor processing of curd with organic acids changed the distribution of AFM_1 in curds when compared to results obtained with a constant amount of rennet at a constant temperature. However, changing the renneting temperature while using a constant amount of rennet decreased the percentage of AFM_1 in the curd. Increasing the temperature when using starter cultures also lead to a decrease of toxin in the curd.

2. Aflatoxins B_1 and G_1, in natural and processed cheese: Lie and Marth[100] were among the earliest investigators to show that aflatoxigenic molds can produce aflatoxins in cheese. Aflatoxin contents of inoculated cheese ranged from 4.8 to 49,600 µg of AFB_1 or aflatoxin G, (AFG_1) per kg of cheese, depending on where the sample was taken and what strain of mold had grown on the cheese. Later others found that contaminating brick[161] or Tilsit[53] cheese with aflatoxigenic fungi resulted in only about 4% of the amount of aflatoxin found by Lie and Marth in Cheddar cheese. Production of AFB_1 and AFG_1 can also occur in Emmentaler[154] but apparently not in Camembert or Romadur cheese.[88]

Several surveys have been conducted to determine if AFB_1 could be found in commercial cheese samples and to quantify the toxin in positive samples. One survey[139] showed that no AFB_1 was found in 169 samples of 19 varieties of cheese. However, another survey[86] of the same 19 varieties revealed that 34 of 222 samples were positive for aflatoxin. Corbion and Fremy[38] analyzed 100 samples of Camembert cheese from 40 factories in Northern France and found no AFB_1. In addition to natural cheese, AFB_1 can also occur in process cheese; Kiermeier and Rumpf[90] examined 115 samples of process cheese and found two positive samples containing AFB_1 or AFG_1.

F. Amines:

Ingestion of foods containing tyramine together with drugs of the monoamine oxidase inhibitor (MAOI) type can lead to hypertension attacks in some persons. Drugs of the MAOI type include iproniazid, nialamide, tranylcypromine, tranylcypromine sulfate, isocarboxazid and phenelzine sulfate. Some of these continue to be used as antidepressants.[146]

Tyramine acts by releasing norepinephrine from tissue stores, which in turn causes an increase in blood pressure.[146] Tyramine has $\frac{1}{20}$ to $\frac{1}{50}$ of the ability of epinephrine to increase blood pressure. MAOI-type drugs increase the tissue stores of norepinephrine and thus potentiate the action of tyramine. Symptoms of a hypertensive crisis prompted by tyramine include high blood pressure, headache, fever and sometimes perspiration and vomiting.[146]

Histamine has been implicated in several outbreaks of food poisoning, including at least one that resulted from eating 2-year-old Gouda cheese.[43] Symptoms of histamine poisoning include nausea, vomiting, facial flushing, intense headache, epigastric pain, burning sensation in the throat, dysphagia, thirst, swelling of the lips and urticaria.[32]

Tyramine and histamine are commonly found in fermented foods, including cheese. Some lactic acid bacteria, as well as other bacteria likely to be in cheese, possess the enzymes needed to decarboxylate tyrosine or histidine and form tyramine and histamine, respectively.[146] This could account for the presence of these compounds in cheese. Typical values for some cheeses are: histamine ($\mu g/g$) and tryramine ($\mu g/g$) 0–2300 and 27–1100 for blue, 0–480 and 20–2000 for Camembert, 0–1300 and 0–1500 for Cheddar, 0–500 and 100–560 for Colby, 0 and 0–410 for mozzarella, 0–58 and 4–290 for Parmesan, and 0 and 0–1800 for Swiss.

3.5 Cultured Milks

Cultured milks (e.g. yogurt, kefir, cultured buttermilk) are made from skim milk, partially skimmed milk or whole milk. Nonfat dry milk is commonly added to milk used for making yogurt. The type of milk chosen, with or without added nonfat dry milk, is commonly heated at 82 to 84°C

3.5 Cultured Milks

for 30 min, both to pasteurize the milk and to insure that the desired body will develop in the fermented product. Cooled milk is inoculated (1 to 2.5% inoculum) with the appropriate lactic acid bacteria (commonly used are *S. cremoris, S. lactis, Leuconostoc cremoris, S. thermophilus* or *L. bulgaricus*, depending on the product being made), incubated (21 to 48°C) until the desired amount (usually ca. 0.85%) of lactic acid is produced. The coagulum may or may not be broken, the product is cooled and then is ready to enter marketing channels. Fruit and cereals may be added to yogurt before or after the fermentation, depending on the type of yogurt that is being produced.

The rather extreme pasteurization given milk used for these products, addition of an abundance of lactic acid bacteria in the form of starter culture, the low pH (ca. 3.9 to 4.6) of the finished product, and refrigerated storage of the finished product all serve to control foodborne pathogens in cultured milks. As a consequence, these foods rarely, if ever, are associated with outbreaks of foodborne illness.

That salmonellae do poorly in milk that also contains lactic acid bacteria was demonstrated by Park and Marth.[130] They inoculated skim milk with *S. typhimurium* (ca. 10^3/mL) and with *S. cremoris, S. lactis*, mixtures of *S. cremoris* and *S. lactis, Streptococcus diacetilactis, S. thermophilus, L. bulgaricus*, mixtures of *S. thermophilus* and *L. bulgaricus*, a mixture of *Lactobacillus helveticus* and *S. thermophilus*, or *L. cremoris*. Inocula of lactic acid bacteria ranged from 0.25 to 5.0% and incubation temperatures from 21 to 42°C. *S. cremoris, S. lactis* and mixtures of the two repressed growth but did not inactivate *S. typhimurium* during 18 h of incubation at 21° or 30°C when the lactic inoculum was 0.25%. An increase in inoculum to 1% resulted in inactivation of *S. typhimurium* at 30°C by some of the mixed cultures. Both *S. diacetilactis* and *S. cremoris* were less inhibitory to *S. typhimurium* than were *S. cremoris* or *S. lactis*. When added at the 1% level, *S. thermophilus* was more detrimental to *S. typhimurium* at 42°C than was *L. bulgaricus*. Mixtures of these two lactic acid bacteria, when added at levels of 1.0 or 5.0%, caused virtually complete inactivation of *S. typhimurium* during the interval between 8 and 18 h of incubation at 42°C.

According to data by Frank and Marth,[54,55] growth of enteropathogenic strains of *E. coli* in milk also is inhibited by lactic acid bacteria. They added ca. 10^3 *E. coli*/mL of milk, then inoculated the milk with 0.25 or 2.0% lactic starter culture and incubated the mixture at 21° or 32°C for 15 h. With no lactic acid bacteria present, generation times for several strains of *E. coli* ranged from 28 to 35 min at 32°C and from 66 to 109 min at 21°C. At 32°C, after 6 to 9 h of incubation and a 1- to 3-log increase in numbers, *E. coli* was completely inhibited by both concentrations of starter culture. Complete inhibition of *E. coli* occurred earlier at 32° than at 21°C, but smaller numbers of *E. coli* were obtained at 21°C; some strains virtually did not grow at this temperature in the presence of lactic acid bacteria.

If pathogens enter cultured milks, growth does not occur and survival often is of a short duration. Park and Marth[132] made fermented milks that contained *S. typhimurium* and then refrigerated the fermented milks. Survival of *S. typhimurium* varied somewhat depending on conditions under which the milks were fermented, but often the pathogen was not detectable after 3 to 6 days of storage. Frank and Marth[54,55] did similar work with enteropathogenic strains of *E. coli*. Again, there was no growth of the pathogens in the fermented milks during refrigerated storage, but *E. coli* survived longer than *S. typhimurium*.

Adding *E. coli* (pathogenicity of strains tested is unknown) directly to a finished cultured milk seems to be more detrimental to the bacterium than is cultured milk when the bacterium was in it during the entire fermentation. Goel et al.[65] added *E. coli* to yogurt (pH 3.65 to 4.35) and the bacterium could not be recovered from 4, 5, 6, and 8 of 10 samples after 1, 2, 3 and 4 days, respectively, of refrigerated storage. A similar trial was made with cultured buttermilk and 10 days of refrigerated storage were needed before 6 of 10 samples were free of detectable *E. coli*. The demise of *E. coli* in cultured sour cream was similar to that in cultured buttermilk.

Minor and Marth[114] added *S. aureus* (ca. 10^2 or 10^5/g) to commercially produced cultured buttermilk, yogurt, and sour cream and then held the products at 7 and 23°C. Generally, products with ca. 10^2 *S. aureus*/g were free of the bacterium, or nearly so, within 24 h regardless of storage temperature. When ca. 10^5 *S. aureus*/g were added, yogurt samples were free of viable staphylococci in 2 to 4 d, sour cream in 4 to 7 d and cultured buttermilk in 3 to 5 d of storage, regardless of storage temperature.

If cultured milks are made from milk that contains AFM_1, it is unlikely that the fermentation or subsequent storage of the product will act to eliminate or even appreciably reduce the amount of aflatoxin in the product. This was demonstrated by Wiseman and Marth[186] for yogurt, cultured buttermilk and kefir. Applebaum and Marth[16] found that AFM_1 was stable during the manufacture and subsequent storage of cottage cheese.

3.6 Butter

Butter, when made from adequately pasteurized cream, is rarely associated with outbreaks of foodborne illness. However, in 1970, 24 people became ill after consuming whipped butter made by blending approximately 113 g of milk with 2.7 kg of butter. Both staphylococci and enterotoxin A were recovered from the butter but not from the milk used to prepare the whipped butter.[124] During the same year a major food company voluntarily recalled 33,750 kg of butter which contained up to 100 million staphylococci/g and was suspected of causing illness in several consumers.[155] Several additional outbreaks of staphylococcal food poisoning associated with butter have occurred more recently.

3.7 Nonfat Dried Milk

The initial incidents prompted Minor and Marth[117] to study the fate of *S. aureus* in cream and butter. *S. aureus* was added to unsalted, salted, and whipped butters which were then stored. When *S. aureus* was added to regular butter with 1.5% sodium choloride, numbers of the bacterium decreased markedly during storage for 60 d at 10°C. Changes in numbers were less dramatic when storage was at 23°C for 14 d. These observations prompted additional studies with butters containing different amounts of salt. At 23°C, staphylococci grew in butters with no salt or with 1% salt and decreased in numbers when butter contained 2.4% salt. Whipping of butter containing 1.5% salt made it more suitable for survival of staphylococci than was true of the unwhipped product.

Minor and Marth[117] also inoculated cream with *S. aureus*, incubated the cream, and then churned it into butter. Their data indicate that cream (about 40% fat) supported both growth of and enterotoxin production by staphylococci. It is also evident that some of the enterotoxin in cream appeared in butter made from that cream. The largest proportion of the enterotoxin, however, appeared in the buttermilk. Hence, buttermilk could be more toxic than the butter, and could cause illness even when butter made from the same lot of cream would contain insufficient enterotoxin to produce food poisoning. This suggests that producers of dried buttermilk, which sometimes serves as an ingredient in other foods, must be careful to insure that the product is safe. It is evident from the observations of Minor and Marth[117] that butter is likely to be a problem only if the cream (or milk) from which the butter is made is abused sufficiently to permit staphylococcal growth and enterotoxin production. Production of enterotoxin in cream and appearance of enterotoxin in the various fractions resulting from churning of butter also were demonstrated by Maercklin.[102]

When butter was made from cream that contained AFM_1, only about 4 to 15% of the aflatoxin in cream appeared in butter, according to data by Wiseman and Marth.[187] Most of the AFM_1 (61 to 87%) in the cream appeared in the buttermilk with 3 to 26% also present in the water used to rinse the butter.

3.7 Nonfat Dried Milk

Mistakes made in the manufacture of nonfat dried milk caused some noteworthy outbreaks of staphylococcal food poisoning and salmonellosis. Although the most recent of these problems, the outbreak of salmonellosis, occurred nearly 20 y ago, it is well to briefly review the circumstances under which the problems developed so the mistakes need not be repeated.

A. *Staphylococcal poisoning:*

Eight outbreaks of staphylococcal food poisoning which involved approximately 1,200 children occurred in England in 1953, and were caused by toxic nonfat dry milk.[4] Another series of outbreaks occurred several years later in Puerto Rico when toxic nonfat dry milk was used in a school lunch program.[19] These outbreaks probably resulted because staphylococci grew in milk after it was preheated and before it was dried.

Undoubtedly these incidents prompted George et al.[61] to study some of the conditions associated with production of nonfat dry milk which could lead to problems. They observed that the conventionally used pre-condensing heat treatment of 85°C for 30 min was not detrimental to subsequent growth of *S. aureus* in condensed skim milk which contained 40% solids. Furthermore, *S. aureus* could grow in condensed skim milk at temperatures between 32 and 45°C, although one strain of *S. aureus* also grew in the condensed skim milk at 46.7 and 47.8°C. Finally, vacuum as found in a condensing pan, was only minimally inhibitory to growth of *S. aureus*.

George et al.[61] concluded that to prevent further outbreaks of staphylococcal food poisoning, milk drying factories must operate so that: (a) underpasteurized milk does not reach or leave the condensing apparatus, (b) condensed milk is not held for an extended period at temperatures which permit staphylococcal growth, and (c) condensed milk is protected from direct contamination by staphylococci.

Outbreaks of staphylococcal food poisoning have not been associated with nonfat dry milk in recent years. It appears that the precautions recommended by George et al.,[61] and possibly others practiced by the industry, serve to essentially eliminate nonfat dry milk as a cause of this form of foodborne illness. This is verified by the report of Meyer.[112] He examined 170 samples of dried milk products commercially available in the Federal Republic of Germany and found 0.04–0.1 µg staphylococcal enterotoxin/g in three samples. Two of the three "positive" samples were feed-grade products and the other was roller-dried skim milk.

B. *Salmonellosis:*

Contamination of instant nonfat dried milk made newspaper headlines in the U.S. during 1966 and 1967. This, however, was not the first time these organisms were reported as present in dried milks. The British Ministry of Health[113] reported that 12 cases of salmonellosis in 1950 were associated with consumption of dried milk contaminated with *S. typhimurium*. In 1954, according to Cockburn and Vernon,[36] another outbreak of salmonellosis resulting from *S. typhimurium* was caused by contaminated dried milk. Similar outbreaks in Bulgaria during 1945–1950 were recorded by Iordanov et al.[78]

The United States experience with salmonellae in dried milk began in January, 1966, when the Division of Epidemiology of the Michigan State Department of Public Health investigated two cases of gastroenteritis resulting from *Salmonella new-brunswick*.[8,9] The cases occurred in males less than 6 months old, residing in different areas of the state; each had received a formula made from instant nonfat dry milk. Following this outbreak, it was noted that *S. new-brunswick* had been reported only rarely

3.7 Nonfat Dried Milk

in the United States between 1947 and 1964 (0.02% of all isolations). Between April, 1965, and January, 1966, there were 29 reported isolations of this serotype (as compared to 13 from 1947 to 1964) from humans, a situation which suggested a common source of infection.

As a consequence, state health departments that reported isolation of this organism were asked to submit epidemiologic information about the cases. Of the 29 persons from whom *S. new-brunswick* was isolated, 25 were available for further studies. All of these persons had symptoms characteristic of salmonellosis, but the cases were distributed throughout the United States. Although dietary histories were impossible to obtain, it was determined that dried milk was the only item consumed with greater frequency than was expected. Twenty of the 25 persons had ingested this food within 30 days of their illness.

After dried milk was implicated, bacteriological examinations were made on hundreds of shelf samples of many brands of dried milk. The same rare serotype, *S. new-brunswick,* was subsequently isolated from many samples of instant nonfat dry milk produced by a single factory in the midwestern United States. The organism was isolated within the factory and from other milk products on the premises. Schroeder[158] detailed conditions in the factory where the contaminated powder was produced. The plant handled about 11 million lb of milk per day, most of which was instantized. Raw whole milk was obtained from about 250 farms and raw skimmilk came from four dairies. During processing, the skimmilk passed through a plate heater, where it was supposed to attain a temperature of 72.2°C and be held at that temperature for 10 to 15 sec. Thermostatic controls, a flow diversion valve, and time controls were lacking and, hence, there was no assurance of adequate pasteurization. No heat treatments later in the process were sufficient to destroy salmonellae, if they were present. After heating, milk was concentrated in a series of three vacuum pan evaporators at temperatures which may have favored growth of microorganisms present. The concentrated milk was spray dried and, with the aid of a pneumatic system, transported to a sifting apparatus where coarse aggregates were removed. Dried milk was then subjected to the instantizing or agglomerating process in which moisture in the form of steam is added to the powder. The moistened powder was dried and larger particles obtained as a result of this treatment.

Although much of the equipment used to handle the milk was easily cleaned, inspection of the plant revealed: (a) a baffle plate inside the dryer and a trough and screw used to remove the powder had open seams and crevices, (b) the instantizing system had several welded joints with rough surfaces, numerous crevices and open seams, together with a series of rods and dividers in the after-dryer, all of which permitted accumulation of caked material, (c) the spray dryer was dry-cleaned (with brushes) daily and wet-cleaned monthly, (d) the agglomerator was wet-cleaned weekly but product deposits accumulated in much less than a week, and (e) air intakes for the dryer and after-dryer were located in the same room with sifting and bagging operations, where the atmosphere contained a high level of dried milk solids.

Schroeder's[158] study on the occurrence of *Salmonella* in dried milk led him to the following conclusions: (a) since a large number of producers supply a processing plant, it is difficult to identify the original source of infection, but it is important to guard against such contamination, (b) temperature requirements for pasteurization should be followed carefully, (c) equipment used to dry and to instantize milk is very difficult to clean and, hence, once salmonellae are introduced into a plant great efforts are required to eradicate them, and (d) dried milk is distributed extensively at the wholesale level and hence, contaminated product can receive widespread distribution.

The product from this plant was recalled from the market in April, 1966, and a careful cleanup and remodeling of the plant were instituted. Additional recalls of instant dry milk produced at other locations occurred during 1966 and 1967.

A large-scale testing program was instituted by the USDA during 1966. During the initial phase of the program, April through August, 1966, 2,741 samples from 156 plants in 23 states were tested. Thirty-four samples of nonfat dried milk and 27 environmental samples were found to contain salmonellae. Dried milks contaminated with salmonellae were recovered from factories in Idaho, Iowa, Minnesota, South Dakota, Washington and Wisconsin. Additionally, environmental samples from plants in the following states were found to contain salmonellae: Iowa, Minnesota, Missouri, North Dakota, Oklahoma, South Dakota, Vermont, Washington and Wisconsin[8,9]

Salmonellae recovered from the dried products include: *S. montevideo, S. oranienburg, S. heidelberg, S. tennessee, S. senftenberg, Salmonella alachua, S. cubana,* and *Salmonella orion*[8,9,10,11] A lactose-fermenting strain of *S. newington* also was isolated from instant nonfat milk powder after the product had given rise to nine cases of salmonellosis.

Tests on environmental samples from dried milk plants revealed the presence of the following serotypes: *Salmonella schwarzengrund, S. tennessee, S. newport, S. cubana, S. typhimurium, S. anatum, Salmonella orthington, Salmonella kentucky, S. infantis, S. oranienburg, S. heidelburg* and *S. corion.*[8,9]

Meister[9] reported results obtained in the USDA survey during the first six months of 1967. Product and environmental samples were taken from approximately 200 dry milk samples in 19 states. Results indicated that 34 (1%) of 3,315 product samples and 121 (8.2%) of 1,475 environmental samples were contaminated with salmonellae.

Final survey data[10,11,12] showed that 0.2% of all products tested contained salmonellae and these were manufactured in 13% of the factories.

According to Severs et al.,[160] the Newfoundland Public Health Labora-

3.7 Nonfat Dried Milk

tories reported 20 isolations of *S. newport* from stool specimens examined during February 1 to March 20, 1968. The 20 patients, who came from 17 different households in the St. John's district, gave a history of diarrhea and vomiting; five preschool children and two adults required hospitalization. At least 12 other persons in the families of confirmed cases of *S. newport* infection gave a history of diarrhea and vomiting. All the families had purchased and consumed a particular brand of nonfat dried milk. Opened 3-lb boxes of product, all with the same batch number, were obtained from three of the homes and contained *S. newport*. Dried milk from eight unopened 3-lb boxes with the same batch numbers as those taken from homes of infected persons were obtained in St. John's supermarkets and also contained *S. newport*.

C. *Aflatoxin:*

Purchase et al.[141] studied the effect of concentration (before drying) on AFM_1 in naturally contaminated milk. Milk was reduced to half its volume using reduced pressure (ca. 45 mm Hg abs.) at 40°C. During this processing the content of AFM_1 was reduced by 64%.

Occasionally dried milk has been found to contain AFM_1.[125,136,183] Milk has been dried by several methods to test the effect that these might have on stability of AFM_1. The earliest report on roller drying showed it caused no change in toxicity of AFM_1.[1] However, the duckling bio-assay used in this study is not senitive to small changes in aflatoxin content.[140] Purchase et al.[141] studied roller drying under two conditions. At reduced pressure and 40°C, 61% of the AFM_1 was lost. With steam heating, 75% of the AFM_1 was lost. Some of this degradation occurred during the concentration process. When they spray-dried the milk, a reduction of 86% occurred. During freeze-drying experiments with naturally and artificially contaminated milks, Kiermeier and Mashaley[89] found that AFM_1 decreased up to 12% over a 15-day storage period.

Wiseman and Marth[187] made nonfat dried milk by freeze-drying skimmilk that contained AFM_1. There was no apparent loss of aflatoxin through freeze-drying or during 4 months of storage of the dried product at 22°C. Similar observations were made when buttermilk containing AFM_1 was dried and stored.

Aflatoxins can be produced through growth of mold on milk powders that are moistened or on concentrated milk products. Kiermeier and associates[84,91] found aflatoxin in artifically inoculated and moistened milk powders. The amounts ranged from 0.06 to 19.2 mg/100 g of milk powder. The amounts depended on moisture content of the substrate and strain of mold that was used. The pH of the substrate also influenced the concentrations of aflatoxins produced.[85] This may be of significance because Paul et al.,[136] in India, found that indigenous milk products with the consistency of pastes were contaminated with aflatoxins. Hence there may be concern in areas where these products are consumed.

3.8 Frozen Desserts

Commercial ice cream and related products are commonly made from a "mix" (a mixture of several ingredients) that is pasteurized. Eggs, if they are used at all, are also pasteurized. Certain ingredients (e.g. fruits and nuts) are not a part of the pasteurized "mix", and so could serve as a source of contamination unless care is exercised in the way they are prepared and in how they are used. In the last several decades ice cream has been associated with some outbreaks of salmonellosis and so further comments about this problem will follow. The fate of AFM_1 in frozen desserts also will be mentioned.

Wundt and Voss[190] studied an outbreak of gatroenteritis caused by commercial ice cream containing 10,000 to 100,000 cells of *S. typhimurium* and *Salmonella bareilly* per milliliter. These authors also inoculated vanilla, strawberry, and lemon ice creams with *S. typhimurium* and then subjected them to heat treatments of 60 to 80°C. Ten to 15 min of exposure at the lowest temperature destroyed the organisms in all types of ice cream, but from 20 to 60 min of additional exposure (depending on the type of container) was needed before the desired temperature was attained.

During 1967, 14 outbreaks of salmonellosis occurred in New York, New Jersey, Connecticut, and Maine, and were attributable to a kosher imitation ice cream.[13] Although the imitation ice cream did not contain dairy ingredients, it its the type of product which could easily be manufactured in a dairy plant. Each outbreak occurred 1 to 3 d after a catered banquet was held at which kosher food was served. Of approximately 3,300 persons who attended the banquets, an estimated 1,800 (54%) developed diarrhea, abdominal pain or cramps, fever, chills, headache, nausea, and vomiting. The median duration of the illness was 3 d, and there were no deaths. Examination of stool specimens from 12 persons who represented six of the banquets revealed *S. typhimurium*. Product left over from the banquets also yielded *S. typhimurium*. Studies were then done at the plant which manufactured the ice cream. All environmental samples from the plant and employee stool cultures were free of salmonellae. Ingredients used in the product were also examined and all were free of salmonellae except the frozen egg yolks, which yielded *S. typhimurium*. The frozen egg yolks were unpasteurized and were produced at a New York City egg-breaking plant. Environmental samples taken at the egg-breaking plant revealed several serotypes, including *S. typhimurium*.

As a consequence of this incident, action was taken by the New York City Health Department which resulted in the temporary suspension of operations by the frozen dessert and egg-breaking plants. All imitation ice cream manufacturers in the city have been placed under the milk code, which will require production of a certified pathogen-free product. The incriminated egg-breaking plant agreed to pasteurize its raw egg products. Joint action by the city health department, state officials, and U.S. Food and Drug Administration officials resulted in recovery or destruction of several thousand servings of contaminated product.

Wiseman and Marth[188] made ice cream and orange sherbet from milk (and other suitable ingredients) that was naturally contaminated with AFM_1. The aflatoxin concentration remained unchanged in the ice cream during 8 mo of frozen storage. Results obtained with sherbet were somewhat variable but here too aflatoxin was in the product that had been held frozen for 8 mo.

3.9 References

1. ALLCROFT, R.; CARNAGHAN, R.B.A. Groundnut toxicity: An examination for toxin in human food products from animals fed toxic groundnut meal. Vet. Rec. **75**:259–263; 1963.
2. ALLCROFT, R.; ROBERTS, B.A. Toxic groundnut meal - the relationship between aflatoxin B_1 intake by cows and excretion of aflatoxin M_1 in milk. Vet. Rec. **82**:116–118; 1968.
3. ALLEN, V.D.; STOVALL, W.D. Laboratory aspects of staphylococcal food poisoning from Colby cheese. J. Milk Food Technol. **23**:271–274; 1960.
4. ANDERSON, P.H.R.; STONE, D.M. Staphylococcal food poisonings associated with spray-dried milk. J. Hyg. **53**:387-397; 1955.
5. ANGUS, K.W.; BARR, W.M. An outbreak of *Salmonella typhimurim* infection in a dairy herd. Vet. Rec. **75**:731; 1963.
6. ANONYMOUS. Food poisoning. Lancet **253**:522; 1947.
7. ANONYMOUS. Milk facts. Washington, D.C.: Milk Industry Foundation; 1984.
8. ANONYMOUS. Progress report—Interstate outbreak of salmonellosis related to nonfat dry milk. Salmonella Surveillance No. 53, p. 7; 1966.
9. ANONYMOUS. Progress report—Interstate outbreak of salmonellosis related to nonfat dry milk. Salmonella Surveillance No. 55, p. 2; 1966.
10. ANONYMOUS. Progress report—Interstate outbreak of salmonellosis related to nonfat dry milk. Salmonella Surveillance No. 57, p. 2; 1967.
11. ANONYMOUS. Progress report on *Salmonella* problem. Dairy Div. Consumer and Marketing Service, U.S.D.A.; 1967.
12. ANONYMOUS. Salmonellosis associated with nonfat dry milk. Morbidity Mortality Weekly Report **15**:385; 1966.
13. ANONYMOUS. Salmonellosis - New York, New Jersey, Connecticut, Maine. Morbidity Mortality Weekly Report **16**:141; 1967.
14. ANONYMOUS. Thermally processed low-acid foods in hermetically sealed containers. Code of Federal Regulations No. 21 (revised Apr. 1). Washington, D.C.: U.S. Gov. Printing Office; 1977.
15. ANONYMOUS. Wisconsin dairy facts. Madison: Wisconsin Dept. of Agriculture; 1975.
16. APPLEBAUM, R.S.; MARTH, E.H. Fate of aflatoxin M_1 in cottage cheese. J. Food Prot. **45**:903–904; 1982.
17. APPLEBAUM, R.S.; BRACKETT, R.E.; WISEMAN, D.W.; MARTH, E.H. Aflatoxin: toxicity to dairy cattle and occurrence in milk and milk products—a review. J. Food Prot. **45**:752–777; 1982.
18. APPLEBAUM, R.S.; BRACKETT, R.E.; WISEMAN, D.W.; MARTH, E.H. Responses of dairy cows to dietary aflatoxin: feed intake and yield, toxin content and quality of milk of cows treated with pure and impure aflatoxin. J. Dairy Sci. **65**:1503–1508; 1982.
19. ARMIJO, R.; HENDERSON, D.A.; TIMOTHEE, R.; ROBINSON, H.B. Food poisoning outbreaks associated with spray-dried milk—an epidemiologic study. Am. J. Public Health **47**:1093–1100; 1957.

20. BELIKOV, L.A. A case of milk-borne food poisoning caused by an enteritis bacillus. J. Microbiol. Epidemiol. Immunobiol. **3**:354; 1957.
21. BELL, W.B.; VELIZ, M.O. Production of enterotoxin by staphylococci recovered from the bovine mammary gland. Vet. Med. **47**:321–322; 1952.
22. BONITZ, K. On the epidemiology of *Salmonella bareilly* from infected cheese. Zentralbl. Bakteriol. I. Orig. **168**:244–256; 1957.
23. BOYD, J.M. An outbreak of *Salmonella newport* in Fife. Health Bull., Edinburgh **16**:31; 1958.
24. BRACKETT, R.E.; MARTH, E.H. Association of aflatoxin M_1 with casein. Z. Lebensm. Unters. Forsch. **174**:439–441; 1982.
25. BRACKETT, R.E.; MARTH, E.H. Fate of aflatoxin M_1 in Cheddar cheese and in process cheese spread. J. Food Prot. **45**:549–552; 1982.
26. BRACKETT, R.E.; MARTH, E.H. Fate of aflatoxin M_1 in Parmesan and mozzarella cheese. J. Food Prot. **45**:597–600; 1982.
27. BRACKETT, R.E.; APPLEBAUM, R.S.; WISEMAN, D.W.; MARTH, E.H. Fate of aflatoxin M_1 in brick and Limburger-like cheese. J. Food Prot. **45**:553–556; 1982.
28. BRUHN, P.A.; MOLLER-MADSEN, A.; PEDERSEN, A.H.; JENSEN, H. Thermal-resistant pathogenic bacteria and their survival during aging of cheese. Beret. Forsoegzm. Kbh. **124**:1–85; 1960.
29. BRYAN, F.L. Epidemiology of milk-borne diseases. J. Food Prot. **46**:637–649; 1983.
30. CAMPBELL, A.G.; GIBBARD, J. The survival of *E. typhosa* in Cheddar cheese manufactured from infected raw milk. Can. J. Public health **35**:158–164; 1944.
31. CASMAN, E.P. Staphylococcal enterotoxin. Ann. N.Y. Acad. Sci. **128**:124–131; 1965.
32. CENTER FOR DISEASE CONTROL. Follow-up on scombroid fish poisoning in canned tuna fish—United States. Morbidity Mortality Weekly Report. Rep. **22**:78; 1973.
33. CLARENBURG, A.; HUININK, A.B. *Salmonella dublin* infection by milk. Neth. Milk Dairy J. **2**:5; 1948.
34. CLARK, W.S., JR.; NELSON, F.E. Multiplication of coagulase-positive staphylococci in grade A milk samples. J. Dairy Sci. **44**:232–236; 1961.
35. COCKBURN, W.C.; SIMPSON, E. Food poisoning in England and Wales, 1951 and 1952. Monthly Bull., Ministry Health Lab. Service **13**:12; 1954.
36. COCKBURN, W.C.; VERNON, E. Food poisoning in England and Wales, 1954. A report of the Public Health Laboratory Service. Monthly Bull., Ministry of Health Lab. Service **14**:203; 1955.
37. COLLINS - THOMPSON, D.L.; ERDMAN, I.E.; MILLING, M.E.; BURGNER, D.M.; PURRIS, V.T.; LOIT, A.; COULTER, R.M. Microbiological standards for cheese: survey and viewpoint of the Canadian Health Protection Branch. J. Food Prot. **40**:411–414; 1977.
38. CORBION, B.; FREMY, J.M. Recherche des aflatoxins B_1 et M_1 dan les fromages de type "Camembert". Lait **53**:133–140; 1978.
39. CROMB, LE.; MURDOCK, C.R. An outbreak of food poisoning due to *Salmonella dublin*. Med. Off. **82**:267; 1949.
40. DAVIES, E.T.; VENN, J.A.J. The detection of a bovine carrier of *Salmonella heidelberg*. J. Hyg. **60**:495; 1962.
41. DE IONGH, H.; VLES, R.O.; VAN PELT, J.G. Milk of mammals fed an aflatoxin-containing diet. Nature **202**:466–467; 1964.
42. DOBRIER, I.B. An outbreak of food poisoning following the consumption of milk infected with *Salmonella* bacteria on a farm. Zhur. Mikrobiol. Epidemiol. Immunol. **40**:101; 1963.
43. DOEGLAS, H.M.G.; HUISMAN, J; NATER, J.P. Histamine intoxication after cheese consumption. Lancet **ii**:1361–1362; 1967.
44. DOMMET, T.W. Studies on coliform organisms in Cheddar cheese. Aust. J. Dairy Technol. **25**:54–61; 1970.

3.9 References

45. DONNELLY, C.B.; LESLIE, J.E.; BLACK, L.A. Production of enterotoxin A in milk. Appl Microbiol. **16**:917–924; 1968.
46. DOYLE, M.P. *Campylobacter fetus* subsp. *jejuni*: An old pathogen of new concern. J. Food Prot. **44**:480–488; 1981.
47. DUPONT, H.I.; FORMAL, S.B.; HORNECK, R.B; SNYDER, M.J.; LIBONATI, J.P.; SHEAHAN, D.G.; LABRIE, E.H.; KALAS, J.P. Pathogenesis of *Escherichia coli* diarrhea. N. Engl. J. Med. **285**:3–11; 1971.
48. ELLIOTT, J.A.; MILLARD, G.E. A. comparison of two methods for counting coliforms in cheese. Can. Inst. Food Sci. Technol. J. **9**:95–97; 1976.
49. ERNSTROM, C.A. An early gas defect in pasteurized milk Cheddar cheese. Milk Prod. J. **45**:21, 42; 1954.
50. FANTASIA, L.D.; MESTRANDREA, L.; SCHRADE, J.P.; YATER, J. Detection and growth of enteropathogenic *Escherichia coli* in soft ripened cheese. Appl. Microbiol. **29**:179–185; 1975.
51. FIELD, H.J. A survey of bovine salmonellosis in Mid and West Wales. Vet. J. **104**:251; 1948.
52. FOLEY, A.R.; POISSON, E. A cheese-borne outbreak of typhoid fever, 1944. Can J. Public health **36**:116–118; 1945.
53. FRANK, H.K. Diffusion of aflatoxins in foodstuffs. J. Food Sci.**33**:98–100; 1968.
54. FRANK, J.F.; MARTH, E.H. Inhibition of enteropathogenic *Escherichia coli* by homofermentative lactic acid bacteria in skimmilk. I. Comparison of strains of *Escherichia coli*. J. Food Prot. **40**:749–753; 1977.
55. FRANK, J.F.; MARTH, E.H. Inhibition of enteropathogenic *Escherichia coli* by homofermentative lactic acid bacteria in skimmilk. II. Comparison of lactic acid bacteria and enumeration methods. J. Food Prot. **40**:754–759; 1977.
56. FRANK, J.F.; MARTH, E.H. Survey of soft and semisoft cheese for presence of fecal coliforms and serotypes of enteropathogenic *Escherichia coli*. J. Food Prot. **41**:198–200; 1978.
57. FRANK, J.F.; MARTH, E.H.; OLSON, N.F. Survival of enteropathogenic and nonpathogenic *Escherichia coli* during the manufacture of Camembert cheese. J. Food Prot. **40**:835–842; 1977.
58. FRANK, J.F.; MARTH, E.H.; OLSON, N.F. Behavior of enteropathogenic *Escherichia coli* during manufacture and ripening of brick cheese. J. Food Prot. **41**:111–115; 1978.
59. GANDHI, N.R.; RICHARDSON, G.H. Staphylococcal enterotoxin A development in Cheddar cheese slurries. J. Dairy Sci. **56**:1004–1010; 1973.
60. GAUTHIER, J.; FOLEY, A.R. A cheese-borne outbreak of typhoid fever. Can. J. Public Health **34**:543–556; 1943.
61. GEORGE, E., JR.; OLSON, J.C.; JR.; COULTER, S.T. The growth of staphylococci in condensed milk. J. Dairy Sci. **42**:816–823; 1959.
62. GIBSON, E.A. An outbreak of *Salmonella dublin* infection in goats. Vet. Rec. **69**:1026; 1957.
63. GILLILAND, S.E.; SPECK, M.L. Interactions of food starter cultures and foodborne pathogens. J. Milk Food Technol. **35**:307–310; 1972.
64. GLATZ, B.A.; BRUDVIG, S.A. Survey of commerically available cheese for enterotoxigenic *Escherichia coli*. J. Food Prot. **43**:395–398; 1980.
65. GOEL, M.C.; KULSHRESTHA, D.C.; MARTH, E.H.; FRANCIS, D.W.; BRADSHAW, J.G.; READ, R.B., JR. Fate of coliforms in yogurt, buttermilk, sour cream and cottage cheese during refrigerated storage. J Milk Food Technol. **34**:54–58; 1971.
66. GOEPFERT, J.M.; OLSON, N.F.; MARTH, E.H. Behavior of *Salmonella typhimurium* during manufacture and curing of Cheddar cheese. Appl. Microbiol. **16**:862–866; 1968.
67. GRUNSELL, C.S.; OSBORNE, A.D. *Salmonella dublin* infection of adult cattle. Vet. Rec. **60**:85; 1948.

68. GUINEE, P.A.M.; KAMPELMACHER, E.H; VAN KEULEN, A.; HOFSTRA, K. Salmonellae in healthy cows and calves in the Netherlands. Zentralbl. Veterinärmed. **11**:728; 1964.
69. HAINES, W.C.; HARMON, L.G. Effect of variations in conditions of incubation upon inhibition of *staphylococcus aureaus* by *Pediococcus cerevisiae* and *Streptococcus lactis*. Appl. Microbiol. **25**:169–172; 1973.
70. HALVERSON, W.L. Preliminary report of typhoid fever due to unpasteurized cheese. Calif. Health **1**:171–173; 1944.
71. HARMS, F. Salmonellen-Dauerausscheider und Milchhygiene. Deut. tierärztl. Wochenschr. **66**:131; 1959.
72. HARRISON, F.C. Gas producing bacteria and their effect on milk and its products. Ont. Agric. Coll. Bull. 141; 1905.
73. HAUSLER, W.J., JR.; BYERS, E.J., JR.; SCARBOROUGH, L.C., JR.; HENDRICKS, S.L. Staphylococcal food intoxication due to Cheddar cheese. II. Laboratory evaluation. J. Milk Food Technol. **23**:1–6; 1960.
74. HEINEMANN, B. Growth and thermal destruction of *Micrococcus pyogenes* var. *aureus* in heated and raw milk. J. Dairy Sci. **40**:1585–1589; 1957.
75. HEMMES, G.D. Transmission of typhoid through cheese. Ned. Tijdschr. Geneesk. **86**:3159–3161; 1942.
76. HOLTERMAN, A; MATHER, J. Outbreak of *Salmonella typhimurium* due to contaminated raw milk. Salmonella Surveillance No. **53**:5; 1966.
77. HUTCHINSON, R.I. Milk-borne outbreak of *Salmonella heidelberg*. Brit. Med J. 1964-I: 479; 1964.
78. IORDANOV, M.; ZLATEV, I.; SLAVKOV, I.; DELCHEV, K. Vurku salmonelozite u was. Izv. Mikrobiol. Inst., Sofia **5**:77; 1954.
79. JEZESKI, J.J.; TATINI, S.R.; DEGARCIA, P.C.; OLSON, J.C. JR. Influence of *Streptococcus lactis* on growth and enterotoxin production by *Staphylococcus aureus*. Bacteriol. Proc. **1967**:12; 1967.
80. JONES, G.A.; GIBSON, D.L.; CHENG, K.J. Coliform bacteria in Canadian pasteurized dairy products. Can. J. Public Health **58**:257–264; 1967.
81. KAUTTER, D.A.; LILLY, T., JR.; LYNT, R.K.; SOLOMON, H.M. Toxin production by *Clostridium botulinum* in shelf-stable pasteurized process cheese spreads. J. Food Prot. **42**:784–786; 1979.
82. KEYL, A.C.; BOOTH, A.N. Aflatoxin effects in livestock. J. Amer. Oil Chem. Soc. **48**:599–600; 1971.
83. KIERMEIER, F. Significance of aflatoxins in the dairy industry. Annu. Bull. Intern. Dairy. Fed. Document No. 98; 1977.
84. KIERMEIER, F. Zur Aflatoxinbildung in Milch and Milchprodukten. IV. Modellversuchen mit Milchpulver. Z. Lebensm. Unters. Forsch. **146**:262–265; 1971.
85. KIERMEIER, F.; BEHRINGER, G. Einfluss der pH Wertes auf die Aflatoxinbildung in Modellversuchen mit Milchpulver. Z. Lebensm. Unters. Forsch. **151**:392–394; 1973.
86. KIERMEIER, F.; BÖHM, S. Aflatoxin formation in milk and milk products. V. Application of hen embryo test for confirmation of thin layer chromatographic determination of aflatoxin in cheese. Z. Lebensm. Unters. Forsch. **147**:61–64; 1971.
87. KIERMEIER, F.; BUCHNER, J. Verhalten von Aflatoxin M$_1$ während der Reifung und Lagerung von Käse. Z. Lebensm. Unters. Forsch. **164**:87–91; 1977.
88. KIERMEIER, F.; GROLL, D. Zur Bestimmung von Aflatoxin B$_1$ in Käse. Z. Lebensm. Unters. Forsch. **142**:120–123; 1970.
89. KIERMEIER, F.; MASHALEY, R. Einfluss der molkereitechnischen Behandlung der Rohmilch auf des Aflatoxin M$_1$ Gehalt der daraus hergestellten Produkte. Z. Lebensm. Unters. Forsch. **164**:183–187; 1977.
90. KIERMEIER, F.; RUMPF, S. Behavior of aflatoxin during production of processed cheese. Z. Lebensm. Unters. Forsch. **157**:211–216; 1975.

3.9 References

91. KIERMEIER, F.; WEISS, G. Zur Untersuchung von Milch und Milchprodukten auf die Aflatoxine B_1, B_2, G_1, G_2, und M_1. Z. Lebensm. Unters. Forsch. **160**:337–344; 1976.
92. KIERMEIER, F.; WEISS, G.; BEHRINGER, G; MILLER, M. Ueber das Vorkommen der Gehalt von Aflatoxin M_1 in Käsen den Handels. Z. Lebensm. Unters. Forsch. **163**:268–271; 1977.
93. KNOX, W.A.; GALBRAITH, N.S.; LEWIS, M.J.; HICKIE, G.C.; JOHNSTON, H.H. A milk-borne outbreak of food poisoning due to *Salmonella heidelberg*. J. Hyg. **61**:175; 1963.
94. KORNACKI, J.L.; MARTH, E.H. Fate of nonpathogenic and enteropathogenic *Escherichia coli* during the manufacture of Colby-like cheese. J. Food Prot. **45**:310–316; 1982.
95. KORNACKI, J.L.; MARTH, E.H. Foodborne illness caused by *Escherichia coli:* a review. J. Food Prot. **45**:1051–1067; 1982.
96. KOSIKOWSKI, F. Cheese and fermented milk foods, 2nd ed. Ann Arbor, Mich.: Edwards Brothers; 1977.
97. KRISTOFFERSEN, T.; MIKOLAJCIK, E.M.; GOULD, I.A. Cheddar cheese flavor. IV. Directed and accelerated ripening process. J. Dairy Sci. **50**:292–297; 1967.
98. LANCASTER, M.C. Mycotoxins in ruminants. Proc. Nutr. Soc. **28**:2303–212; 1969.
99. LEWIN, W.; ROUX, P. Bacteriology aspects of recent food-poisoning outbreaks in Witwatersrand. S. Afric. Med. J. **19**:185; 1945.
100. LIE, J.L.; MARTH, E.H. Formation of aflatoxin in Cheddar cheese by *Aspergillus flavus* and *Aspergillus parasiticus*. J. Dairy Sci. **50**:1708–1710; 1967.
100a. LIGHTBODY, L.G. Coliform organisms in Queensland Cheddar cheese. Queensl. J. Agric. Sci. **19**:305–307; 1962.
101. LYNCH, G.P. Mycotoxins in feedstuffs and their effect on dairy cattle. J. Dairy Sci. **55**:1243–1255; 1972.
102. MAERCKLIN, A. Untersuchungen zum Vorkommen von Staphylokokken-Enterotoxin in Butter. Thesis, Tierärztliche Hochschule, Hannover; 1972.
103. MARRIER, R.; WELLS, J.G.; SWANSON R.C.; CALLAHAN, W.; MEHLMAN, I.J. An outbreak of enteropathogenic *Escherichia coli* foodborne disease traced to imported French cheese. Lancet **ii**:1376–1378; 1973.
104. MARSHALL, C.E. Gassy curd and cheese. Mich. Agric. Coll. Bull. 183; 1900.
105. MARTH, E.H. Assuring the quality of milk. J. Dairy Sci. **64**:1017–1022; 1981.
106. MARTH, E.H. Salmonellae and salmonellosis associated with milk and milk products. A review. J. Dairy Sci. **52**:283–315; 1969.
107. MASRI, M.S.; GARCIA, V.C.; PAGE, J.R. Aflatoxin M content of milk from cows fed known amounts of aflatoxin. Vet. Rec. **84**:146–147; 1969.
108. McCALL, A.M. An explosive outbreak of food poisoning caused by *Salmonella dublin*. Lancet **264**:1302; 1953.
109. McFARLANE, D.; RENNIE, J.C. The behavior of substances introduced into the bovine mammary gland via the streak canal. Austral. Vet. J. **26**:53; 1950.
110. McKINNEY, J.D.; CAVANAGH, G.C. Extraction of "bound" aflatoxin. Zeszyty Problemowe Postepow Nauk Rolniczyeh **189**:247–253; 1977.
111. MENZIES, D.B. An outbreak of typhoid fever in Alberta traceable to infected Cheddar cheese. Can. J. Public Health **35**:431–438; 1944.
112. MEYER, J. Zum Vorkomomen von Staphylokokken-Enterotoxinen in Trockenmilchprodukten. Thesis, Tierärztliche Hochschule, Hannover; 1973
113. MINISTRY OF HEALTH. Food poisoning in England and Wales, 1950. Monthly Bull., Ministry of Health Lab. Service **10**:228; 1950.
114. MINOR, T.E.: MARTH, E.H. Fate of *Staphylococcus aureus* in cultured buttermilk, sour cream and yogurt during storage. J. Milk Food Technol. **35**:302–306; 1972.
115. MINOR, T.E.; MARTH, E.H. Growth of *Staphylococcus aureus* in acidified pasteurized milk. J. Milk Food Technol. **33**:516–520; 1970
116. MINOR, T.E.; MARTH, E.H. Staphylococci and their significance in foods. Amsterdam: Elsevier Scientific Pubishing Co.; 1976.

117. MINOR, T.E.; MARTH, E.H. Staphylococcus aureus and enterotoxin A in butter. J. Dairy Sci. 55:1410–1414; 1972.
118. MITSCHERLICH, E; MARTH, E.H. Microbial survival in the environment. Heidelberg: Springer Verlag; 1984.
119. MOCQUOT, G.; LAFONT, P.; VASSAL, L. Further observations on the survival of Salmonella in cheese. Ann. Inst. Pasteur (Paris) 104:570–583; 1963.
120. MOUSTAFA, M.K.; AHMED, A.A-H.; MARTH, E.H. Occurrence of Yersinia enterocolitica in raw and pasteurized milk. J. Food Prot. 46:276–278; 1983.
121. MOUTSY, A.W.; NASR, S. Studies on the sanitary condition of fresh Kareish cheese with special reference to the incidence of some food poisoning organisms. J. Arab. Vet. Med. Assoc. 24:99–106; 1964.
122. MURDOCK, C.R.; GORDON, W.A.M. The incidence of Salmonella in healthy cattle in Northern Ireland. Monthly Bull., Ministry Health Lab. Service 12:72; 1953
123. MURRAY, J.G. Salmonellae in farm milk. 17th Intern. Dairy Congr. B: 427; 1966.
124. NATIONAL CENTER FOR DISEASE CONTROL. Staphylococcal food poisoning traced to butter-Alabama. Morbidity Mortality Weekly Report 19:271; 1970.
125. NEUMAN-KLEINPAUL, A.; TERPLAN, G. Occurrence of aflatoxin M_1 in dried milk products. Arch. Lebensmittelhyg. 23:128–132; 1972.
126. NEVOT, A.; MOCQUOT, G.; LAFONT, P.; PLOMMET, M. Studies on the survival of pathogenic bacteria in soft ripened cheese. Ann. Inst. Pasteur 103:128–134; 1962.
127. NORTON, R.; ARMSTRONG, E.C. A milk-borne outbreak due to Salmonella typhimurium. Monthy Bull., Ministry Health Lab. Service 13:90; 1954.
128. NYSTRÖM, K.G.; KARLSON, K.A.; MENTZING, L.O. An outbreak of Salmonella dublin infection in cattle and people in Northern Kalmar County, 1963. Nord. Vet. Med. 16:697; 1964.
129. OLSON, J.C., JR.; CASMAN, E.P.; BAER, E.F.; STONE, J.E. Enterotoxigenicity of Staphylococcus aureus cultures isolated from acute cases of bovine mastitis. Appl. Microbiol. 20:605–607; 1970.
130. PARK, H.S.; MARTH, E.H. Behavior of Salmonella typhimurium in skimmilk during fermentation by lactic acid bacteria. J. Milk Food Technol. 35:482–488; 1972.
131. PARK, H.S.; MARTH, E.H. Inactivation of Salmonella typhimurium by sorbic acid. J. Milk Food Technol. 35:532–539; 1972.
132. PARK, H.S.; MARTH, E.H. Survival of Salmonella typhimurium in refrigerated cultured milks. J. Milk Food Technol. 35:489–495; 1972.
133. PARK, H.S.; MARTH, E.H.; OLSON, N.F. Survival of Salmonella typhimurium in cold-pack cheese food during refrigerated storage. J. Milk Food Technol. 33:383–388; 1970.
134. PARK, H.S.; MARTH, E.H.; OLSON, N.F. Fate of enteropathogenic strains of Escherichia coli during the manufacture and ripening of Camembert cheese. J. Milk Food Technol. 36:543–546; 1973.
135. PARK, H.S.; MARTH, E.H.; GOEPFERT, J.M.; OLSON, N.F. The fact of Salmonella typhimurium in the manufacture and ripening of low-acid Cheddar cheese. J. Milk Food Technol. 33:280–284; 1970.
136. PAUL, R.; KALRA, M.S.; SINGH, A. Incidence of aflatoxins in milk and milk products. Ind. J. Dairy Sci. 29:318–321; 1976.
137. POLAN, C.E.; HAYES, J.R.; CAMPBELL, T.C. Consumption and fate of aflatoxin B_1 by lactating cows. J. Agr. Food Chem. 22:635–638; 1974.
138. POLZHOFER, K. Aflatoxinbestimmung in Milch und Milchprodukten. Z. Lebensm. Unters. Forsch. 163:175–177; 1977.
139. POLZHOFER, K.P. Hitzestabilität von Aflatoxin M_1. Z. Lebensm. Unters. Forsch. 164:80–81; 1977.
140. PURCHASE, I.F. H. Aflatoxin residues in food of animal origin. Food Cosmet. Toxicol. 10:531–544; 1967.

3.9 References

141. PURCHASE, I.F. H.; STEYN, M.; RINSMA, R,; TUSTIN, R. C. Reduction of aflatoxin M content of milk by processing. Food Cosmet. Toxicol. **10**:383–387; 1972.
142. RANTA, L.E.; DOLMAN, C.E. Preliminary observations on the survival of *S. typhi* in Canadian Cheddar-type cheese. Can J. Public Health **33**:73–74; 1941.
143. RASCH, K. Die Bekämpfung der Rindersalmonellose in Blickpunkt der Lebensmittelhygiene. Berlin u. Münch. tierärztl. Wochenschr. **70**:161; 1957.
144. RASH, K.E.; KOSIKOWSKI, F.V. Influence of lactic acid starter bacteria on enteropathogenic *Escherichia coli* in ultrafiltration prepared Camembert cheese. J. Dairy Sci **65**:537–543; 1982.
145. REITER, B.; FEWINS, B.G.; FRYER, T.F.; SHARPE, M. E. Factors affecting the multiplication and survival of coagulase-positive staphylococci in Cheddar cheese. J. Dairy Res. **31**:261–272; 1964.
146. RICE, S.L.; EITENMILLER, R.R.; KOEHLER, P. E. Biologically active amines in food: a review, J. Milk Food Technol. **39**:353–358; 1976.
147. RICE, T.B. Typhoid epidemic in Northern Indiana. Ind. Board Health Monthly Bull. **47**:29, 42; 1944.
148. RICHARDSON, G.H.; DIVATIA, M.A. Lactic culture inocula required to inhibit staphylococci in sterile milk. J. Dairy Sci. **56**:706–709; 1973.
149. RIDDELL, J.; NORVAL, J.; ANDERSON, G.K. An outbreak of salmonellosis in a cattle court dairy farm. Health Bull., Edinburgh **17**:67; 1959.
150. RITCHIE, J.M.; CLAYTON, N.M. An incidence of *Salmonella dublin* in healthy cattle. Monthly Bull., Ministry Health Lab. Service **10**:272; 1951.
151. ROBINSON, D.A. Infective dose of *Campylobacter jejuni* in milk. Br. Med. J. **282**:1584; 1981.
152. ROSS, P.W. Ecology of group B streptococci. Skinner, F.A.; Quesnel, L.B. eds. Streptococci. London: Academic Pres; **1978**:127–142.
153. ROTHENBACHER, H. Mortality and morbidity in calves with salmonellosis. J. Amer. Vet. Med. Assoc. **147**:1211; 1965.
154. ROTHENBÜLER, E.; BACHMANN, M. Untersuchungen zum Problem der Aflatoxinbildung in Käse. Schweiz. Milchzeitung **96**(74) (Wissenschaftliche Beilage **123**):1053–1056; 1970.
155. ROTHSCHILD, L., JR. (ED.). Firm voluntarily recalls butter because of contamination. Food Chem. News **12**:33; 1970.
156. RUSSELL, H.L. Gas producing bacteria and the relation of the same to cheese. Wis. Agric. Exp. Stn. 12th Annu. Rep.: 139–150; 1895.
157. RYDER, R.W.; WACHSMUTH, I.K.; BUSTON A.E.; BARRETT, F.F. Infantile diarrhea produced by heat-stable enterotoxigenic *Escherichia coli*. N. Engl. J. Med. **295**:849–853; 1976.
158. SCHROEDER, S.A. What the sanitarian should know about salmonellae and staphylococci in milk and milk products. J. Milk Food Technol. **30**:376–380; 1967.
159. SCOTT, W.M.; MINETT, F.C. Experiments on infection of cows with typhoid bacilli. J. Hyg. **45**:159.
160. SEVERS, D., FARDY, P.; BUTLER, R. Salmonellosis from powdered milk. Samonella Surveillance No. 73, p. 12; 1968.
161. SHIH, C.N.; MARTH, E.H. Experimental production of aflatoxin in brick cheese. J. Milk Food Technol. **35**:585–587; 1972.
162. SKELTON, W.R.; HARMON, L.G. Growth rate of coliform organisms in cottage cheese and reconstituted nonfat dry milk. J. Milk Food Technol. **27**:197–201; 1964.
163. SKIRROW, M.B. *Campylobacter* enteritis: the first five years. J. Hyg. **89**:175–184; 1982.
164. SMITH, H.W.; BUXTON, A. Isolation of salmonellae from feces of domestic animals. Brit. Med. J. **1951–I**:1478; 1951.
165. SUTHERLAND, P.L.; BERGER, F.M. A milk-borne outbreak of gastroenteritis due to *Salmonella dublin*. Brit. Med. J. **1944**:488; 1944.

166. TAKAHASHI, I.; JOHNS, C.K. *Staphylococcus aureus* in Cheddar cheese. J. Dairy Sci. **42**:1032–1037; 1959.
167. TANAKA, N.; GOEPFERT, J.M.; TRAISMAN, E.; HOFFBECK, W. M. A challenge of pasteurized process cheese spread with *Clostridium botulinum* spores. J. Food Prot. **42**:787–789; 1979.
168. TATINI, S.R.; JEZESKI, J.J.; MORRIS, H.A.; OLSON, J.C., JR.; CASMAN, E.P. Production of staphylococcal enterotoxin A in Cheddar and Colby cheese. J. Dairy Sci. **54**:815–825; 1971.
169. TATINI, S.R.; WESALA, W.D.; JEZESKI, J.J.; MORRIS, H.A. Production of staphylococcal enterotoxin A in brick, mozzarella and Swiss cheeses. J. Dairy Sci. **56**:429–435; 1973.
170. THAL, E.; HOLMQUIST, H. Observations on in vivo dissociation of *Salmonella dublin*. Nord. Vet. Med. **9**:421; 1957.
171. THOMAS, C.T.; WHITE, J.C.; LONGREE, K. Thermal resistance of salmonellae and staphylococci in foods. Appl. Microbiol. **14**:815–820; 1966.
172. THOMASSON, H.L. A typhoid epidemic from cheese as experienced by a milk sanitarian. Ind. Board Health Monthly Bull. **48**:283, 295–296, 1944.
173. TODOROV, D. Resistance of *Salmonella typhimurium* and *Salmonella enteritidis* in hard cheese (Kachkaval). Proc. 17th Intern. Dairy Congr. D, Intern, Dairy Fed., Munich; 1966.
174. TUCKER, C.B.; CAMERON, G. M.; HENDERSON, M.P.; BEYER, M.R. *Salmonella typhimurium* food infection from Colby cheese. J. Am Med. Assoc. **131**:1119–1120; 1946.
175. TUCKEY, S.L.; STILES, M.E.; ORDAL, Z.J.; WITTER, L. D. Relation of cheese-making operations to survival of *Staphylococcus aureus* in different varieties of cheese. J. Dairy Sci. **47**:604–611; 1964.
176. TULLOCK, W.J. Observations concerning bacillary food infection in Dundee during the period 1923-38. J. Hyg. **39**:324; 1959.
177. VAN DER LINDE, J.A.; VAN DER FRENS, A.M.; VAN ESCH, G.J. Experiments with cows fed groundnut meal containing aflatoxin. Wogan, G.N. ed. Mycotoxins in foodstuffs. Cambridge: MIT Press; **1965**:247–249.
178. VARELA, G.; OLARTE, J. Classification and distribution of 1,075 cultures of *Salmonella* isoated in the City of Mexico. J. Lab. Clin. Med. **40**:73; 1952.
179. VERNON, E. Food poisoning in England and Wales 1965, Monthy Bull., Ministry of Health **25**:194; 1966.
180. VIZIR, P.E. The bacterial contamination of brynza cheese, methods for counteracting it, and indexes for judging the cheese. V. The effect of degree of acidity and salt concentration on the viability of *B. breslau*. Mikrobiol. Zh. Kiev **7**:65–78; 1940.
181. WAHBY, A.M.; ROUSHDY, A. Viability of enteric fever organisms in some Egyptian dairy products. Zentralbl. Veterinaermed. **2**:57–65; 1955.
182. WALKER, G.C.; HARMON, L.G.; STINE, C.M. Staphylococci in Colby cheese. J. Dairy Sci. **44**:1272–1282; 1961.
183. WEISS, G.; MILLER, M.; BEHRINGER, G. Vorkommen und Gehalt an Aflatoxin M_1, in Molkereiprodukten. Störungs möglichkeiten beim Nachweis von Aflatoxin M_1. Milchwissenschaft **33**:409–412; 1978.
184. WILLIAMS, W.L. *Staphylococcus aureus* contamination of a grade "A" milk supply. J. Milk Technol. **4**:311–313.
185. WILSON; C.D.; SALT, G.F.H. Streptococci in animal disease. Skinner, F.A.; Quesnel, L.G. eds. Streptococci, London: Academic Press; **1978**:143–156.
186. WISEMAN, D.W.; MARTH, E.H. Behavior of aflatoxin M_1 in yogurt, buttermilk and kefir. J. Food Prot. **46**:115–118; 1983.
187. WISEMAN, D.W.; MARTH, E.H. Stability of aflatoxin M_1 during manufacture and storage of a butter-like spread, non-fat dried milk and dried buttermilk. J. Food Prot. **46**:633–636; 1983.

3.9 References

188. WISEMAN, D.W.; MARTH, E. H. Stability of aflatoxin M_1 during manufacture and storage of ice-cream and sherbet. Z. Lebensm. Unters. Forsch. **177**:22–24; 1983.
189. WRIGHT, H.A.; NORVAL, J.; ORR, A. *Salmonella thompson* gastroenteritis. Report of two outbreaks. Brit. Med. J. **1957-II**:69; 1957.
190. WUNDT, W.; VOSS, J. On the prevention of infections through ice cream, Arch. Hyg. Bakteriol. **147**:358–368; 1963.
191. YALE, M.W.; MARQUARDT, J.C. Coliform bacteria in Cheddar cheese. N.Y. Agric. Exp. Stn. Tech. Bull. 270; 1943.
192. ZAGAEVSKII, I.S. Sources of contamination of milk with *Salmonella* on dairy farms. Voprosy Pitaniya **21**:86; 1962.
193. ZAGAEVSKII, I.S. The hygiene of milk production on farms infected with paratyphoid. Veterinariya **40**:58–62; 1963.
194. ZEHREN, V.L.; ZEHREN, V.F. Examination of large quantities of cheese for staphylococcal enterotoxin. J. Dairy Sci. **51**:635–644; 1968.
195. ZEHREN, V.L.; ZEHREN, V.F. Relation of acid development during cheese making to development of staphylococcal enterotoxin A. J. Dairy Sci. **51**:645–649; 1968.
196. ZOTTOLA, E.A.; AL-DULAIMI, A.N.; JEZESKI, J. J. Heat resistance of *Staphylococcus aureus* isolated from milk and cheese. J. Dairy Sci. **48**:774; 1965.
197. ZOTTOLA, E.A.; JEZESKI, J.J.; AL-DULAIMI, A. N. Effect of short-time pasteurization treatments on the destruction of *Staphylococcus aureus* in milk for cheese manufacture. J. Dairy Sci. **52**:1707–1714; 1969

CHAPTER 4

SAMPLING DAIRY AND RELATED PRODUCTS

V. Grace, G. A. Houghtby, S. E. Barnard, and J. Lindamood
(J. Messer, Tech Comm.)

4.1 Fluid Milk and Cream Samples

The sample must be representative of the mass of material being examined. It must be collected in such a way that it is not contaminated microbiologically or chemically. Protection must be provided against change in the sample during the period between collection and analysis.

A. Equipment for collecting samples:

Regardless of the type of sample being collected and/or the designated laboratory procedure, the following items of equipment are basic and required.

1. Thermometer: Mercury-filled, dial, or electronic type, graduation intervals not to exceed 1°C. A chart to convert units of measurement for some temperatures, weights, volumes, and lengths is provided in Table 4.1. Accuracy of a mercury-filled thermometer must be checked at least biennially with either a thermometer certified by the National Bureau of Standards [2.4O] or one of equivalent accuracy. When a dial-type thermometer or electronic thermometer is used, its accuracy must be checked at least once during each 6-mo period. A record of the accuracy of the test thermometer shall be maintained by the sample collector or by the laboratory where checked, and the thermometer is to be tagged, showing date checked and correction factor. Every thermometer must be accurate to plus or minus 1 division. A dial-type thermometer with an appropriate range is suggested when samples are collected from storage or balance tanks, vats, cans, or farm bulk tanks, or wherever glass thermometers might be broken.
2. Sample transfer instruments: Individually wrapped, sterile or pre-sterilized, plastic, single-service sampling tubes are preferred. However, sanitized tubes or stainless steel metal dippers with long handles (capacity 10 mL or greater) are commonly used. Other transfer means may be used, but must be demonstrably equivalent.

Table 4.1. Conversion Factors. The following are provided to aid in converting data from one measurement system to another.

1.0 oz (ounce) (weight)	=	28.3 g (grams)
1.5 oz	=	42.5 g
4.0 oz	=	113.4 g
1.0 lb (pound)	=	453.6 g
1.0 oz (ounces) (volume)	=	29.6 mL (milliliter)
1.0 pt (pint)	=	473.2 mL (0.4732 L)
1.0 qt (quart)	=	946.3 mL (0.9463 L)
1.0 gal (gallon)	=	3785.0 mL (3.785 mL)
32°F (Farenheit)	=	0.0°C (Celcius)
40°F	=	4.4°C
45°F	=	7.2°C
104°F	=	40.0°C
180°F	=	82.2°C
1.0" (inch)	=	2.54 cm (centimeters)
1.0' (foot)	=	30.48 cm

For line sampling, the needle of a disposable, plastic, sterile hypodermic syringe may be inserted through a sanitized gasket into a pipeline, or into a sterile, rubber, serum cap fitted to a stainless steel nipple that is welded onto a standard line fitting,[4] Collection of an initial sample from appropriate sampling cocks, properly sanitized and purged (2L) before sampling, is satisfactory. All sampling instruments must be protected from contamination before use.

3. Sample containers: a) Multiple-use containers holding a minimum of 15 mL but of sufficient capacity to permit thorough mixing of contents may be used. These containers must be sterilized by dry heat [5.10B] or steam [5.10A], and have leakproof closures that have been demonstrated to be nontoxic. b) Sterile, evacuated sampling equipment may be used for collecting at least 10-mL portions. c) Presterilized polyethylene or other suitable nontoxic plastic containers of adequate size are acceptable. d) Single-service, nonsterile, leak-proof vials[13] may be used for samples of raw milk and cream assuming certain conditions are met. 1) The plant(s) in which containers are manufactured and prepared for shipment have been inspected and certified by State Milk Sanitation Rating Authorities as being in compliance with established fabrication procedures[15] and are listed in the FDA Quarterly Publication.[17] 2) Containers made according to each different formulation are nontoxic, not bacteriostatic or bactericidal, and nonabsorbent. 3) The closure is designed so the container can be opened and closed easily without contami-

nating the lip of the vial or the inner surface of the closure. 4) The sampling plan includes the testing of 60 randomly selected vials from each lot. (The lot is to be accepted only if all vials tested are in compliance with the standard of one viable organism or less/mL of capacity, each lot is tested and certified, and results of tests are included with the lot containers when it is purchased.) 5) The sampling plan includes the testing of 24 randomly selected vials from each lot, and the lot is accepted only if all vials tested are in compliance with the standard of no detected leakage. Each lot must be tested and certified, and the results of tests are to be included with the lot of containers when purchased. Suggested methods of testing are:

Method I (Class C)

Select 24 vials at random from the lot of vials to be tested. Fill each vial three-quarters full with powdered sugar or a color-indicating desiccant that has been dried. Cap the vials, making certain that the closure is properly positioned.

Secure the vials upright in racks by using a wire cover or other means. Place the rack into a case containing ice-water and a dye or indicator solution (0 to 1°C) and submerge the vials for 6 h or overnight.

Remove the racks from ice-water solution and examine vials for leakage. Using paper towels or other absorbent material, carefully remove water on the outside of the vials, especially around the rim of the closure to prevent water from accidentally entering vials when opened. Carefully open vials, inspect the contents of each for signs of water entry and record the number of vials that leaked. The lot will be accepted if no defective vials are observed.

Method II (Class C)

Select 24 vials at random from the lot of vials to be tested. Fill each vial three-quarters full with raw milk. Cap the vials, making certain that the closure is firmly and completely positioned.

Place inverted vials in racks and put racks in a refrigerator (0 to 4.4°C) overnight ($\geq$ 15 h). Remove racks and transfer inverted vials to an incubator and warm the vials and contents to 40°C. If any vials leak, the lot is rejected.

If no vials leak, return the vials to the upright position and shake 25 times in 7 sec with a 30-cm movement. If no vials leak, the lot is accepted.

4. Sample case: Should be of rigid metal or plastic construction, completely insulated, with a tight cover, and with ample space for cracked ice or other refrigerant to cool and maintain samples at 0 to 4.4°C until removed at the laboratory. When a mixture of ice and water is used for rapid cooling, the sample case shall be provided

with racks, compartments or baffles to: a) hold sample containers vertically, b) keep necks of containers above the surface of the cooling medium at all times, and c) maintain the cooling medium at a level slightly above the level of the liquid in sample containers. To protect sample(s) from moisture, the sample container may be inserted into a moisture-proof metal box or can, or a properly closed plastic bag.
5. Shipping case: When samples are transferred to the laboratory from a receiving station, plant, or other point of collection via a common carrier, the shipping case must be provided with a tamper-proof sealing and locking device and with handles. Attach "This Side Up" labels on top of the shipping case.
6. Inner shipping case: When inner shipping cases are needed to protect samples in small vials or containers, provide boxes with tightly fitted covers, just large enough to hold bottles upright and to prevent container breakage when the box is placed in a larger shipping case [4.1A5].
7. Agitators: The 3-A Sanitary Standards (3-A)[8,9] specify types of equipment or mechanical agitators acceptable for use in bulk milk operations, weigh-tanks, vats, and storage tanks. Air systems used for agitation must comply with the applicable 3-A Sanitary Standards[7] for odor-free, filtered air. To sample from cans or weigh-tanks, use: a) a stainless steel disc scoop (diam. about 15.24 cm) on the end of a stainless steel rod long enough to reach the bottom of the container, or b) a bowl-type agitator-sampler combination consisting of a solid handle welded to the outside of a bowl with a pouring spout.

B. *Cleaning, sterilizing and sanitizing equipment:*

All multiple-use equipment for collecting samples must be cleaned thoroughly before it is sanitized or sterilized. Single-service containers and sampling devices must be properly protected before use. Rinse all instruments in tap water immediately after use and clean with a suitable detergent solution. Avoid excessive exposure to strong alkali and, immediately after treatment, rinse thoroughly with tap water to remove detergent residues.
1. Hot-air sterilization: Glassware and metal equipment may be sterilized by holding at a temperature not lower than 170°C for not less than 1 h [5.10B].
2. Autoclave (steam) sterilization: Glassware and metal equipment may be steam-sterilized by holding at 121°C for not less than 15 min [5.10A]. Equipment must be dry before use.
3. Hot water sanitization: Use two stainless steel containers of approximately 40-L capacity (or two 37.85 L milk cans in good condition) equipped with threaded fittings to provide metal inlets (do not substitute rubber hose) for steam and water connections. The first container is for a cold water rinse of the agitator and

4.1 Fluid Milk and Cream Samples

sampling instrument. The second container, with a bottom inlet for steam under pressure, is for a hot water (not less than 82°C) treatment of instruments. Equip the second container with a thermometer that is immersed in the water. Minimal time for exposure in hot water should be 1 min. Minimize the risk of burns and excessive noise by installing a steam coil with a condensate trap on the exhaust side. To insure against possible back-siphonage into potable water lines, the cold water lines should terminate not less than 2.54 cm above the lip of the can or be equipped with an approved anti-siphon device.

4. Chemical sanitization: Use two containers as in 4.1B3, except that in the second container, the hot water is replaced with a solution of an approved germicide[16] that is bactericidally equivalent at all times to at least 100 mg/L (100 ppm) of available chlorine as hypochlorite. Submerge sampling equipment for at least 30 sec in the chemical sanitizer. Sampling dippers carried on bulk tank trucks shall be stored in a tightly stoppered tube of sanitizer solution. Wash dipper and tube-holding rack in sanitizer solution each day after use. Prepare sanitizing solution fresh daily and check with applicable test kit.

C. *Sampling procedures (Class 0):*

Perishability of raw and pasteurized milk and milk products requires special handling to prevent their direct contamination with bacteria and the subsequent growth of such contaminants during sampling, transportation and storage of samples before analysis. Collection procedures are a fundamental consideration when samples are to be used as a basis for payment or to determine compliance with legal bacteriological or chemical standards.

1. General requirements: Use the application items of equipment [4.1A] to collect samples. Thoroughly agitate or mix the volume of milk being sampled or randomly select products to be sampled.

    Meaningful results cannot be obtained unless representative samples are collected without contamination and tested within 36 h. Cool samples immediately to and maintain at 0 to 4.4°C; do not freeze samples. Promptly identify each sample legibly with a number, label or tag corresponding to inspection records. Record time and date of sampling. Record temperature of milk at all sampling locations. If milkfat or other chemical tests are to be done on the samples, first remove portions of samples for microbiological analysis after suitable agitation [4.1D]. The temperature of all samples received at the testing laboratory must be between 0 and 4.4°C [4.1D].

    Always include an extra sample of milk or dairy product in the sample case or shipping case for use as a temperature control. The temperature control sample shall be collected at the first sampling point. Subject the temperature control sample to the same storage conditions as other samples during completion of the sampling schedule and subsequent transit to the laboratory. The temperature control sample shall be identified with the following information: Temperature Control (TC); date of collection, time of collection, the temperature of the product collected, collection site (farm, producer, permit number, or tank number), and the identity of the sample collector by initials or a number.

Do not use the temperature control sample for microbial tests, but otherwise treat it exactly as are all other samples [4.1D]. When samples are to be examined by the Direct Microscopic Count Method for bacteria within 36 h after collection and cooling is impractical, preserve samples by adding 0.08% formaldehyde to the milk.[2,11]

2. Raw milk:
   a. Farm bulk-tank milk: The farm bulk-milk hauler is critically important in the current structure of milk marketing. When the hauler's obligation includes collection and delivery of samples to the laboratory for analysis, he/she becomes a vital part of quality control and regulatory programs. Deviation from the following acceptable sample collection practices could cause a milk producer to lose his market or payment for his product.
      1. Preparations for collecting the milk sample:
         a. Before measuring and sampling milk, wash and dry hands after connecting the milk transfer hose. Keep hands clean during the sampling operation. Do not open the bulk tank outlet valve until the milk has been measured and sampled.
         b. Observe milk and do not sample frozen, partially frozen, lumpy, curdled, or churned milk for regulatory purposes. If such milk is sampled for other purposes, record any abnormalities that are present.
         c. Dry the measuring stick with a clean, single-service paper towel. Measuring sticks stored outside the bulk tank should first be sanitized, then wiped dry at the milk level area with a paper towel.
         d. Agitate milk stored in a farm holding cooling tank at least 5 min immediately before sampling.[9] Agitate contents of farm storage tanks over 5677.9 L (1,500 gal) for 10 min or according to tank specifications.[8]
         e. Check temperature of milk. Use an accurate thermometer [4.1A1]. Sanitize the test thermometer before insertion into milk. See 4.1B4.
         f. At least once a month, compare accuracy of bulk-tank thermometer with that of the test thermometer.
      2. Preparations for collecting the milk sample with an in-line mechanical sampler: Follow instructions in 1 above except

4.1 Fluid Milk and Cream Samples

that in (d) it is not necessary to agitate for 5 or 10 min. Instead, turn on the agitator and proceed to 3.
3. Identification and information needed for each sample:
   a. Determine and record temperature of milk sampled at all sampling locations.
   b. Record time in hours and date at each sampling location. (Knowledge of the time the bulk milk was picked up is helpful when planning milk-hauler inspection schedules.)
   c. Legibly and indelibly identify each sample container with producer name, number, official tag, or label at each sampling location.
   d. Record all data on either laboratory report forms or weigh slips and submit to the laboratory with samples.
4. Special precautions:
   a. Protect sampling instruments from exposure to contamination before and during use.
   b. When using a dipper, remove it from the sanitizing solution, completely drain, and rinse at least twice in milk before transferring the sample.
   c. Handle sample container or caps aseptically. Do not drop, lay down, or otherwise contaminate containers or caps.
   d. Do not carry presterilized plastic bags or plastic vials in pockets of clothing. Clipboards, covered containers, or boxes may be used to carry empty single-service sample containers.
   e. Pre-cool multi-use sterile glass containers in sample case as needed.
   f. Do not hold or fill sample container over top of tank while transferring sample.
   g. Fill sample container not more than ¾ full to permit proper mixing of sample at the laboratory. Do not expel air when folding or whirling plastic bags in order to leave ¼ of the volume of the bag as air space.
   h. Collect a representative sample of each producer's total milk supply. Sample all tanks individually and label each sample.
   i. Provide a temperature control by collecting two samples at the first collection point (one sample to be used as the temperature control). Identify the temperature control (TC) sample and show producer number, date, time of collection, and temperature of that sample on the container. If there is no temperature control sample, the laboratory may use the first official sample collected [4.1D] or reject the set of samples.

j. Place sample in sample case immediately. If temperature of sample is above 4.4°C, use ice water to reduce the temperature.

k. Rinse dipper in tap water immediately after use and return it to sanitizing solution.

5. Sampling with an in-line mechanical device:
   a. Use the Isolok series MSA or MS-6 Mobile Service Sampler (Bristol Engineering Co., PO Box 696, Yorkville, IL 60560), or equivalent, installed as per manufacturer's instructions.
   b. Follow preparatory steps [4.1C3 and 4].
   c. The sanitized sampling device must have a sterile sample container placed on it immediately after sanitization, both initially and between stops. Because sanitizer may drip into this container, replace it with an identified sterile container at each farm just prior to pickup.
   d. Set the select switch on the sampler to collect at a rate that will provide 40 to 60 mL of sample.
   e. When the sampling is complete, remove the sample container, sanitize the container holder, and replace the container with the one removed in c, or a fresh one. Sanitize the container holder by flushing with clean water then applying sanitizer by immersing, spraying, or brushing.
   f. Controlled experiments[12] have indicated the risks of a carryover of milk from the previously sampled tank into the sample are low. The following practices considerably reduce such risks:
      1. Position the milk hose to minimize the amount of residual milk when pumping is stopped.
      2. Require that samples for microbiological tests be collected manually when tank volumes are less than 757L. (When quantities collected per tank are 757L or more, selection of the proper sampling rate ensures that the built-in time delay will be long enough to allow practically all of the milk from the previous tank to be flushed from the hose prior to sample collection.)

6. Storage and transportation of sample:
   a. Use ice or mechanical refrigerant at all times to cool and hold samples at 0 to 4.4°C. Keep ice level in sample case slightly above the milk level. Do not freeze fluid samples before or during shipment to the laboratory. Do not rely on winter air temperatures to keep samples cold.
   b. Protect samples from contamination. Contamination by ice water in the sample case may be prevented by using racks, compartments, or other means [4.1A3 and 4]. Ice water

4.1 Fluid Milk and Cream Samples

shall be only slightly higher than the level of milk in sample containers. Do not bury tops of containers in ice.
c. Promptly deliver samples to the laboratory. Microbiological and somatic cell examinations must be started within 36 h after collection [4.1D].
d. Samples shipped to a laboratory by a common carrier shall be packed in a shipping case with a tamper-proof lock or seal [4.1A5].
b. Milk cans and weigh tanks: After agitation [4.1A6], sample milk with a sanitized dipper or sterile sampling tube [4.1B]; then cool to and hold sample [4.1C1] at 0 to 4.4°C. Place sample in a sample case [4.1A4] and deliver promptly to the laboratory [4.1C1]. If the weigh tank is not large enough for the producer's total volume of milk, collect into a single sample container proportionate amounts of milk from each filling.
c. Mobile tank trucks (farm pickup and over-the-road tankers) and dairy plant storage tanks: Collection of representative samples from a transport tank or storage tank requires adequate agitation of the milk. The time interval required to agitate a tank of milk until it is homogeneous is determined by the size and shape of the tank; volume of product held; type, location, and force of agitator; and time allowed for creaming before starting agitation. Because of the importance of these variables, it is necesaary to determine for each given tank how much agitation time is needed to assure homogeneity of its contents.

The agitation time for milk storage tanks and milk tank trucks may be determined by taking a series of milkfat samples at specified intervals (e.g., 3 min, 4 min, etc.) during mixing until at least five milkfat tests stabilize at a definite value. From a full tank, take samples from the top and bottom, and from points near to and far from the source of agitation. If construction of a tank does not permit direct removal of subsamples, draw samples through the outlet valve while the agitator is in operation. Adequate agitation[8] is that degree of agitation which, at full tank capacity, will result in the milkfat content of the product in the tank varying by not more than ± 0.1% as determined by Babcock methods, ± 0.5% by automated methods, or 0.03% by the Roese-Gottlieb method [18.8E]. When the appropriate time interval has been determined for each tank installation or truck, record and use this time as the guide for proper agitation time for taking samples.

Studies show that inadequate agitation of bulk tank milk (farm tank, tank truck, and/or plant storage tanks) before top sampling is accompanied by the presence of more bacteria, somatic cells, and milkfat in the sample than occur in the milk lower in the tank.[1a] This increase in bacteria is not caused by growth but by the rising of fat globules, which serve to "sweep" microorganisms and somatic cells toward the surface, thus concentrating them in the cream layer.

The following methods may be used to agitate milk in tankers or storage tanks:

1. A mechanical agitator built into the tank and driven by an electric motor.
2. A propeller or agitator driven by an electric motor or by an air-driven motor and placed on the manhole with the agitator suspended in the milk.
3. An oil-free, filtered, 3-A quality air supply[7] injected through a quick coupling device on the rear of the tanker or through a sanitary air hose inserted through the manhold of the tanker.
4. An approved sanitary tank washer/air agitator system, permanently installed in the tank, through which 3-A-quality air[7] enters and agitates the milk. This method of agitation is not recommended unless an approved, cleaned in place (CIP) system has been installed at the receiving plant.
5. The milk transfer hose, when attached to the tanker unloading pump and inserted through the manhole, permits recirculation and mixing of the milk. A well-agitated sample may also be obtained by pumping milk directly from the tank truck into an empty, plant-storage tank. The sample may be collected through the sampling cock located in the tank access door. Do not collect the sample until the sampling cock has been adequately sanitized on the outside and at least 2 L of milk have been purged through it [4.1A2].

An automatic line-sampling device [4.1C5] may be used to provide a representative sample from a tanker as milk is unloaded. Samples collected by this method for bacteriological analysis must be cooled to and maintained at 0 to 4.4°C during the collection period.

To sample over-the-road tankers, collect two samples through the tank truck dome immediately after filling. The first sample is to be used for milkfat, bacteriological, or chemical analysis by the shipper; the second sample, held at 0 to 4.4°C, should accompany the over-the-road tank truck to the final destination for delivery to the receiving laboratory for analysis. Transport samples to the laboratory so that testing can be started within 36 h after collection [4.1C3].

To sample commingled milk in dairy plant storage tanks, collect the sample directly from the weigh tank before pas-

teurization or at some other processing step. When collecting samples of raw milk through the sampling cock in the storage tank door, use proper sanitization procedures to prevent contamination of the sample. [4.1A2].
   d. Line samples: To collect representative samples of raw milk from silo and/or dairy plant storage tanks, or to locate sources of contamination for raw or pasteurized milk, collect samples aseptically at suitable locations between the storage tank and the first processing step, or at different points after pasteurization. Sterile or sanitized metal tubes or dippers may be suitable transfer instruments at certain dairy plant locations. A special line-sampling technique involves use of a disposable, sterile, hypodermic syringe, the needle of which is inserted into a pipeline gasket or the rubber closure of a stainless steel nipple [4.1A2]. The nipple can be clamped on or permanently located as desired. Cool samples to 0 to 4.4°C and deliver promptly to the laboratory [4.1C1].
3. Pasteurized and ultrapasteurized package products, dispensers, coffee creamers, and pressurized containers: Select representative samples of all products, regardless of age, while they are still in possession of the processor. Note processing code date of sample on form submitted to the laboratory. Preferably submit an unopened container of retail product to the laboratory. If necessary, after thoroughly mixing the product in its container, aseptically transfer a representative portion to a sterile sample container. Store samples at 0 to 4.4°C and rapidly transfer samples to the laboratory so that testing will be started within 36 h after collection. Collect and examine acidified cultured milk, and milk product samples for coliforms within 24 h after processing [4.2D2].

When shipping samples to the laboratory by a common carrier, protect containers from contamination by using tight-fitting waterproof bags. Do not crush or damage plastic and paper containers [4.1A5].

Always transmit a temperature control (TC) sample [4.1C1] to the laboratory.
   a. Dispenser milk: Collect an official sample from the dispenser tube into sample container [4.1A3] without flushing milk through tube. Optionally, collect a second sample, after drawing off about 0.5L of milk, to determine if the tube is contaminated. Keep samples between 0 to 4.4°C until tested [4.1C1].
   b. Coffee creamers-pasteurized, ultrapasteurized, dried or frozen: Follow the general procedures for milk and cream [4.1A, 4.1B]. Since a single sample (container of product) is usually of insufficient quantity for analytical procedures, randomly select suffi-

cient samples to satisfy testing requirements and be representative of a lot [4.2]. Submit original unopened containers to the laboratory.

1. Pasteurized and ultrapasteurized products: Place samples in a bag in sample case, and maintain at 0 to 4.4°C until tested.
2. Dried creamers: Place dried product in a bag or box to avoid contamination. Do not carry in clothing. Hold and transport samples at room temperatures. Samples of dried products may be transferred to the laboratory without refrigeration.
3. Frozen creamers: Follow general sampling procedures for ice cream [4.2E]. Place in plastic bag or container to avoid contamination. Keep sample frozen until analyzed.
4. Pressurized containers: Randomly select commercial-size pressurized containers. Shake containers through twenty-five complete inversions within 30 sec. Aseptically dispense contents of container into sterile sample container. Cool samples to 0 to 4.4°C, and deliver promptly to the laboratory [4.1C1]. Consumer-size pressurized containers should be randomly selected and delivered directly to the laboratory.
5. Milk from the udder to be tested for microbial infections: Collect samples immediately before regular milking. Avoid dusty, windy areas. Preferably remove loose soil from animals before they enter sampling facility.

    Thoroughly wash teats and adjacent floor of udder with warm water containing a disinfectant (e.g., 25 mg/L of iodophor). Avoid wetting large portions of udder because of difficulties in drying. Dry thoroughly with disposable towels. Disinfect the teat canal by swabbing thoroughly with a cotton ball saturated with 70% alcohol after opening it with pressure with two fingers. Alternatively, dip teats in iodophor teat dip, wait 10 sec, and swab off the excess with a clean ball of cotton. To prevent recontamination of teat ends, first swab teats farthest from the collector. Commence collection of samples at the proximal (closest) teats, removing two streams of milk into a separate vessel before collecting samples in sterile containers. Promptly cool collected samples to 0 to 4.4°C and transfer to the laboratory within 12 h.

D. *Responsibilities of laboratory receiving samples:*

As samples arrive at the laboratory, determine temperature of each group of samples by inserting a precooled (0 to 4.4°C) thermometer [4.1C1] into temperature control (TC) samples [4.1C1]. If a TC sample is not received, determine the temperature of one of the official samples (preferably the first sample collected) and label it as the TC, or reject the set of samples.

4.2 Other Dairy Products

The temperature of all samples received at the testing laboratory should be between 0 and 4.4°C [4.1C1].

Requirements defined in this section occasionally are unattainable for pasteurized products. For these occasions, the following is recommended: When the elapsed time between collection and arrival of pasteurized samples at the laboratory is less than 3 h, samples should be accepted by the testing laboratory, provided cooling was provided as soon as the sample was collected.

Record temperature of sample(s), time received, and date of receipt on sample identification form. Store samples at 0 to 4.4°C until tested. When samples are tested, record time and date of analysis and temperature of sample control for each set of samples. If chemical tests are also to be done on samples, first agitate and then aseptically remove a portion of each sample for microbiological analyses [6.6B].

Record and transmit with each series of samples a record of the time, date, and temperature of the control sample when the series is transferred at interim sample transfer points, from the time of collection until receipt at the testing laboratory. (Interim transfer points are locations where refrigerators are maintained at 0 to 4.4°C.) Temperature records must be maintained.

4.2 Other Dairy Products [Chapter 10]

When several chemical and microbiological analyses are to be done, large samples may be required. Specific procedures should be reviewed before sampling so that adequate and proper samples may be taken for the required analyses.

In sampling bulk or large quantities where numerous units compose a lot, it should also be considered that while a single sample may not be representative, it may be practical under certain conditions. It is necessary to consider the condition or type of product to be sampled, e.g., dry powder, solid, semi-solid, or viscous, before sampling so that the number of units will be representative and/or statistically significant.

Unfortunately, it is not possible to present and use one set of tables, graphs, or instructions for all sampling situations. Each set of physical conditions may require the use of different techniques. Consequently, a number of diversified plans for particular applications are presented in the literature.[10] The mode of measuring the items determines choice between two general set of plans. An item may be examined by taking analytical data on its physical, chemical, or biological characteristics (variable sampling), or it may be inspected (attribute sampling) and classified into different groups. More than one measurement of attributes or variables can be obtained per item.

Sampling plans[3,10] can be arranged in several ways to assure adequate consumer and producer protection. The required number of samples can be taken all at once (single sampling), in groups (multiple sampling), or one item or unit may be sampled at a time (sequential sampling). Economic considerations often indicate the advantage of one or the other of these arrangements. Some measurements destroy samples, and others leave them unharmed. The cost added by destructive sampling is an example of an economic consideration.

An individual wishing to devise a sampling plan for inspection of items or units should obtain the services of a statistician to evaluate specific situations.

Current practice is a compromise between the desire to achieve a high level of consumer protection and the cost of providing the service. Available experimental data indicate that at least 5 to 10 units are necessary for representative sampling when a large lot or large number of units is to be sampled.[1a] A lot may be defined as an identifiable unit of production such as one day's production.[10]

An alternative procedure that reduces laboratory workload, if substantiated by background data, would be composite groups of samples within a given lot or series of units.

A. *Evaporated, concentrated, and condensed milks (Class 0):*
  1. Equipment and supplies: Follow the general procedures for milk and cream [4.1A to 4.1B and 4.2A2].
  2. Procedure: Collect samples in original, unopened, hermeticallysealed containers. When necessary to take samples from opened containers, e.g., bulk tanks, or industrial-size containers, provide equipment for transferring representative portions to sample containers [4.1A3]. Promptly identify all samples legibly and indelibly with an official number or tag corresponding to the inspection record. Cool perishable samples to 0 to 4.4°C and deliver promptly to the laboratory [4.1C].
     a. Retail-size cans: Where sterility of contents of retail cans is to be determined, take the original unopened package to the laboratory.
     b. Bulk tanks and industrial-size containers: To determine the bacterial count of evaporated or concentrated milk in bulk tanks or industrial-size containers, thoroughly agitate the product and aseptically transfer a representative portion (not less than 30 mL) to a sterile sample container [4.1A, 4.1B], cool to and maintain at 0 to 4.4°C [4.1C].

B. *Dry milk, whey, casein, dry whip topping, and related products (Class 0):*
  1. Equipment and supplies: Provide sterile or sanitized trier(s), spoons, or spatulas for sample collection.
  2. Procedure: Follow the general procedures and the equipment specified for milk [4.1A, 4.1B]. With sterile tube or trier, transfer at least

4.2 Other Dairy Products

three plugs or columns from each large bulk-type container; or with sterile spoon or spatula, transfer not less than 30 g of product from retail-type containers to sterile sample container. Since dry milk absorbs moisture quickly, make transfers rapidly after opening container, and seal container immediately after completing transfer. Sample contents of stitched bags by opening them at the top. Minimize contamination when untying bags; collect a sample and reseal the bag. Where necessary, aseptically prepare a composite sample in a large sample container and, after proper mixing, transfer representative portions to sample container. If some portions of the package appear to be damp, caked or soiled, sample such portions separately, noting the conditions in the sampling record and the bacteriological report.

C. Butter, margarine, and related products (Class 0):

Follow the general procedures for sampling milk and cream [4.1A, 4.1B] and butter [4.2C1 and 2]. Cool samples at once to 0 to 4.4°C, record temperature, time and date of collection, and deliver as soon as possible to the laboratory [4.1B]. Although butter is specifically mentioned in the discussion that follows, procedures are applicable to margarine and related products.

1. Equipment and supplies:
    a. Butter trier, sanitized: Under commercial conditions, when numerous samples must be taken daily, use of a sterile trier for each sample may be assured if a single, polished, stainless steel trier is wiped thoroughly after each sample with clean tissue paper, dipped in 70% alcohol, ignited to remove excess alcohol, and tempered by twice inserting it into the butter to be sampled before removing the official sample. For composition or chemical examination, clean and dry the trier between each sample.
    b. Triers, spatulas, or spoons, sterilized: Wrap stainless steel triers, spatulas, or spoons in kraft paper or place them in metal or glass containers before steam sterilization.
    c. Sample jars or containers, sterilized: Size 240 mL or larger, with screw tops; or pre-sterilized plastic containers.
2. Procedure:
    a. Butter from churns: Transfer three 120-g portions of finished butter, with sanitized stainless steel trier, from center and end locations of a batch of butter or from successive sampling points in the stream of butter from continuous churns.
    b. Butter in boxes or bulk packages: With a stainless steel trier, remove at least 120 g of butter from three different corners of the container by inserting trier diagonally through the butter. Take plugs at least 75 mm long, including surface portions. With a sanitized spatula or spoon, transfer butter to sterile sample

container. Use a small portion of the sample plug to seal the hole from which the plug was removed. (NOTE: A sample from the corner of the butter box, which includes the outside surface on three sides, is more effective for detecting surface mold contamination than is a subsurface sample.)

c. Print butter: Because of differences in the ratio between surface area and weights of 113.5, 227, and 454 g quantities (prints), preferably remove samples from print with trier to insure uniformity in surface area per sample. Use a No. 8 sanitized cheeze trier to take at least a 75-mm plug (approximately 120 g) from end of print and transfer it (including surface portion) with sanitized spatula or spoon to the sample container. To sample a 113.5 g print, send entire print to the laboratory.

D. *Cheese and other cultured products* (Class 0):
1. Equipment and supplies: Follow general procedures for milk and cream [4.1A-4.1B] and cheese [4.2D 1a and b].
    a. Sampling instruments, sterile: Cheese trier, cork borer, spoon, wooden tongue depressors, knife, or spatula to remove sample. Wrap the sampling equipment in kraft paper, parchment, or foil, or place it in covered metal or glass containers and sterilize [4.1B 1 and 2]. Where insufficient laboratory-sterilized sampling equipment is available, sanitize instruments [4.1B 3 and 4 and 4.2c 1a].
    b. Sample container: Sterile glass or presterilized plastic containers, clean, wide-mouth, dry, leakproof, with a suitable space for sample identification [4.1A 3].
2. Procedure: Collect samples in original unopened containers, or otherwise transfer representative portions from opened containers, using apparatus as outlined for milk and cream [4.1A and 4.1B] and cheese [4.2D 1a]. Submit cottage cheese and cultured products to the laboratory in the original containers except where container size makes this impractical. Since numbers of coliform organisms tend to decrease rapidly in cultured products because of the acid environment, examine samples within 24 h after products are manufactured. Protect samples from potential contaminants at all times. Promptly cool samples to 0 to 4.4°C and transport or ship to the laboratory for examination within 24 h after processing for all acid-treated or cultured products.
    a. Surface samples: Surface portions of cheese are subject to diverse forms of contamination. Surface microorganisms may differ from these in the interior because of the different environment. Surface layers should be sampled only as required to establish presence or absence of surface contaminants, or as necessary for other special purposes.

b. Natural, processed, dehydrated, and imitation cheeses: Sample these types of cheese with sterile or sanitized sampling instruments [4.2D 1a]. Transfer at least five samples, each consisting of not less than 5 g, from different parts of the individual cheese to sample container. Discard the outer 2-cm portions of each sample. Remove the surface portions of soft cheese as noted. With a sterile or sanitized instrument, transfer at least five samples, each consisting of not less than 5 g from different parts of the cheese, from freshly exposed areas to the sample container. Cheese can be sampled by the following techniques depending upon the size and shape of cheese. If the cheese has a circular base, make two cuts radiating from the center of cheese using a knife with a pointed blade. If the cheese is rectangular, make two cuts parallel to the sides. Discard outer surfaces and place resultant piece of cheese into sample container.

Sampling can be done using a cheese trier and one of the following techniques depending upon the shape, size and type cheese: 1) insert trier obliquely towards center of cheese once or several times (depending upon the sample size desired) into one of the flat surfaces at a point ≤ 10 cm from edge, 2) insert trier perpendicularly into one face and pass through center of cheese to opposite face, 3) insert trier horizontally into vertical face of cheese midway between two plane faces, toward center of cheese, and 4) with cheese transported in barrels, boxes or other bulk containers, or cheese that is formed into large compact blocks, sample by passing trier obliquely through contents of container from top to base. Use outer 2 cm or more of plug containing rind to close the hole made in cheese. The remainder of the plug constitutes the sample. Close plug holes and seal over with suitable sealing compound such as one made by mixing molten paraffin, beeswax and white petrolatum (1 + 1 + 2) or white petrolatum and paraffin (1 + 1). Alternatively, reseal cheese in a vacuum-heat sealable plastic pouch.

Small cheeses and cheese packed in small containers are normally sampled by taking the entire cheese. Crumbled, grated, and dehydrated cheese can be sampled by transferring at least five samples, each consisting of not less than 5 g, from a single lot of cheese to a sample container.

c. In a barrel container the mass of cheese (225 to 250 kg) cools slowly from the outside toward the center. By the time cooling is complete, a moisture pattern in the the cheese is fixed with the highest on the outside, lowest in the center and the average somewhere in between.

Studies were supervised by the National Cheese Institute (NCI) and the University of Wisconsin to determine the most representative sampling areas.[1] Over nine thousand moisture analyses were made on current, medium cured, and aged cheese from representative geographic regions. The results were statistically analyzed to identify a moisture profile in barrel cheese.

The sample procedure resulting from this study indicates that a 30.5 cm blade length trier be inserted 7 cm from the edge of the cheese and toward the nearest outside edge of the barrel at 11 degrees from vertical. The center of the plug hole shall be 7 cm from the edge of the cheese. A trier guide fixed at the 11 degree angle is available and may be used as an aid. (For information on guide, contact NCI, 699 Prince St., Alexandria, VA 22314.)

If the cheese barrel is full, i.e., with not more than a 2 to 3 cm headspace, it is possible to draw a reliable sample through a bung or sample port in the cover which permits insertion of the trier at the 7 cm point. If the head space is more than 3 cm, the cover should be removed as the point of trier insertion will otherwise be distorted. In no instance should a barrel containing cheese from more than one vat be selected as a sample for moisture analysis.

It is desirable for the reliability of the sample that the trier be inserted to draw a full 27.9 to 30.5 cm (11 to 12 in) plug from a full container. If a plug breaks short of 25.5 cm, (10 in) draw another plug from a different location 7 cm from the edge.

For a plug between 25.5 and 30.5 cm remove the top 11.43 cm (4.5 in) for sealing the plug hole. Transfer the next 10.16 cm (4 in) portion to the sample container. The remaining bottom portion of the plug should be discarded.

E. *Ice cream and related frozen products (Class 0):*

Observe the general sampling procedures for milk and cream [4.1A, or 4.1B]. Ice cream ingredients such as evaporated milk, condensed milk, dry milk, whey, butter, fluid milk, and cream are sampled according to applicable procedures [4.2A to 4.2C].

Keep frozen product samples frozen until they arrive at the laboratory and the analysis is begun. Dry ice is the refrigerant of choice. Store samples of semi-frozen products (mix, soft-serve) at 0 to 4.4°C. Start examinations within 36 h after collection for mix and soft-serve products. Samples of dry materials may be transported to the laboratory without refrigeration.

1. Equipment and supplies [4.1A to 4.1B].
    a. Sample cases for transporting samples [4.1A4].
    b. Sampling instruments: Sterilize in the laboratory or sanitize immediately before use [4.1B1 to 4]. Collect samples of semi-frozen products (mix or soft-serve), with a sanitized dry spoon

4.2 Other Dairy Products

or sampling tube, or a stainless steel dipper with a capacity of approximately 50 mL. For frozen ingredients, use a sanitized dry knife, spoon, butter trier, or spatula. Samples of eggs or fruits, frozen solid in large blocks or containers, may consist of composite borings taken with a sanitized boring device such as a bit or hollow tube driven by an appropriate electric drill. Where cans or other large-volume containers or unfrozen or melted products are to be sampled, agitate the product until contents are thoroughly mixed [4.1A7, 4.1C2].

c. Sample containers: These should be clean, sterile, dry and leakproof, have a wide mouth, and suitable space for sample identification [4.1A3].

2. Sampling procedures:
   a. Packaged samples of ice cream and frozen desserts: If possible collect and submit samples in their original containers, keeping the samples frozen at all times until they are to be analyzed. Otherwise, follow the procedure [4.2A2] for samples not in original containers.
   b. Bulk samples of ice cream and frozen desserts: Aseptically transfer representative portions of not less than 50 g to sterile sample containers [4.1A3] for bacteriological examination. If the container has been opened previously, use sterile instruments to remove and discard the surface portion to a depth of 2 cm around the area from which the sample is to be taken. With a second sterile instrument, aseptically transfer to the sample container at least 50 g from the newly exposed surface.

   If the object is to determine the care used by the retailer in vending operations, sample surface portions only, to a depth of 1 cm of the sample, using a minimum of 50 g of material. Sample individual servings of frozen dairy products as dipped by the retailer by taking representative portions of not less than 50 g with a service dipper or other dispensing device, and treat such samples as regular bulk samples. When high bacterial counts have been observed previously on samples from the retailer's cabinet, take additional samples from unopened packages to determine whether the responsibility for high counts rests with supplier or retailer.
   c. Process samples: When a particular part of the processing operation is being evaluated, take samples of unfrozen or partially frozen mixes by passing a sterile sample bottle or dipper under properly sanitized orifices at intervals or by withdrawing samples with sterilized sampling instruments. Thoroughly agitate contents of cans or vats before aseptically removing 50-g portions to sterile containers.

d. Coloring and flavoring materials: Mix individual solutions of coloring materials, extracts, flavors, etc., until they are homogeneous and pour directly and aseptically into a sample container; or transfer with a sterile sampling tube or pipette not less than 30 g from the original container into a sterile sample container. With a sterile spoon, aseptically transfer not less than 30 g of well-mixed coloring powders, fruits, fruit preparations, nuts or nut preparations, cakes, pastries, or confections to a sample container. Where cold-pack or frozen fruits are not thawed, use a high speed electric drill with sanitized auger, hollow coring tube, or other suitable sharp, sanitized instrument to obtain enough material for the sample; assure a representative cross section of material from outer edge to center of block. Aseptically transfer representative portions to the sample container. Cool perishable unfrozen samples to 0 to 4.4°C and place under refrigeration. Maintain frozen products in frozen condition and transport as soon as possible to the laboratory.
e. Stabilizers and emulsifiers: With sterile or sanitized trier, transfer at least three plugs or columns (total not less than 120 g) from each container or with sterile or sanitized spoon or spatula remove not less than 120 g of powder from various places and transfer to sterile sample container. Make transfers as rapidly as possible after opening containers and immediately seal the sample container after transfer is completed. If some portions of package contents are believed to be more highly contaminated than others, sample such portions separately.
f. Sweetening agents: Transfer not less than 30 g of sweetener to sterile sample container, cool to 0 to 4.4°C, and transfer to the laboratory.
g. Eggs and egg products[6]: Take samples from a representative number of containers in the lot. Sterilize all sampling equipment (sampling tubes, dippers, ladles, augers, borers, corers, spoons, triers, chisels, and hatchets) by sponging off surfaces with alcohol and flaming over alcohol lamp or similar burner. Between samplings, thoroughly wash instruments, dry, and resterilize. Open all containers under as nearly aseptic conditions as possible.
  1. Liquid eggs: With a sterile tube or dipper mix contents of container until homogeneous. With a sterile long-handle dipper or ladle aseptically transfer about 400 mL to a sterile sample container. Place sample in sample case at 0 to 4.4°C and deliver promptly to laboratory. Preferably examine upon receipt or store at 0 to 4.4°C until tested. After removing sample,

record odor of each container sampled as normal, abnormal, musty, or other obvious deviation from expected.
2. Frozen eggs: Remove top layer of egg with a sterilized hatchet or chisel. Take three cores with a sterile auger or corer from the top to the bottom of the container; first core in center, second core midway between center and periphery, and third core near the edge. With a sterile spoon transfer cores to a sterile sample container. Pack sample in an insulated box with dry ice for transport to the laboratory. Keep frozen until ready for analysis. Examine organoleptically by smelling sample after removal of the bacteriological sample.
3. Dried eggs: For small packages, take entire package as sample. For boxes and barrels, remove top layer with a sterile spoon or other sterile instrument. With a sterile trier remove ≥ 3 cores as in [4.2E2g2] above. Sample should consist of about 400 g. Aseptically transfer core to sample container with sterile spoon or other suitable instrument. Keep samples at 0 to 4.4°C until tested.

h. Sample handling: Maintain frozen samples in frozen condition until examined. Samples collected locally that are to be tested promptly may be packed in cracked ice and water when no other refrigeration is available. Store perishable, semi-frozen, or unfrozen samples at 0 to 4.4°C until examined. Keep records of the time when samples are collected and shipped or delivered to the laboratory. Test unfrozen samples as soon as possible and always within 36 h after collection. Test frozen samples within 72 h after collection. Prevent the thawing and refreezing of frozen samples. Avoid freezing frozen dessert mixes and semi-frozen soft-serve products.

F. Aseptically (UHT) processed and packaged milk and milk products:
The routine sampling and testing of aseptically (UHT) processed and packaged milk and milk products is generally done at the processing location, to meet low-acid, canned food regulations (21CFR 113). If samples are to be shipped to a laboratory, collect them in their original, unopened, hermetically sealed containers. Identify all samples legibly and indelibly with the time and date of collection and the lot number, have the official number or tag correspond to the inspection record. Samples should be rapidly delivered to the laboratory with the original storage conditions maintained if possible. The time and date the samples arrive at the laboratory should also be recorded.

The product need not be refrigerated if unopened. Transport the samples in a sturdy box or container to avoid crushing or damaging the package.

4.3 Sampling for Specific Laboratory Procedures

A. Direct microscopic count methods (Class 0) [Chapter 11]:

Procedure for collecting samples: To collect samples for direct microscopic examination, follow the general procedures for milk and cream [4.1A, 4.1B, and 4.1C2].

B. Reduction methods (Class 0) [Chapter 13]:
  1. Equipment and supplies: See 4.1A and 4.1B; use transfer dipper or pipette to deliver 10-mL portions.
     a. Tubes, vials, or suitable nontoxic plastic bags: Use tubes without lips, preferred length 150 mm × 15 to 18 mm inside diameter, or special tubes of similar diameter with 10-mL graduation. Uniformity in diameter and length of tubes or vials permits interchangeable use of closures and possible use of a rigid cover to hold firmly over closures when inverting a rack of tubes. Samples may be collected in plastic bags [4.1A3] and transferred later to tubes or vials.
     b. Racks, sheet metal or wire: Corrosion-resistant, for holding tubes vertically in single or multiple rows.
  2. Procedure: Using sterile equipment [4.1B1 and 2], transfer 10-mL portion of well-stirred or otherwise adequately mixed sample directly from can, weigh tank, or other container to the tube in which the test is to be made. Optionally, take samples as in 4.1C1 and 4.1C2a and b, but before subsequent transfer of 10-mL portions to tubes or vials, shake samples by rapidly inverting each container 25 times [6.5B]. Use a sterile pipette for each sample, or sanitize the dipper or pipette between samples [4.1B3 and 4]. Test sample(s) within 36 h after collection.

C. Microbiological tests of containers: closures, equipment, water, and air (Class 0) [Chapter 15]:
  1. Equipment and supplies: Same as 4.1A, plus the following:
     a. Tubes or vials, sterile: Same as tubes and vials under 4.32B1a.
     b. Scalpels or scissors, sterile.
     c. Kraft paper or envelopes, sterile.
     d. Forceps, sterile.
  2. Procedure for paper stock[14]: Paper stock shall be sampled before it enters into any converting operation. Partially converted items may also be sampled in intermediate stages of conversion. Sampling methods for these items shall, in general, comply with sampling methods for paper stock or finished products. Items fabricated from plastic or plastic-laminated paper or paperboard, dry-waxed paper, wet-waxed paper, metal, and/or combinations of these materials are usually not sampled before conversion; therefore, sampling methods for finished products only would apply to them.

4.3 Sampling for Specific Laboratory Procedures

a. Roll stock sampling: Either of the following two methods is acceptable.
   1. Roll stock shall be sampled as a butt roll, with a minimum radial thickness of 2.54 cm of paper on the core and a minimum roll width 30.48 cm. With a lesser mill roll width, the full width of the mill roll would constitute the sample. The cut web shall be firmly taped before removal of the butt roll from the roll stand. If the width of the roll is over 30.48 cm, for convenience the butt roll may be cut to a minimum of 30.48 cm, by using either a handsaw or a power band saw, with care to minimize contamination. The roll stock sample shall be wrapped carefully in wrapping paper and sealed with tape.
   2. Samples shall be cut from the roll of paper with a clean sharp knife as follows: Cut into roll two laps deep and discard these first two layers. Then cut into roll six laps deep to a size of approximately 20.22 × 25.4C cm, grasp sample by top and bottom sheets near corners, remove and tape in three or four places to keep sheets in alignment, thus protecting inner sheets for testing. When cutting samples from the roll of paper, three sides of the sample may be cut first, thus forming a flap. This flap may then be stapled or taped to prevent undue exposure and riffling, and the hinge side of the sample is finally cut as assembled sample is removed. Place in a clean paper envelope, seal, and insert into a stout manila or brown kraft envelope. For sample protection, envelopes should be of a type that can be sealed, preferably with a pressure-sensitive adhesive rather than one that must be moistened.
b. Sheeted stock sampling: Final samples shall consist of a minimum of six sheets, each approximately 20.32 × 25.40 cm. Sheeted paper selected for sampling shall be taken from any portion of the stock except the bottom or top 5.08 cm. Large sheets shall be reduced to a final sample size, using a guillotine-type cutter or equivalent means not involving direct handling of any sample sheets. Stacks of cut sample sheets shall be taped and packaged as in paragraph 2 above. Top and bottom sheets constitute protective covers during sampling and shipping, and shall be discarded before testing.
3. Procedure for single-service product samples[14]: Samples shall consist of a regular unit package, when the unit package contains at least 50 individual items. If a unit package contains fewer than 50 individual items, multiples of the unit package shall be used to make up one sample. The sample unit package(s) shall be taken from regular stock produced under normal operating procedures. The sample unit packages shall not be opened. The ends of each sample unit shall

be sealed with tape to protect the sample from contamination during shipment to the laboratory. Upon arrival at the laboratory, samples should be examined for transit damage. If damage is found, replacement samples should be submitted.
4. Sampling procedures for rinse testing:
   a. Firm-walled capped items: Remove four containers from the conveyor line (without touching lips or interiors) before containers reach the filler valves. Insert the containers to be capped mechanically into the conveyor lines beyond the filler valves. Should inspection show that mechanical cappers are insanitary, apply laboratory-sterilized closures to four other containers taken from the conveyor before reaching the filler valves. Transport to laboratory unrefrigerated.
   b. Firm-walled uncapped items: Paper, laminate, or plastic containers that are formed, filled, and sealed in a single device must be obtained after sanitization but before the product to be packaged is introduced. Collect four sealed containers from line before product is introduced to filler. Transport to laboratory unrefrigerated.
   c. Flexible-walled items: Collect four containers, transport to laboratory unrefrigerated.
   d. Modification for CIP equipment: The rinse solution method may be employed on individual items of equipment or on systems designed for "cleaning-in-place." This method is not practical for other equipment because the volume of rinse solution required would be equivalent to the capacity of the entire unit and adequate agitation to flush surfaces thoroughly might not be available.

   The minimum amount of solution needed to thoroughly flush the product-contact surface of CIP items is much greater than can be autoclaved conveniently. Instead, determine the volume of water needed to completely fill the CIP system[16] and sanitize the rinse solution by susperchlorination (25 mg chlorine per L and hold for 10 min). Then neutralize the hypochlorite by adding an excess of sterile 10% sodium thiosulfate solution (0.2 mL per 100 mL of rinse solution). Tap water may also be sterilized by membrane filtration followed by addition of sodium thiosulfate to inactivate residual disinfectant [15.3A1F].

   Use a suitable wide-mouth container of sufficient size (at least 1 L) made of borosilicate glass or polypropylene plastic. Keep 1 L of the chlorine-treated water that has been neutralized for use as a control sample.

   Introduce neutralized chlorine-treated water into the "ready for use" CIP unit. Start circulation and continue for 20 times the period required to move the same volume of water with the pump.

4.3 Sampling for Specific Laboratory Procedures

Stop the pump and open the outlet value at the lowest point in the system. Fill separate sterile sample containers at the beginning, middle, and end of the unit-emptying step.

For large volumes of neutralized chlorine-treated rinses, the membrane filter procedure may be used to analyze both rinse and control samples.

Cool rinse and control samples to 0 to 4.4°C and transport to the laboratory.

e. Modification for line sampling: The sanitary condition of components of various assemblies can be determined by sampling milk progressively at specific sanitized access points [4.1A2]. Samples may be analyzed using the Standard Plate Count to determine where significant changes in SPC counts occur. Specific flora can be determined by employing appropriate differential media [Chapter 9] and incubation temperatures. To determine the presence of thermoduric bacteria, samples may be laboratory-pasteurized and then plated.

f. Modification of rinse test for ice cream freezers: The sanitary condition of scrapers and interior surfaces of "batch" freezers may be determined by a modified rinse method. Prepare a sterile rinse solution [16.3A1e] in 100-mL quantities per 9.46-L capacity of batch freezer barrels to be tested. With the outlet valve of the barrel closed, pour sterile rinse into inlet. Operate the scraper for 2 min, collect sample through outlet valve into a sterile bottle, cool to 0 to 4.4°C, and transport to laboratory.

5. Swab test procedure: Use the swab test method to determine the sanitary condition of equipment and large containers that can not be tested by the rinse method. Open a sterile swab container, grasp end of stick (being careful not to touch any portion that might be inserted into the vial), and remove aseptically. Open a vial of buffered rinse solution [15.3B1b], moisten swab head, and press out excess solution against the interior of the vial with a rotational motion. Hold the swab handle to make a 30°-angle contact with the surface. Rub the swab head slowly and thoroughly over approximately 50 cm^2 of surface. Rub the swab three times over this surface, reversing direction between successive strokes. Return the swab head to the solution vial, rinse briefly in solution, and then press out excess. Swab four more 50-cm^2 areas of surface of the item, as above; rinse swab in solution after each swabbing and remove excess moisture. After the fifth area has been swabbed, position head in vial and break or cut with sterile scissors or other device, leaving swab head in vial. Replace screw cap, put vial in waterproof container, cool to 0 to 4.4°C and deliver to the laboratory. Begin tests within 36 h of collection.

6. Procedures for water supplies: Procedures and equipment for sampling water supplies are detailed in the current edition of *Standard Methods for the Examination of Water and Wastewater*.[5]
7. Procedures for air supplies: Sampling for airborne microorganisms in dairy plants involves collecting samples in air pathways by which organisms may be introduced onto processing equipment. Samples may be taken: a) at openings in equipment subject to potential contamination from organisms transported by air currents, b) at selected points for testing quality of room air, (e.g., where product is filled into containers), and c) in areas where employees are concentrated. Because of air turbulence during operating hours, sampling by volumetric methods will be more effective and dependable than use of the sedimentation technique.

   Sampling time for all methods of collection usually is standardized at 15, 30, and 60 min, with duration determined by experience or estimation [Chapter 15]. Samples may be taken monthly to detect specific microorganisms unless more frequent samples are necessitated by emergency conditions or suspected contamination.

D. *Sediment in fluid milk (Class 0) [Chapter 16]*:
  1. Mixed-sample method: For retail containers, 18.92- to 37.85-L cans, and bulk storage tanks, a 500-mL, 120-mL, 60-mL, or 30-mL sample may be used depending on the diameter of the filter disc being used. Before mixing the sample, with a small strainer transfer any floating extraneous contaminants such as flies, hair or debris to a separate disc and properly identify. Agitate contents of bulk storage tanks for at least 5 min or according to manufacturer's recommendation before test portions are removed. Avoid contamination of samples with foreign matter.
  2. Off-bottom method: For 18.92- to 37.85-L cans, take a 500-mL sample with an off-bottom tester from an unstirred can of milk. Before withdrawing sample, remove any floating extraneous matter with a small strainer as directed in 4.3D1.

E. *Phosphatase method (Class 0) [Chapter 17]*:
  1. Equipment and supplies:
     a. Trier: must be thoroughly clean, phenol-free and sterile for sampling solid and semisolid products.
     b. Tubes, bottles, plastic bags: clean and phenol-free.
  2. Procedure: Collect not less than 10 mL of fluid milk products or not less than 10 g of solid or semisolid dairy products. Obtain samples of solid products from the interior with a clean trier. Avoid taking samples from exposed surfaces of solid or semisolid products. Take all samples in phenol-free containers [4.1A3] and refrigerate at 0 to 4.4°C until tested. Examine the sample within 36 h. Preferably take samples for phosphatase before addition of nuts, fruits, candy, etc.

4.3 Sampling for Specific Laboratory Procedures

When testing flavored products, however, strain out nuts, fruits, etc., before testing. If a preservative is necessary, add 1 to 3% chloroform.

F. *Other chemical methods (Class 0) [Chapter 18]:*
1. Equipment and supplies: Tube, vial, or plastic bag: Airtight, nonabsorbent.
2. Procedure: Size of sample needed varies with analyses required. For the usual analyses, collect approximately 500 mL of sample; for milkfat determination, about 120 mL. Collect one or more containers of bottled or packaged milk as processed for sale. Thoroughly agitate milk until homogeneous by pouring back and forth from one container to another. If a cream layer has formed, dislodge from sides of the container and stir until evenly distributed.

Place sample(s) in nonabsorbent, airtight containers(s). Cool to and store at 0 to 4.4°C until examined. Use preservative as needed. Tablets containing potassium dichloride or other suitable preservative, with at least 0.5 g of active ingredient per tablet, should be used for each 240 mL of milk, but the total weight of such a tablet should not be more than 1 g. A 36% solution of HCHO, 0.1 mL (2 drops) per 30 mL, may be used unless the presence of the preservative is objectionable in physical or chemical tests to be made in addition to the determination of milkfat.

G. *Radionuclides in milk (Class 0) [18.15]:*

Sampling of milk for analysis depends on whether samples are intended for determination of radionuclides: 1) as affected by farm practices, 2) in the vicinity of nuclear power generating stations, or 3) at the consumer level. To study the effect of farm practices on radionuclides in milk, samples are collected from individual cows or a portion of the pooled milk is taken from the storage tank on the farm. Milk samples in the vicinity of nuclear power generating stations are taken from within a discrete geographical area depending on meteorological conditions and the number of farms affected. The sample may be from an individual farm or from groups of farms. To study market milk, samples at each location are composites of subsamples from each milk processing plant, the quantity being proportional to the plant's average sales in the community served.

Frequency of sampling depends on the seriousness of the situation and the radionuclide considered for analysis. With iodine-131 (^{131}I, whose half-life is 8.08 days) it is necessary to collect milk samples every day or at least once a week. With strontium-90 (^{90}Sr, whose half-life is 28 years) it is necessary to collect at least once a month.

At least 3.78 L of milk (raw or pasteurized) are collected, preserved with formaldehyde (H_2CO) (10 mL of 40% H_2CO per 3.18 L), and shipped to the laboratory for analysis. H_2CO does not interfere with analysis of ^{131}I, ^{140}Ba, ^{137}Ba, and ^{137}Cs by gamma spectroscopy or ^{89}Sr and ^{90}Sr by ion-exchange and/or by chemical methods. Addition of H_2CO to raw milk does, however, cause transformation of ^{131}I to protein-bound ^{131}I. The enzyme, xanthine oxidase, in raw milk, oxidizes H_2CO to H_2O_2 in the presence of oxygen, and H_2O_2 in the presence of oxygen and H_2O_2 oxidizes ^{131}I to ^{131}I2, which binds to protein. In pasteurized milk or dried milk, the enzyme is absent, consequently, addition of H_2CO to processed milk has no effect on ^{131}I. Processed milk containing ^{131}I from fallout (or labeled in vitro) can be preserved with H_2CO and successfully used for determination of ^{131}I by ion exchange methods. However, if raw milk is the sample of choice and ^{131}I is the only radionuclide of concern, no preservative should be added to the milk. Analysis can be simplified by passing milk through an anion-exchange resin column at the sampling station and then shipping the column to the laboratory for further processing. This presupposes that all the ^{131}I in the milk is in the inorganic form as iodide.

Due to the complex nature of these determinations, The Association of Official Analytical Chemists, *Official Methods of Analysis*[6] should be consulted for detail concerning methods.

4.4 References

1. BLATTNER, T.M.; OLSON, N.F.; WICHERN, D.W. Sampling barrel cheese: comparison of methods. J. Assoc. Offic. Anal. Chem. (submitted); 1985.
1a. BRAZIS, A.R.; BLACK, L.A. Bacterial counts of bulk milk for interstate shipment. 1. Effect of sampling procedures. J. Milk Food Technol. **25**:172–175; 1962.
2. CLAIBORNE, F.B.; COX, K.E. The direct microscopic count on preserved milk samples; an effective measure for uniform state-wide control. J. Milk Food Technol. **14**:105–108, 1951.
3. DODGE, H.F.; ROMIG, H.G. Sampling inspection tables, 2nd ed. New York, NY: John Wiley and Sons; 1959.
4. ELLIKER, P.R.; SING, E.L.; CHRISTENSEN, L.J.; SANDINE, W.E. Psychrophilic bacteria and keeping quality of pasteurized dairy products. J. Milk Food Technol. **27**:69–75; 1964.
5. GREENBERG, A.E.; CONNORS, J.J.; JENKINS, D., EDS. Standard methods for the examination of water and wastewater, 15th ed. Washington, DC: Amer. Pub. Health Assoc.; 1981.
6. HORWITZ, W., ED. Official methods of analysis, 13th ed. Washington, DC: Assoc. Offic. Anal. Chem.; 1980.
7. INTERN. ASSOC. MILK FOOD ENVIRON. Sanitarians; U.S. Pub. Health Serv.; Dairy Ind. Committee. 3-A accepted practices for supplying air under pressure in contact with milk, milk products and product contact surfaces, serial #60403. J. Milk Food Technol. **35**:378–382; 1972.
8. INTERN. ASSOC. MILK FOOD ENVIRON. Sanitarians; U.S. PHS; DIC. 3-A sanitary standards for storage tanks for milk and milk products, 01–06. J. Milk Food technol. **37**:36–61; 1974.
9. INTERN. ASSOC. MILK FOOD ENVIRON. Sanitarians; U.S. PHS; DIC. 3-A sanitary standards for farm milk cooling and holding tanks, no. 13–06. J. Milk Food Technol. **38**:646–653; 1975.
10. INTERNATIONAL COMMISSION ON MICROBIOLOGICAL SPECIFICATIONS FOR FOODS OF THE INTERNATIONAL ASSOCIATION OF MICROBIOLOGICAL SOCIETIES. Microorganisms in Foods. 2. Sampling for microbiological analysis: principles and specific applications. Toronto Canada: Univ. Toronto Press; 1974.
11. LEVINE, B.S. Effect of formaldehyde on the direct microscopic count of raw milk. Pub. Health Rep. **65**:931–938; 1950.
12. MARSHALL, R.T.; SHELLEY, D.S.: Comparisons of tests of milk samples taken conventionally and with an automatic in-line sampler. J. Food Prot. **44**:257–262; 1981.
13. MARTH, E.H.; REINBOLD, G.W.; MARSHALL, R.T.; MATHER, D.W., CLARK, JR., W.S.; DIZIKES,

4.4 References

J.L.; HAUSLER, JR., W.J.; RICHARDSON, G.H.; ULLMANN, W.W. An addition to the thirteenth edition of *Standard Methods for the Examination of Dairy Products*. Single-service plastic vials for samples of raw milk and cream. J. Milk Food Technol. 37:159; 1974.

14. PARROW, T. Manual of recommended laboratory methods for sampling and microbiological testing of single service food packages and their components. Syracuse, NY: Syracuse Research Corporation; 1978.
15. U.S. DEPT. OF HEALTH AND HUMAN SERV. Fabrication of single-service containers and closures for milk and milk products. Washington, DC: PHS/FDA.
16. U.S. DEPT. OF HEALTH AND HUMAN SERV. Grade A pasteurized milk ordinance. No. 017-001-00419-7. Washington, DC: U.S. Gov. Printing Office; 1980.
17. U.S. DEPT. OF HEALTH AND HUMAN SERV. Sanitation compliance and enforcement ratings of interstate milk shippers, Quarterly publication. Washington, DC: PHS/FDA.

CHAPTER 5

MEDIA

R. H. Bell, D. A. Power, and G. H. Richardson
(G. H. Richardson, Tech Comm.)

5.1 Introduction

Accuracy and precision of microbial data obtained from testing of dairy foods depend upon the culture media employed and the way these media are used in the laboratory. Media, reagents, and dilution water must be prepared and handled properly.

The culture media have been given generic names throughout this book. Commercially available dehydrated media should be used exclusively. Composition information and preparation directions are on the labels. Only information on Standard Methods Agar (SMA) and media not readily available commercially is included in this edition [5.9].

5.2 Suitability of Water for Microbiological Applications

Only water that has been treated to free it from traces of dissolved metals, and bactericidal and inhibitory compounds should be used to prepare culture media, reagents, and dilution blanks. Inhibitor-free water will be referred to as Microbiologically Suitable (MS) water. For routine use in dairy product analysis, it is the growth medium or dilution system prepared with the water that is of concern, not the water itself. Culture media provide considerable protection from toxic agents present in the water;[22] therefore, the primary interest in water quality for routine applications is in its use as a dilution fluid. A procedure to determine if dilution water may have a toxic effect is given in Section 5.4A.

Specific resistance in ohms may be determined as a rough indicator of distilled water quality.[26] Generally, between 400,000 and 500,000 ohms specific resistance is the breakpoint between acceptable and unacceptable distilled water. However, only ionic concentration is measured by specific resistance; the presence of organic substances is not detected.

Portions of the following are paraphrased from *Standard Methods for the Examination of Water and Wastewater, 15th edition.*[5]

Distilled water may contain ions from the materials used in the still. Stills fabricated of borosilicate glass, fused quartz, silver, or block tin are preferred. Distillation will not remove ammonia (NH_3) or carbon dioxide (CO_2). Distilled water is often supersaturated with carbon dioxide because of the decomposition of raw-water bicarbonates to carbonates in the boiler. Remove ammonia by distillation from acid solution or by passing the water through a column of mixed anionic and cationic resins. Remove carbon dioxide by distillation from a solution containing an excess of alkali hydroxide, by boiling for a few minutes, by vigorously aerating the water with a stream of inert gas for a sufficient period, or by passing the water through a column of strong anion-exchange resin in the hydroxide form. Distilled water kept in glass containers slowly leaches the more soluble material from the glass, and the concentration of total dissolved solids increases with time of storage.

Four types of laboratory water are described:

A. Distilled water:

Prepare reagent-quality water by distillation. Distilled water quality depends on the type of still and quality of feedwater. Deionized feedwater is preferred. Commercial resin purification trains can produce equivalent or superior water for some applications. Confirm toxicity of distilled water or equivalent for its intended specific analysis [5.4].

Distillate from chlorinated water occasionally contains significant amounts of free chlorine, even when passed through an ion exchange column before use. Residual chlorine in water should be neutralized with sodium thiosulphate ($Na_2SO_2O_3 \cdot 5H_2O$) before use in milk dilution blanks. Addition of 0.1 mL of a 10% solution of sodium thiosulphate to every 120 mL of distilled water will neutralize about 15 mg of residual chlorine.

B. Carbon dioxide-free distilled water:

Prepare fresh as needed by boiling distilled water for 15 min and cooling rapidly to room temperature. Cap the flask or bottle in which the water has been boiled with a slightly oversized, inverted breaker to minimize entry of atmospheric CO_2 during the cooling process. For best results, exclude CO_2 entry with a drying tube containing soda lime.

C. Redistilled water:

Prepare by redistilling single-distilled water from an all-borosilicate glass still.

D. Deionized distilled water:

Slowly pass distilled water through a 25-cm column of glass tubing (1- to 2.5-cm in diameter) that has been charged with 2 parts, by volume, of a strongly basic anion-exchange resin in the hydroxyl form and 1 part by volume of a strongly acidic cation-exchange resin in the hydrogen form. Use ion-exchange resins of a quality suitable for analytical work. Units using such resins in flow-through in sequence can produce water of very high quality and are available commercially.[5]

Deionized or demineralized water from a mixed-bed ion exchanger is satisfactory for standard method applications. This will be referred to as Laboratory Grade (LG) water [18.1]. Because ion exchange units do not always remove nonelectrolytes, colloids, plankton, nonionic organic materials, and dissolved air, however, LG water is not suitable for determinations in which such constitutents interfere. Some ion exchangers leave traces of formic acid or other organic matter that make the demineralized water unsuitable for use in certain microbiological methods. A water toxicity test [5.4] is suggested when the water supply is questionable. Any water passing the water toxicity tests would be considered Microbiologically Suitable (MS) water.

5.3 Preparation of Dilution Water

Dilution blank preparations vary considerably in dairy and food laboratories around the world.[3,4,5,8,11,12,13,15,16,17,19,22,24] Janus[12] summarized the results of an International Dairy Federation/International Organization for Standardization (IDF/ISO) study involving eight different diluents. The study group recommended quarter-strength Ringers solution as a reference standard, with peptone-salt diluent, peptone solution, and phosphate buffer being suitable alternatives. The phosphate solution is retained in *SMEDP* because of its equivalence to other diluents,[12] and the lack of sufficient evidence to dictate a change.[11,19] The $MgCl_2$ modification[8] is now included as a Class C option. Perhaps MS water can be selected in future *SMEDP* editions if proven adequate in high dilutions. Distilled water performed as well as phosphate buffer (without $MgCl_2$)[11,12] and would be useable where raw milk dilutions are low. Class O emphasis of one diluent is essential in this edition, however, to ensure standardization.

A. *Stock phosphate buffer and magnesium chloride solutions:*
  1. *Phosphate buffer stock solution:* Dissolve 34 g of potassium dihydrogen phosphate (KH_2PO_4) in 500 mL of MS water, adjust to pH 7.2 with 1 N NaOH solution, and make up to 1 L with MS water. If desired, place in small vials. Sterilize at 121°C for 15 min. After sterilization, tightly seal the vials and store at 0 to 4.4°C.
  2. *Magnesium chloride stock solution:* To prepare stock solutions, measure 38 g of $MgCl_2$ into a 1-L volumetric flask and add MS water to make 1 L of solution. Small vials may be filled with the stock solution, sterilized, and stored [5.3A1].

B. *Phosphate dilution water (Class O):*
Add 1.25 mL of stock phosphate buffer solution to MS water and make up to 1 L. Dispense in 99 ± 2 mL [2.4D] or 9 ± 0.2 mL quantities or other volumes as required and autoclave at 121°C for 15 min [5.12]. Bacteria should not be suspended in any dilution water for more than 30 min at room temperature to minimize death or multiplication of the bacteria.

C. Phosphate and magnesium chloride dilution water (Class C):

Add 1.25 mL of stock phosphate buffer solution and 5 mL of stock magnesium chloride solution to MS water and make up to 1 L. Dispense in 99 ± 2 mL [2.4D], 9 ± 0.2 mL quantities or other volumes as required and autoclave at 121°C for 15 min [5.12]. The addition of magnesium chloride improves recovery of organisms with metabolic injuries, which may have been induced by toxic properties in the dilution water.[4,8,16]

5.4 Water Toxicity Tests

A. Dilution water toxicity test (Class O):

A screening test for dilution-water toxicity can be conducted by preparing a dilution of the appropriate dairy food to provide 100 to 300 colonies per plate when 1 mL is plated. The diluted sample(s) is then plated in duplicate after 0, 15, 30, and 45 min by the SPC procedure [Chapter 6]. Distinct trends toward lower counts among successively plated samples suggest probable toxicity. Decreases of over 20% in 45 min are indicative of dilution water toxicity.

B. Interval plating procedure (Class D):

Because of its limited usage and the debatable results obtained by using this technique, it is not presented in this edition.[9,20] See [4.7C] of the previous edition.[18]

C. Distilled water quality test (Class O):

An elaborate method is described in the water standard methods book.[5]

5.5 Cleaning Glassware and Testing for Detergent Residues

Modern detergents are very effective for cleaning laboratory glassware. Most of these are of the anionic type, usually with alkaline substances such as phosphates, carbonates, or silicates. Some detergents, especially the cationic type (quaternary ammonium compounds or "quats"), are highly bactericidal and great care must be excercised to ensure their removal. Because detergents and soaps have a great affinity for all surfaces, they displace dirt and allow it to be washed away, but they are difficult to remove completely.

A. Cleaning glassware:

Deposits of "milk stone" or calcium salts, which are resistant to ordinary detergents, are occasionally encountered. These salts must be removed by rinsing glassware several minutes in acid solutions before effective rinsing can be achieved. Commercial dairy-acid cleaners designed to remove milk stone from milk processing surfaces, can be used on laboratory glassware. Proper usage levels should be obtained from the supplier or label.

The detergent wash is best done with hot water after preliminary rinsing with warm water to remove most of the dirt. Soaking aids in the removal of stubborn residues. Traces of acid on glassware can be detected with pH indicator paper or brom thymol blue (BTB) indicator [5.5C] after the last rinse. Six to 12 rinses with running tap water followed by several rinses with MS water are sometimes necessary to completely remove a detergent.

B. Detergent residues:

Laboratories should check for bactericidal detergent residues whenever changing brands or lots of detergents. Evaluate by preparing three sets of glass petri dishes: one set of three dishes is washed and sterilized by the method routinely used; the second set is washed in acid-detergent solution, then in castile soap. This set is then rinsed four times in tap water and six times in MS water before sterilization. A third set is dipped in the presently-used detergent solution and sterilized without rinsing.

A sample of milk is plated in triplicate in these dishes and colonies are counted after two days at 32 ± 1°C. There should be no significant difference in the counts between the first and second sets of plates. A reduction in bacterial count of over 20%, or a diminished size of colonies may be apparent in plates of group three if the detergent in use is bactericidal or bacteriostatic.

C. Glassware pH check:

Acid and alkaline residues can be detected on cleaned labware by adding and dispersing a few drops of 0.04% bromthymol blue (BTB) indicator. Localized blue (alkaline) or yellow (acid) zones in the dispersing BTB solution demonstrate incomplete removal of cleaning solutions. The BTB solution is prepared by dissolving 0.1 g of BTB in 16 mL of 0.01 N NaOH and diluting to 250 mL with MS water; the BTB solution should be green.

5.6 Media

A. Dehydrated:

Dehydrated media should be stored in tightly closed containers at below 30°C and protected from light. If the environment is hot and humid, media may be stored in a refrigerator or freezer. Properly stored, most media should remain stable or at least three years; however, purchases should be planned to encourage a complete turnover within one year.

After its first usage, when the seal has been broken, the subsequent quality of the medium may depend upon the storage environment. Use of a desiccator after the seal has been broken will extend shelf life to about one year. Otherwise, the media must be used or discarded within six months after breaking the seal. Suitable package sizes should be purchased to minimize repeated use of material from unsealed containers. Unsealed containers may admit air and moisture, which can initiate reactions that result in reduced efficiency of media. Any medium that has absorbed moisture and become caked should be discarded. Contamination may take place if unclean spatulas are used to transfer material from containers.

B. Prepared:

The "life" of any prepared medium, whether in tubes, bottles, or plates, depends upon storage conditions and the type of medium. Prepared media should not be stored unless protected against water loss. This may be accomplished by using screw-capped tubes and bottles instead of cotton-plugged or loosely covered containers. Prepared plates should be used within one week, unless stored in moisture-proof containers such as plastic bags. Media in screw-capped tubes should be used within approximately six months. Commercially prepared media should be used prior to the expiration date. If dehydration has occurred, media should be discarded. Prepared plated or tubed media should be stored at 2 to 8°C.

5.7 Basic Steps in Media Preparation

Weigh carefully the proper amount of dehydrated medium.

Place the requisite amount of MS water into a suitable container (e.g., borosilicate glass, stainless steel or heat-resistant plastic such as polypropylene, polycarbonate, or polysulfone).

Add the weighed, dehydrated medium to part of the water. Mix with a stirring rod or magnetic stirrer. Add remaining water and mix again.

If necessary to effect complete solution, heat to boiling on wire gauze over a free flame or on a magnetic stirrer-hot plate. Stir frequently to prevent burning of the medium. A non-pressurized, free-flowing steam unit may also be used. Media containing agar should be boiled 1 min to insure solution of the agar. Prolonged boiling may cause undesirable foaming; this can be reduced by holding the flask in cold water for a few seconds after initial boiling has been accomplished. If necessary, restore water lost by evaporation. For example, weigh the container and media before and after heating. Add sufficient water to restore the original weight before distributing and sterilizing the rehydrated medium. A microwave oven might be used to dissolve dried medium if compatible with the container in use. Adjust the medium pH if necessary [5.8].

Distribute medium into a suitable container. This can be most easily accomplished for a tubed liquid medium by using automatic pipettors. A hand-held adjustable pipettor is most suitable for small numbers of test tubes, whereas an electricity operated machine is more satisfactory for large numbers. In each instance, before use, the machine must be flushed with MS water, followed by medium until the water has been completely removed from the system. After use, the apparatus must be flushed with warm water, detergent solution, tap water and MS water. If any residues accumulate, the apparatus should be disassembled and scrubbed. A small amount of water should be left in the syringe portion to prevent drying out and subsequent "freezing" of the plunger and barrel [2.4E].

The amount of medium per container should be limited so that no point within the volume medium is more than 2.5 cm from the top surface of the medium or area of medium interfacing with the container (this is to assure rapid equilibration of temperature when placed in a waterbath).

Sterilize at 12°C for 15 min or according to recommended procedures for each medium. Checking of pH before use is recommended [5.13G].

Sterilizing efficiency is greatly reduced by overloading the autoclave or by improper spacing of containers. For best results, separate containers by at least 1 cm in all directions. Do not use volumes of medium in excess of 30 to 40% of the volume of the autoclave. If larger volumes are used, sterilizing times should be increased [5.12A].

Melt and hold solid media at 44 to 46°C until used, but not exceeding 3 h. Hold inhibitory-substances media at 55 or 64°C depending upon the test requirements. Solid media may be melted in boiling water, flowing steam not under pressure, or in a microwave oven. The time to assure complete solution will vary with the method and the volume in the container. If a precipitate forms, discard the media. Do not re-melt media.

If it becomes necessary to formulate a dehydrated medium that is not commercially available, use only reagent grade chemicals and USP-approved ingredients prepared by culture media suppliers. Follow manufacturer's instructions for storing stock materials. Discontinue use of materials showing any evidence of contamination, decomposition or dehydration.

Machines for the automatic preparation of agar media are now available and are of significant value where laboratory volume warrants their application.

5.8 Adjustment of pH

Prior to use, determine the hydrogen-ion concentration (pH) of culture media at 25°C electrometrically. Determinations made at 45°C to take advantage of the fluid state of agar will be low! For example, SMA [5.9E] showed a pH = 7.04 at 25°C and pH = 6.80 at 45°C.[25] Meters only measure pH accurately if samples and buffers are at the same temperature. Commercially prepared reference buffers should be used.

A. Electrometric procedure (potentiometer):

Prepare the pH meter as described in 18.2. Macerate solid medium thoroughly with a glass rod before inserting electrodes. Be sure electrodes and sample are at 25°C. Alternatively, flat surface electrodes are now available that can test agar surfaces without prior mixing and breaking of the medium.

5.9 Formulae and Directions for Preparation of Culture Media

This section contains formulae and/or preparation directions for culture media mentioned in this edition that are not readily available commercially. Details for SMA [5.9E] are also included since *SMEDP* is the original reference for this medium. Additional media formulae are included in the Compendium.[24]

A. *Citrate azide agar:*[21,23]

Yeast extract	10 g
Pancreatic digest of casein	10 g
Sodium citrate	20 g
Sodium azide	0.4 g
Ditrazolium chloride (tetrazolium blue)	0.01 g
Agar	15 g
MS water to make	1 L

To prepare the basic medium, do not add sodium azide or tetrazolium blue. Heat to boiling to dissolve other ingredients. Disperse in 100-mL amounts and autoclave at 121°C for 20 min.

Before adding the final ingredients, temper the medium to 48°C and add 1 mL of 0.1% sterile (121°C for 20 min) aqueous solution of tetrazolium blue and 1 mL of 4% sterile (121°C for 20 min) aqueous solution of sodium azide to each batch of 100 mL. Final pH should be 7.0 ± 0.2 at 25°C.

Pour about 15 mL of the final prepared medium per plate. After solidification, overlay the plates with a thin layer of the same medium.

For enumeration of colonies, place a thin sheet of white paper (facial tissue) underneath the petri dish on the illuminated colony counter to enhance color contrast between colonies and background.

B. *Levine's eosin methylene blue agar:*

Avoid overheating, as the agar is likely to become soft and unsuitable for streaking. When it is to be used as a streaking medium, melt agar in flowing steam or boiling water. Gently swirl contents of flask to distribute flocculant precipitate and pour a thick layer into petri plates. Set cover aside to leave about one-third of the plate exposed (for escape of moisture vapor and to dry surface of agar after solidification). A dry surface promotes growth of distinct individual colonies and gives better type differentiation. Plates can be stored inverted in a refrigerator and should be used within a week. After storage, incubate noninoculated plates as a check for contamination.

C. *Lactose broth:*

When fermentation tubes or other containers are prepared to examine 10-mL portions of samples, the lactose broth medium must be of such strength that addition of that volume of sample to the medium will not reduce the concentration of ingredients below that in the standard medium. When 10-mL portions are to be inoculated into 10 mL of medium, the medium must be formulated at double strength. When 10-mL portions are inoculated into 20 mL of medium, the medium must be formulated at 1.5 times the strength of the standard ingredients. The pH should be 6.7 ± 0.1 for double- and 6.9 ± 0.1 for single-strength broth.

5.9 Formulae and Directions

D. Lactobacillus selection (LBS) agar:

Mix thoroughly 84 g of commercially prepared LBS agar in 1 L of MS water. Add 1.32 mL of glacial acetic acid. Heat with frequent agitation and boil for 2 min. Cool to 45°C and pour plates without autoclaving. If storage is necessary, autoclave the medium at 118°C for 15 min.

E. Standard methods agar (SMA):

Formulation of SMA shall conform to that described in *SMEDP*, 14th ed.[18] This is the same formulation described for Plate Count Agar, Dehydrated (Tryptone Glucose Yeast Agar) in *Standard Methods for the Examination of Water and Wastewater*, 15th ed.[5] and in the *Compendium of Methods for the Microbiological Examination of Foods*.[24] It is also the same formulation as for Tryptone Glucose Yeast Agar of the AOAC[10] and consists of the following ingredients and quantities per liter of medium:

Pancreatic digest of casein	5 g
Yeast extract	2.5 g
Glucose	1 g
Agar	15 g
MS water top make	1 L

Final pH should be 7.0 ± 0.2 at 25°C after sterilization at 121°C for 15 min.

Ingredients used in formulating SMA shall conform to the following specifications:

1. Pancreatic digest of casein—Shall be of USP quality as given in the *U.S. Pharmacopoeia XX*, p 1078.[2]
2. Yeast extract—Shall conform to specifications for yeast extract as given in the *U.S. Pharmacopoeia XX*, p 1098.[2]
3. Glucose—Shall conform to specifications for glucose as given in the *U.S. Pharmacopoeia XX*, p 218,[2] and it shall be anhydrous.
4. Agar—Shall conform to the following specifications:
   a. General characteristics: The dry material shall be in granule or flake form, cream-white to pale tan in color. An amount of 2.0 g should be evenly suspendable in 100 mL of cold water without lumping. On application of heat, solution should be complete within 1 min at 100°C.

   After autoclaving the mixture at 121°C for 15 min, the hot 2.0% solution may be clear or hazy, but should be without milkiness and free from extraneous matter or suspended particles that might be confused with bacterial colonies. After autoclaving and cooling to room temperature, the pH should be not less than 6.0.
   b. Temperature of gelation and melting: A 1.5% hot aqueous solution should, upon cooling, gel at not over 43°C, nor less than 33°C. After gelation, the solidified material should not melt at less than 70°C.

c. Viable aerobic mesophilic spore count: The viable spore count should not exceed 50. Add 2 g of agar to 100 mL of sterile soybean casein digest broth USP, and autoclave the mixture at 121°C for 5 min. Rotate the mixture to obtain uniform distribution and, after cooling to 45°C, pour it into three or four sterile petri plates. Incubate the plates for 48 h at 32 to 35°C and count colonies. The total count, divided by 2, should not exceed 50.

F. *Physical Standards for Standard Methods Agar (SMA):*
SMA shall conform to the following requirements:
1. It shall be prepared from ingredients that meet specifications for ingredients of, and according to the formula for, SMA medium as described in [5.9E] of this chapter.
2. It shall be a uniform, free-flowing, beige or light buff colored powder.
3. It shall have a moisture content of less than 5.0%. Moisture is to be determined by using a moisture balance with a 125-W, 115- to 125-V infrared industrial reflector bulb; a powerstat transformer set at 100 is to be in the line to prevent charring. The exposure shall be 15 min.
4. It shall be soluble in a 2.35% solution in MS water when boiled for 1 min. The yield is to be a slightly opalescent-to-clear solution without precipitate.
5. It shall have a final reaction of pH 7.0 ± 0.2 at 25°C after having been autoclaved at 121°C for 15 min.
6. It shall not develop an objectionable precipitate (when held at 50 to 56°C for 2 h after autoclaving) that could interfere with the Standard Plate Count.

G. *Productivity tests for SMA:*
The SMA Reference Standard was replaced in 1981 at the APHA headquarters in Washington, D.C. Those desiring to compare productivity of lots of SMA should obtain a Reference Standard and conduct the study outlined in *SMEDP*, 14th ed. [4.11].[18] The method is omitted from this edition.

H. *Violet red bile agar (VRBA):*
Suspend the dry ingredients in MS water and allow to stand for a few minutes. Mix thoroughly while heating to boiling. Boil for 2 min. Cool to 45 to 46°C and use as a plating medium. After solidification, add a cover layer of approximately 4 mL over the agar to prevent surface growth and spreading of colonies.

Inadequate dispersion of VRBA may cause the medium to appear "grainy", to gel at higher temperatures or solidify incompletely. Since autoclaving reduces VRBA agar productivity,[7,14] only boiling for 2 min maximum should be used. A magnetic mixer system should be used if necessary to assure proper solution of the agar granules.

Although VRBA is stable almost indefinitely when in dehydrated form in sealed containers,[6] the boiled-cooled medium should not be stored over one day before use. The medium pH should be 7.4 ± 0.2 at 25°C.

5.10 Sterilization and Storage

Before sterilization, bring the medium to boiling temperature while stirring frequently. Restore lost water if necessary [5.7] and autoclave. Because the pH of the medium may change during sterilization and because of potential browning reactions, it is important not to exceed the recommended temperature and time. Sugar broths should not be subjected to a sterilization cycle longer than 45 min. Reduce pressure with reasonable promptness (but after no less than 15 min) to prevent undue changes in the nutritional properties of the medium or steam bubbles production. Remove from the autoclave when atmospheric pressure is attained. Preferably use flasks or bottles from which melted medium may be poured into plates. Optionally, test tubes containing 15- to 20-mL amounts may be used for pouring melted agar into plates.

Prepare medium in quantities so that, if stored, it will be used before loss of moisture through evaporation becomes evident. To prevent contamination and excessive evaporation during storage, optionally fit pliable aluminum foil, rubber, plastic, or heavy kraft paper, securely over the closures before autoclaving. Use of screw-cap or crown (cork-and-seal type) closures on containers appreciably reduces such risks. If tubed media are used within a short time, commercially available heat-resistant plastic or stainless-steel closures may be used. Media should be stored at 2 to 8°C in a dry, dust-free area and should not be exposed to direct sunlight.

A. Steam sterilization:

Autoclave media, water, and materials (such as rubber, cork, cotton, paper, and heat-labile plastic tubes and closures), that are likely to be charred in the dry-air sterilizer, by subjection to steam sterilization at 121°C for 15 min [2.4A]. Start timing after the temperature has reached 121°C. Autoclave media and dilution blanks within an hour of preparation. Slightly loosen stoppers to allow passage of steam into and air from closed containers when autoclaved. Make certain that the load is loosely packed [5.7]. Before allowing steam pressure to rise, automatically or manually expell all air from the sterilizer through an exhaust valve of suitable size. If manual means are used, make sure that free steam is being exhausted before pressurization begins. (If air is not exhausted, a 15 psig pressure can mean a temperature of only 100°C!) Because the temperature obtained at a constant pressure of saturated-steam will vary according to atmospheric pressure, rely on upon a properly operating and calibrated thermometer (not a pressure gauge) to insure sterilization.

Carefully review label instructions before heat treating any medium. For example, VRBA agar requires boiling for 2 min rather than autoclaving [5.9H].[14]

Avoid overloading autoclaves so that the rate of air exhaust or heating is not appreciably delayed [5.7]. The autoclave should reach 121°C slowly, but within 15 min. The maximum time from application of steam to unloading the autoclave should not exceed 45 min. A rapid come-up time does not necessarily result in more efficient autoclaving since air is not replaced by steam when the steam enters the autoclave too rapidly. A steam flow that is too slow also results in decreased efficiency because air-steam mixtures form.

When non-liquid materials with slow heat conductance are to be sterilized, or where the packing arrangement or volume of materials otherwise retards penetration of heat, allow extra time for materials to reach 121°C before beginning to time the sterilization period and, if necessary, use longer sterilization periods to insure sterility.

After sterilization, *gradually* reduce pressure within the autoclave (no less than 15 min is recommended) since liquids may be at temperatures above their boiling point at atmospheric pressure. Liquids can be lost through boiling when lowering the pressure too rapidly. When sterilizing dry materials such as sampling equipment or empty sample bottles, pressure may be released rapidly at the end of the 15-min holding period at 121°C. This prevents collection of condensate and speeds up drying of paper-wrapped equipment.

Used plates, pipettes, tubes, etc. must be routinely decontaminated in microbiological laboratories. Dairy products, especially raw milk, may harbor potential pathogens such as staphylococci, streptococci, etc. The same principles apply for loading the autoclave and sterilization as for media preparation. Plastic petri dishes are most conveniently sterilized by placing them in heat-resistant, autoclavable bags or other containers. When large numbers of petri dishes are autoclaved, the sterilization time should be at least 30 min. Plastic dishes become amorphous when properly autoclaved. Sterilization tapes and other indicators are available from scientific suppliers. These can be applied to containers and placed in autoclaves to indicate that sterilization temperatures have been reached.

B. Hot-air sterilization:

Sterilize equipment with dry heat in hot-air sterilizers or ovens so that materials at the center of the load are heated to not less than 170°C for not less than 1 h. (This usually requires exposure for about 2 h at 170°C.) When the oven is loaded to capacity, use a longer period or a slightly higher temperature. Spore packs or thermocouples should be used to check sterilization efficiency.

5.11 Quality Control

General principles are covered in Chapter 2. Contents of this chapter deal with quality control in media preparation. The following procedures will provide minimal but adequate quality control in media production. Many are mentioned elsewhere in this chapter and some have been taken from a DHEW publication.[1]

Date the label of each bottle of dehydrated medium indicating when it was received and when it was first opened.

Store dehydrated media in tightly capped bottles in a cool, dry place protected from light, or in a refrigerator if necessary. Keep no more than a sealed, 6- to 12-months' supply on hand, being sure to use older stocks first. Dehydrated media should be free-flowing powders; if change is noted in this property, or in color, discard.

Whenever possible, commercially dehydrated media should be used.

Complete mixing of medium to form a homogeneous solution in water is necessary before sterilization and dispensing. Stirring that causes foaming should be avoided. Restore any water lost due to evaporation after getting ingredients into a homogeneous solution.

Autoclave performance should be monitored by either a continuous temperature-recording device in combination with properly placed indicator strips or discs, or spore strips, or suspensions. The record for each "run" should be dated, numbered and filed. A maximum registering autoclave thermometer should be used to confirm sterilization temperatures.

Limit heating of the medium to the minimum necessary to insure solution and sterilization. When the autoclave has reached atmospheric pressure following the recommended cycle, it should be opened immediately. Prolonged storage of melted media in a water bath should be avoided.

Check and document the final pH of each lot of medium at 25°C [5.8].

Aseptic technique should be followed strictly during dispensing of sterilized material. Hands should not touch any part of the dispensing tubing that comes in contact with sterile material. During interruptions in a dispensing cycle, the spout of a dispenser train should be placed in a sterile glass container. A dispensing cycle should not be interrupted, except in an emergency. Any leaks should be diverted into a sterile container.

Moving parts of any dispensing apparatus should be oiled or greased as indicated by the manufacturer [2.4E].

Media containing dyes should be protected from light by storage in a dark room, in a dark glass bottle, or by wrapping the container with brown paper.

Each container of autoclaved medium should be labeled with the name of the medium and the autoclave "run number."

Inclusive dates during which each lot and container of dehydrated medium were used should be recorded for possible future trouble shooting.

Each lot should be inspected visually before use for volume, tightness of closure, clarity, color, consistency, and completeness of label.

5.12 References

1. ANONYMOUS. Manual of quality control procedures for microbiological laboratories. Atlanta, GA: Dept. of Health, Education and Welfare.; 1974.
2. ANONYMOUS. U.S. Pharmacopoeia, 20th ed. Easton, PA: Mack Publishing Co.; 1980.

3. BUTTERFIELD, C.T. The selection of a dilution water for bacteriological examinations. J. Bacteriol. 23:355–368; 1932.
4. GELDREICH, E.E. Handbook for evaluating water bacteriological laboratories, 2nd ed. Cincinnati, OH: U.S. Environ. Prot. Agency; 1975.
5. GREENBERG, A.E.; CONNORS, J.J.; JENKINS, D., EDS. Standard methods for the examination of water and wastewater, 15th ed. Washington, D.C.: Amer. Public Health Assoc.; 1981.
6. HARTMAN, P.A.; HARTMAN, P.S.; FREY, K.T. Longevity of dehydrated violet red bile agar. Appl. Microbiol. 29:295–296; 1975.
7. HARTMAN, P.A.; HARTMAN, P.S. Coliform analyses at 30C. J. Milk Food Technol. 39:763–767; 1976.
8. HARTMAN, P.A. Modification of conventional methods for recovery of injured coliforms and salmonellae. J. Food Prot. 42:356–361; 1979.
9. HAUSLER JR., W.J.; JENSEN, J.P.; CREAMER, C.H. Interval plating: A simplified method to determine suitability of distilled water as a dilution fluid. J. Milk Food Technol. 39:848–851; 1976.
10. HORWITZ, W., ED. Official methods of analysis, 13th ed. Washington, D.C.: Assoc. Offic. Anal. Chem.; 1980.
11. HUHTANEN, C.N.; BRAZIS, A.R.; ARLEDGE, W.L.; DONNELLY, C.B.; GINN, R.E.; RANDOLPH, H.E.; KOCH, E.J. A comparison of phosphate buffered and distilled water dilution blanks for the standard plate count of raw-milk bacteria. J. Milk Food Technol. 38:264–268; 1975.
12. JANUS, J.A. Investigations into the bactericidal effect of diluents applied for the enumeration of enterobacteriaceae in food products. The Netherlands: Intern. Dairy Fed. report of Group E 50. ISO/TC 34/SC 9. 1980.
13. JAYNE-WILLIAMS, D.J. Report of a discussion on the effect of the diluent on the recovery of bacteria. J. Appl. Bacteriol. 26:398–404; 1963.
14. JENSEN, J.P.; HAUSLER JR., W.J. Productivity of boiled and autoclaved violet red bile agar. J. Milk Food Prot. 38:279–280; 1975.
15. KING, W.L.; HURST, A. A note on the survival of some bacteria in different diluents. J. Appl. Bacteriol. 26:504–506; 1963.
16. MACLEOD, R.A.; KUO, S.C.; GELINAS, R. Metabolic injury to bacteria. II. Metabolic injury induced by distilled water or Cu^{++} in the plating diluent. J. Bacteriol. 93:961–969; 1967.
17. MALLETTE, M.F. A pH 7 buffer devoid of nitrogen, sulphur, and phosphorus for use in bacteriological systems. J. Bacteriol. 94:283–290; 1967.
18. MARTH, E.H., ED. Standard methods for the examination of dairy products. Washington, D.C.: Amer. Public Health Assoc.; 1978.
19. OBLINGER, J.L.; KENNEDY JR., J.E. Evaluation of diluents used for total counts. J. Milk Food Technol. 39:114–116; 1976.
20. RICHARDSON, G.H.; MARTH, E.H.; MARSHALL, R.T.; MESSER, J.W.; GINN, R.E.; WEHR, H.M.; CASE, R.; BRUHN, J.C. Update of the 14th edition of Standard Methods for the Examination of Dairy Products. J. Food Prot. 43:324–328; 1980.
21. REINBOLD, G.W.; SWERN, M.; HUSSONG, R.V. A plating medium for the isolation and enumeration of enterococci. J. Dairy Sci. 36:1–6; 1953.
22. RONALD, G.W.; MORRIS, R.L. The effect of copper on distilled water quality for use in milk and water laboratories. J. Milk Food Technol. 30:305–309; 1967.
23. SARASWAT, D.S.; CLARK JR., W.S.; REINBOLD, G.W. Selection of a medium for the isolation and enumeration of enterococci in dairy products. J. Milk Food Technol. 26:114–118; 1963.
24. SPECK, M.L., ED. Compendium of methods for the microbiological examination of foods. Washington, D.C.: Amer. Public Health Assoc.; 1976.
25. WALTER, W.G.; KUEHN, K.; BOUGHTON, A. Electrometric pH determinations of different agar media at 25 and 45C. J. Milk Food Technol. 33:343–345; 1970.
26. WINSTEAD, M. Reagent grade water. How, when and why? Austin, TX: Amer. Soc. Med. Technol.; 1967.

Chapter 6

MICROBIOLOGICAL COUNT METHODS

J. W. Messer, H.M. Behney, and L.O. Leudecke
(J.W. Messer, Tech Comm.)

6.1 Introduction

The Standard Plate Count (SPC) is suitable for estimating bacterial populations in most types of dairy products, and it is the reference method specified in the Grade A Pasteurized Milk Ordinance[24] to be used to examine raw and pasteurized milk and milk products. This procedure is also recommended for industry application in detecting sources of contamination by testing line samples taken at successive stages of processing. Cultured dairy products (e.g., cultured buttermilk and fermented acidophilus milk), or dairy products to which a bacterial culture has been added, ordinarily are not tested by the SPC.

While standardization of equipment, materials, and incubation in the SPC has been of considerable value, questions continue to be raised concerning the ability of this procedure to completely reflect sanitary practices used to produce and handle milk. As a result, stipulation of procedures for preliminary incubation of milk in conjunction with determining the SPC may be useful to provide additional information about the bacteriological quality of a sample.[7,16,17,20] A more complete picture of the bacterial content of dairy products may be obtained by incubating additional plates at other temperatures and for various periods of time [Chapter 9].

6.2 Standard Plate Count (Class O)

A. *Sampling procedures:* Are described in Chapter 4.
B. *Equipment and supplies:*
   1. Work area:
      [2.3C, 2.3D, 2.3E, 2.3F and 2.3G]
   2. Storage space:
      [2.3F]
   3. Refrigerator:
      [2.4N]

4. Thermometers:
   [2.4O]
5. Transfer pipettes (glass and plastic):
   [2.4M]
6. Pipette containers:
   [2.4K]
7. Dilution bottles:
   [2.4D]
8. Petri dishes or plates:
   [2.4I]
9. Petri dish containers:
   [2.4J]
10. Balance:
    [2.4B]
11. Water bath:
    [2.4Q]
12. Colony counters, tallies:
    [2.4C]
13. Incubators:
    [2.4H]
14. Autoclave:
    [2.4A]
15. Sterilizing ovens:
    [2.4P]

C. *Materials:*
  1. Standard Methods Agar prepared as described in 5.7.
  2. Microbiologically Suitable (MS) water: MS water as described in 5.2.
  3. Dilution water: Use only buffered MS water for dilution[5] [5.3B]. Make tests semi-annually to determine whether toxic substances are present [5.4A].
  4. Dilution-water blanks: Fill dilution bottles with buffered MS water so that after sterilization each will contain 99 ± 2 mL. Autoclave bottles of buffered MS water at 121°C for 15 min. When autoclave loading conditions restrict heat penetration, 20- to 30-minute sterilization periods may be required. When less than 99-mL amounts are required (e.g., 9 mL, 90 mL, etc.) apply proportionally smaller deviation tolerances for amounts in dilution blanks. Optionally use correctly calibrated automatic water-measuring devices. When bulk-sterilized diluent is used, measure MS water directly into sterile dilution bottles and use prepared blanks promptly.

   Ready-to-use plastic, single-service dilution blanks are available commercially (Gibco Div., The Dexter Corp. Chagrin Falls, OH 44022.)

6.2 Standard Plate Count (Class O)

D. *Sterilization:*
1. Steam:
   [2.4, 5.10A].
2. Hot air:
   [2.14, 5.10B].
3. Recording: Record time at which oven or autoclave reaches sterilization temperature, minimum temperature used, and time of discontinuing heat for each lot of materials sterilized. Recording devices are useful for this purpose.

E. *Examination of samples:*
1. Temperature: After collecting samples, refrigerate and transport to the testing laboratory. Upon arrival at the laboratory, determine the temperature of each set of samples by inserting a precooled thermometer into a separate pilot container treated exactly as the samples; record the temperature. Do not insert a thermometer into any sample intented for microbiological examination before the test portion has been removed.
2. Collection, storage, and examination times: Standard methods to collect samples of milk and milk products are described in Chapter 4. Record times and dates of sample collection and receipt at the laboratory for each set of samples. At the laboratory, store samples at 0 to 4.4°C until tested. Samples must be tested within 36 h after collection and the time of plating recorded. Samples should be at 0 to 4.4°C at the time of plating. Fluid milk samples that have been frozen should not be tested microbiologically because freezing causes a significant change in the viable bacterial count in milk.[20]
3. Chemical tests: If chemical tests are necessary, first remove portions for microbiological analysis [6.2F2].

F. *Preparing samples:*
1. Identifying plates: Identify each plate on its lid with sample number, dilution, and other desired information before shaking samples and making dilutions.
2. Shaking samples and dilutions: Before opening a sample container, remove from the closure all obvious materials that may contaminate the sample. Optionally, wipe the tops of unopened sample containers with a sterile cloth saturated with 70% ethyl alcohol. Invert filled containers (containing air space such as 227 g milk carton) 25 times, or until contents are homogeneous. To insure a homogeneous sample where no air space is present, aseptically open the container and pour the product from the filled carton back and forth into a sterile container three times.) The interval between mixing and removing the test aliquot must not exceed 3 min.

   Immediately before transferring test portions of milk or cream (except from filled containers with no air space, as noted above) and of dilutions thereof, shake container, making 25 complete up-and-down (or back-and-forth) movements of about 30 cm in 7 sec.[12,13] Optionally, a mechanical shaker may be used to shake the dilution blanks uniformly for 15 sec.

G. *Diluting samples:*
   1. Selecting dilutions: For SPCs, select dilutions(s) so that the total number of colonies[23] on a plate will be between 25 and 250 (Fig. 6.1) [Table 10.1]. For example, where a SPC is expected to reach a number between 2,500 and 250,000, prepare plates containing 10^{-2} and 10^{-3} dilutions. Dilutions of $10°$ (no dilution), 10^{-1}, and 10^{-2} should be used for low-count samples.
   2. Measuring sample portions: Use a sterile pipette for initial and subsequent transfers from the same container if the pipette is not

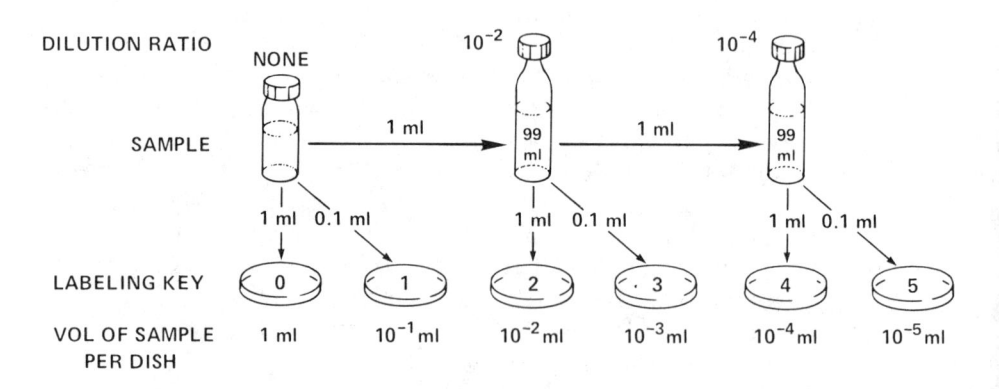

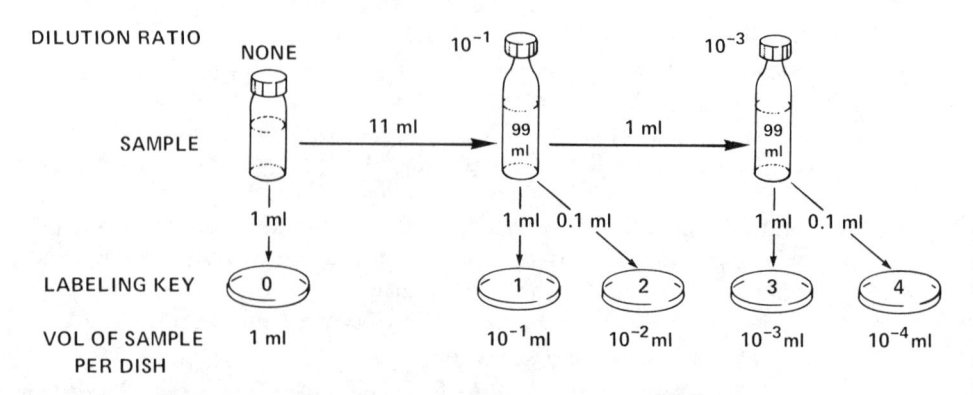

Figure 6.1. Examples for preparing dilutions.

6.2 Standard Plate Count (Class O)

otherwise contaminated. If the pipette does become contaminated before completing transfers, replace it with a sterile pipette. Do not flame to decontaminate. Use a separate sterile pipette for transfers from each different dilution. Add test portions of milk to dilution blanks adjusted to 15 to 25°C.

CAUTION. Do not prepare or dispense dilutions or pour plates in direct sunlight. When removing sterile pipettes from the pipette container, do not drag tips over exposed exteriors of the pipettes remaining in the case because exposed ends of such pipettes are subject to contamination. Do not wipe or drag pipette across lips and necks of vials or dilution bottles. Do not insert pipettes more than 2.5 cm below the surface of the sample or dilution. Draw test portions above the pipette graduation, than raise the pipette tip above the liquid level, and adjust to the desired mark by allowing the lower side of the pipette tip to contact the inside of the container (Fig. 6.2) in such a manner that drainage is complete and excess

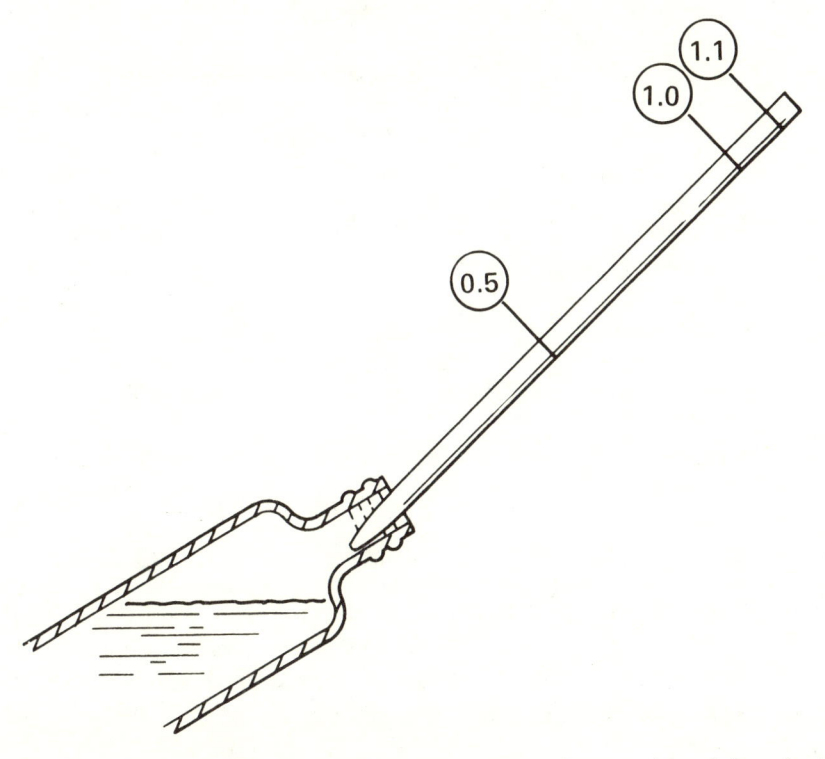

Figure 6.2. Measuring diluted sample or portions. Lower side of the pipette touches the inside surface of the container.

liquid does not adhere when pipettes are removed from sample or dilution bottles,[6]. Do not flame sterile pipettes.

  a. Milk and milk products: When measuring dairy products having a viscosity similar to milk, complete each transfer by letting the column drain from the graduation mark to the rest point of the liquid in the tip of the pipette within 2 to 4 sec. Promptly and gently blow the last drop out. Make subsequent transfers carefully, do not blow out last drop and do not rinse pipettes in dilution water. Dilutions and plates can be prepared according to one of the methods shown in Fig. 6.1.
  b. Concentrated and cultured dairy products: Where the solids content or viscosity of samples appreciably exceeds that of whole milk, prepare the initial 10^{-2} (or 10^{-1}) dilution by weighing a test portion into a warmed dilution blank. Weigh 1 g, preferably 11 g, aseptically into dilution bottles containing 99 mL of sterile, warmed (40 to 45°C), buffered dilution water. Viscosities of these dairy products vary at different temperatures, and differences occur in specific gravities, depending upon their composition. Furthermore, the amount of trapped air will vary, even in properly agitated samples.

3. Measuring dilutions: When measuring a diluted sample of dairy product, hold pipette at an angle of about 45° with tip touching inside bottom of petri dish or inside neck of dilution bottle. Lift cover of petri dish just high enough to insert pipette. Deposit sample away from center of dish to aid in mixing sample with medium. Allow 2 to 4 sec for diluted milk or cream to drain from the graduation mark to the rest point in the tip of the pipette; then, holding the pipette in a vertical position, touch tip once against a dry spot on the petri dish or inside of the dilution bottle neck. *Do not blow out.* When 0.1 mL quantities are measured, hold pipette as directed and let diluted sample drain to the proper graduation point but *do not* retouch pipette to plate.

After depositing test portions in each series of plates, pour the medium promptly [6.2H2].

H. Plating:

1. Melting medium: Melt required amount of medium quickly in boiling water or by exposure to flowing steam in a partially closed container, but avoid prolonged exposure to unnecessarily high temperatures during and after melting. If the medium is melted in two or more batches, use all of each batch in order of melting, provided that contents of separate containers remain fully melted. Discard melted Standard Methods Agar that develops a precipitate. Do not melt more medium than will be used in 3 h. Do not resterilize the medium. Freshly sterilized agar, after tempering to 45°C, also may be used.

6.2 Standard Plate Count (Class O)

Cool melted medium promptly to approximately 45°C and hold in a water bath between 44 and 46°C [5.7I].[15] Place the bulb of a thermometer in an aqueous solution of 1.5% agar in a separate container identical to that used for medium; this temperature-control solution must have been exposed to the same heating and cooling as the medium. Do not depend upon the sense of touch to indicate the proper temperature of the medium when pouring agar.

2. Pouring and mixing agar: Select the number of samples to be plated in any one series so that not more than 20 min (and preferably only 10 min) elapse between diluting the first sample and pouring the last plate in the series.[3,11,13] Should a continuous plating operation be done by a team, plan the work so that the time between initial measurement of a test portion into diluent or directly into a dish and pouring of the last plate for that sample is not more than 20 min. Wipe water from outside of medium bottles before pouring. Introduce 10 to 12 mL of liquefied medium at 44 to 46°C into each plate by gently lifting the cover of the petri dish high enough to pour the medium. Carefully avoid spilling medium on outside of container or on the inside of the plate lid when pouring. As each plate is poured,[22] thoroughly mix medium with the test portions in the petri dish, taking care not to splash the mixture over the edge by: rotating the dish first in one direction and then in the opposite direction, rotating and tilting the dish, or using mechanical rotators. Having thus spread the mixture evenly over the bottom of the plate, allow it to solidify (within 10 min) on a level surface. After solidification, invert plates to prevent spreading colonies from developing due to accumulated moisture [6.2K3] and place them in the incubator within 10 min.

I. Sterility controls of medium, dilutions, and equipment:
1. Material controls: Check sterility of dilution waters and medium by pouring control plates for each sterilization lot of dilution blanks and medium used and, if desired, for each lot of petri dishes and pipettes.
2. Additional controls: Since the number of samples in a series may vary considerably, prepare at least one agar control plate for samples plated in the morning and another agar control for samples plated in the afternoon. If >5 colonies appear on agar control plates, prepare additional control plates to determine if contamination came from the medium, water blanks, plates, pipettes, or air.

J. Incubation:

Incubate plates at 32 ± 1°C for 48 ± 3 h for the SPC.[2,25] Plates must reach the temperature of incubation within 2 h. Incubation times and temperatures for special bacterial groups, and procedural differences, are given in Chapters 7, 8, and 9.

Avoid excessive humidity in the incubator to reduce the tendency toward spreader formation, but prevent excessive drying of the medium by controlling ventilation and air circulation. Agar in plates should not lose more than 15% of its weight during 48 h of incubation.

K. *Counting colonies on plates and recording results:*

Count plates promptly after the incubation period [6.2J].

Record dilutions used and number of colonies counted or estimated on each plate. If impossible to count at once, store plates after the required incubation at 0 to 4.4°C for not more than 24 h,[9] but avoid making this a routine practice. When counting colonies on plates, proceed according to directions given in 6.2K1. For each lot of samples, record the results of sterility tests on materials (dilution of blanks, agar, etc.) used when pouring plates [6.2I2] and the incubation temperature used.

1. Counting of colonies and recording counts:
    a. Manual counting: (Class O) Count colonies with the aid of magnification under uniform and properly controlled artificial illumination, using a tally. Routinely use a colony counter equipped with a guide plate ruled in square centimeters.[1] Plates should be examined in subdued light. Avoid mistaking particles of undissolved medium or sample, or precipitated matter in plates for pinpoint colonies. Examine doubtful objects carefully, using higher magnification where required, to distinguish colonies from foreign matter. Arrange schedules of laboratory analysts to prevent eye fatigue and the inaccuracies that result from eyestrain.
        1. Normal plates (25 to 250 or <25 colonies): Select spreader-free plate(s). Count all colonies, including those of pinpoint size on selected plate(s). Record dilution(s) used and total number of colonies counted.
        2. Crowded plates (more than 250 colonies): Do not record counts on crowded plates from the highest dilution as too numerous to count (TNTC). If the number of colonies per plate exceeds 250, count colonies in those portions of the plate that are representative of colony distribution and calculate the Estimated SPC from these counts. If there are fewer than 10 colonies per square centimeter, count colonies in 12 squares, selecting, if representative, six consecutive squares horizontally across the plate and six consecutive squares at right angles, being careful not to count a square more than once. When there are more than 10 colonies per square centimeter, count colonies in four such representative portions. In both instances, multiply the average number found per square centimeter by the area of the plate used to determine the estimated number of colonies per plate. Each laboratory must determine the area in square centimeters of the plates in use.
        3. Spreaders: Spreading colonies are usually of three distinct types. The first type is a chain of colonies, not too distinctly

separated, that appears to be caused by disintegration of a bacterial clump. The second type develops in the film of water between the agar and the bottom of the dish. The third type forms in the film of water at the edge of or on the surface of the agar. If plates prepared from the sample have spreader growth such that: (a) the area covered by the spreader or the area covered by the spreader plus the repressed area exceeds 50% of the plate area, or (b) the area of repressed growth caused by the spreaders exceeds 25% of the plate area, report the plate as "Spreader." When it is necessary to count plates containing spreaders not eliminated by (a) or (b) above, count each of the three distinct spreader types as one source. If one or more chains appear to originate from separate sources, count each source as one colony. Do not count each individual growth in such chains as a separate colony. The second and third types usually result in distinct colonies and are counted as such. Combine the spreader count and the colony count to compute the plate count. Any laboratory with 5% of plates more than one-fourth covered by spreaders should take immediate steps to eliminate this trouble.
4. Laboratory accidents or bacterial growth inhibitor: When plate(s) from sample are known to be contaminated or are otherwise unsatisfactory, record the plate(s) as Laboratory Accident (LA). If the test for inhibitory substances [Chapter 14] is positive for a sample, record the SPC as "Growth Inhibitor" (GI). The analyst may be inclined to suspect the presence of inhibitory substances in the sample being examined, when plates have no growth or have proportionally less growth in lower dilutions; such developments should not be cited as evidence of inhibition until the presence of inhibitory substances has been confirmed.

b. Automated counting: Automated colony counters (Class B), when determined in individual laboratories to yield counts that 90% of the time are within 10% of those obtained manually, may be used for counting plates. When using colony-counting instruments, exercise the following precautions:[10]

—align petri dish carefully on colony-counter stage
—avoid "counting" stacking ribs or legs of plastic petri dishes
—do not count plates having unsmooth (rippled) agar surfaces
—avoid plates having food particles or air bubbles in the agar
—do not count plates having spreaders or extremely large surface colonies
—avoid scratched plates

—wipe fingerprints and films off petri dish bottom before counting

L. Computing and reporting counts:

To compute the SPC, multiply the total number of colonies or the average number (if duplicate plates of the same dilution) per plate by the reciprocal of the dilution used.

When colonies on duplicate plates of consecutive dilutions are counted, compute the mean number of colonies for each dilution before averaging to define the SPC.

Avoid creating a fictitious impression of precision and accuracy when computing SPC, by reporting only the first two significant digits. Round counts off to two significant figures only at the time of conversion to SPC by raising the second digit to the next highest number only when the third digit from the left is 5, 6, 7, 8, or 9; use zeroes for each successive digit toward the right from the second digit.

1. Guidelines for computing counts (In the examples, underlined figures are used to calculate the SPC.):
   a. Single plate dilutions:
      1. One plate with 25 to 250 colonies: Use the dilution providing 25 to 250 colonies to compute the Standard Plate Count when the other plate is outside of the 25 to 250 range or is excluded by spreaders or laboratory accident.

         Example

Colonies/Dilution		
1:100	1:1000	SPC
_234_	23	23,000 (or 2.3 x 10⁴)
305	_42_	42,000
Spr	_31_	31,000
_243_	LA	24,000

      2. Both plates with 25 to 250 colonies: When two dilutions yield counts of 25 to 250 colonies, average the two dilution counts obtained to compute the SPC, unless the count computed for the higher dilution is more than twice (>2) the count computed for the lower dilution.[6] In the latter case use the lower computed count as the SPC.

         Example

Colonies/Dilution			
1:100	1:1000	Count Ratio	SPC
_240_	_41_	1.7	33,000
_140_	32	2.3	14,000

      3. No plate with 25 to 250 colonies: When there is no plate with 25 to 250 colonies, use the plate having a count nearest 250 to compute the Estimated Standard Plate Count (ESPC).

6.2 Standard Plate Count (Class O)

Example

Colonies/Dilution		
1:100	1:1000	SPC
<u>275</u>	20	28,000 ESPC

4. Both plates with fewer than 25 colonies: When plates from both dilutions yield fewer than 25 colonies each, use the count from the lowest dilution to compute the ESPC.

Example

Colonies/Dilution		
1:100	1:1000	SPC
<u>18</u>	2	1,800 ESPC

5. Both plates with no colonies: When both dilutions yield plates with no colonies and inhibitory substances have not been found, use a count of less than (<) 1 times the lowest dilution plated to compute the ESPC.

Example

Colonies/Dilution		
1:100	1:1000	SPC
<u>0</u>	0	< 100 ESPC

6. Both plates have excessive spreader growth: When both dilutions yield plates with excessive spreader growth, spreader growth is reported for the SPC.

Example

Colonies/Dilution		
1:100	1:1000	SPC
Spr	Spr	Spr

7. Both plates with more than an average of 100 colonies per square centimeter: Estimate the SPC as greater than (>) 100 times the highest dilution plated times the area of the plate. The examples below have an average count of 110 per square centimeter.

Example

Colonies/Dilution		
1:100	1:1000	SPC
TNTC	7150*	> 6,500,000* ESPC
TNTC	6490**	> 5,900,000** ESPC

*based on a plate area of 65 cm^2, **based on a plate area of 59 cm^2

8. Both plates with more than 10 but fewer than 100 colonies per square centimeter: Use the estimated count of the highest dilution to compute the ESPC. The example below has an

average count of 15 colonies per square centimeter when four squares are counted (6.2K1a2).

Example

Colonies/Dilution		
1:100	1:1000	SPC
TNTC	975*	980,000 ESPC

*based on a plate area of 65 cm²

9. Both plates with more than 250 colonies per plate but fewer than 10 colonies per square centimeter: Use the estimated count of the highest dilution to compute the ESPC. The example below has an average count of 8 colonies per square centimeter when 12 squares are counted (6.2K1a2).

Example

Colonies/Dilution		
1:100	1:1000	SPC
TNTC	520*	520,000 ESPC

*based on a plate area of 65 cm²

b. Duplicate plate dilutions:
1. Only one dilution yields plates with 25 to 250 colonies: Compute the mean for that dilution as the basis for the SPC.

Example

Colonies/Dilution		
1:100	1:1000	SPC
175	16	
208	17	19,000

2. Both dilutions yield plates with 25 to 250 colonies: Average the mean count for each dilution as the basis for the SPC unless the count computed for the higher dilution is more than twice the count computed for the lower dilution. In the latter instance, use the lower computed count as the SPC.

Example

Colonies/Dilution		Count	
1:100	1:1000	Ratio	SPC
230	28		
246	36	1.3	28,000
138	42		
162	30	2.4	15,000

3. Neither dilution yields plates with 25 to 250 colonies: Use the mean count of the dilution nearest to 250 as the basis for estimating the SPC.

6.2 Standard Plate Count (Class O)

Example

Colonies/Dilution		
1:100	1:1000	SPC
287	23	
263	19	28,000 ESPC

4. Both dilutions yield only plates with fewer than 25 colonies: Use the mean count of the duplicates of the lowest dilution to estimate the SPC.

Example

Colonies/Dilution		
1:100	1:1000	SPC
18	2	
14	0	1,600 ESPC

5. Both dilutions yield plates with no colonies: Estimate the SPC as less than (<) 1 times the lowest dilution.

Example

Colonies/Dilution		
1:100	1:1000	SPC
0	0	
0	0	< 100 ESPC

6. Only one plate of one dilution contains 25 to 250 colonies: When one plate of one dilution contains 25 to 250 colonies and the duplicate contains more than 250 colonies, use both plates in computing the SPC.

Example

Colonies/Dilution		
1:100	1:1000	SPC
272	23	
248	19	26,000

7. One plate of each dilution contains 25 to 250 colonies and the duplicate contains more than 250 or less than 25 colonies: Use all four in computing the SPC.

Example

Colonies/Dilution		
1:100	1:1000	SPC
275	22	
240	35	27,000

8. Both plates of one dilution contain 25 to 250 colonies and only one duplicate of the other dilution contains 25 to 250 colonies: Use all four plates in computing the SPC unless excluded by spreaders or laboratory accident.

Example Colonies/Dilution		
1:100	1:1000	SPC
258	33	
250	28	28,000
250	33	
236	20	25,000
224	25	
200	Spr	23,000
229	25	
245	LA	24,000

M. Reporting and interpreting counts:

Report counts as the Standard Plate Count (SPC) or the Estimated Standard Plate Count (ESPC) per milliliter or per gram, as applicable. When instrumental colony counters are used, report counts as SPC (I) or ESPC (I).

The matter of interpreting bacterial counts is beyond the scope of this manual. Consult local regulations and the latest edition of the *Grade A Pasteurized Milk Ordinance*.[24]

N. Personnel Errors

Avoid inaccuracies in counting caused by carelessness, impaired vision, or failure to recognize colonies. Laboratory workers should duplicate their own counts on the same plate within 8% and the counts of other analysts within 10% on 90% of samples.[19]

6.3 Preliminary Incubation (PI) for Raw Milk (Class B):

A. *Sampling procedures* [Chapter 4].

B. *Equipment and supplies:*

Same as 6.2B. In addition provide 10-mL pipets, 30-mL sterile plastic or glass vials, and an incubator set at 13 ± 1°C.

C. *Materials:*

Same as 6.2C.

D. *Sterilization:*

Same as 6.2D.

E. *Examination of samples:*

1. Temperature: After collection of samples, refrigerate during transport to the testing laboratory. Upon arrival at the laboratory, determine the temperature of each set of samples by inserting a precooled thermometer into a separate pilot container treated exactly as the samples; record the temperature. Do not insert a thermometer into any sample intended for microbiological examination before removing the test portion.
2. Collection, storage, and examination times: The sample to be analyzed should represent more than two milkings from a given

producer. If only one or two milkings are represented, the sample should be held at 0 to 4.4°C for an additional 24 h before starting the PI at 13°C to obtain full benefit of the test.[21]

F. Procedure:

After thorough mixing [6.2F2], pipette 10 mL into a sterile plastic vial (capacity about 30 mL). Temper samples to 13°C and incubate at that temperature for 18 h; then determine the SPC [6.2F to 6.2L] or use the alternative method [Chapter 7] in the usual manner. When more than four samples are to be plated, cool them to 4.4°C or lower before plating but do not allow samples to freeze.

G. Results and significance:

The value of PI is based on the theory that microorganisms making up the normal flora of the udder do not grow well at 13°C whereas many saprophytic contaminants can grow actively at this temperature.[16] The SPC of milk in a bulk tank might not be affected greatly by use of a milk machine (or other piece of equipment) that is grossly contaminated because the incubation temperature is 32 ± 1°C. Contaminants would, however, grow much better at 13°C than would the normal flora of the milk as it comes from the cow. Thus, a high count after PI suggests careless handling practices, which allowed contamination of the milk. Johns[16] proposed a maximum allowable count of 200,000/mL following PI, and at least one company has successfully used 5000/mL as the basis for bonus payments to its procedures.[17]

6.4 Moseley Keeping Quality Test (Class B)

A. Sampling procedures: [Chapter 4]:

B. Equipment and supplies:
[6.2B] In addition, provide a refrigeration at 7 ± 1°C.

C. Materials:
[6.2C]

D. Sterilization:
[6.2D]

E. Examination of samples:
[6.2E]

F. Procedure:

After making a SPC [6.2F to 6.2L], store containers of freshly processed milk or cream in a refrigerator at 7 ± 1°C for 5 or 7 days, then replate [6.2F to 6.2L]).

G. Results and significance:

Large increases in bacterial count between the first and second plating suggest that keeping-quality problems can be expected during refrigerated storage of the product. For specific illustrations of this test under various conditions of plant sanitation, consult the paper by Elliker *et al.*[8] Many plants have eliminated the initial SPC, and plate samples only after 5 or more days of storage.

6.5 Keeping Quality of Pasteurized Milk, Impedimetric Method (Class B)

A. Principle:

The use of impedimetric detection time in incubated pasteurized whole milk samples has been shown to be superior to traditional keeping quality methods[4]. Results correlated better to actual keeping quality and were available in less than 48 h. A preliminary incubation step is combined with the instrumental method to allow microbial growth.

B. Equipment and supplies:
   1. The basic instrumentation and solid medium are described in 7.7.
   2. Incubator at 18°C.
   3. Broth Medium
      Yeast extract .... 5.0 g
      Pancreatic digest of casein .... 10.0 g
      Dextrose .... 2.0 g
      MS water to make .... 1 L

Dispense 5 mL of broth medium into screw-capped test tubes. Sterilize at 121°C for 15 min. The final pH should be 7.0 ± 0.2.

C. Preparation of samples and analysis:
   1. Shake the sample container as specified [6.2F2].
   2. Add 5 mL of mixed sample into 5 mL of sterile broth medium.
   3. Vortex or otherwise shake tubes 10 times through a 30 cm arc to insure proper mixing.
   4. Place tubes into 18°C incubator for 18 h.
   5. Shake the tubes 25 times as in step 3.
   6. Remove caps from wells of instrument modules containing dried MSMA medium [7.7D].
   7. Pipette 0.1 mL of each sample into an agar-filled (MSMA) module well. Pipette a second 0.1-mL volume into another well as a sample duplicate.
   8. Aseptically place the caps back on all wells.
   9. Adjust the instrument incubation temperature to 21°C.
   10. Initiate the instrument operation using the proper product code and software. Operate the system for up to 48 h. The results are expressed as days of shelf life.

6.6 References

1. ARCHAMBAULT, J.; CUROT, J.; MCCRADY, M.H. The need of uniformity of conditions for counting plates with suggestions for a standard colony counter. Amer. J. Public Health 27:809–812; 1937.
2. BABEL, F.J.; COLLINS, E.B.; OLSON, J.C.; PETERS, I.I.; WATROUS, G.H.; SPECK, M.L. The

6.6 References

standard plate count of milk as affected by the temperature of incubation. J. Dairy Sci. 38:499–503; 1955.
3. BERRY, J.M.; MCNEILL, D.A.; WITTER, L.D. Effect of delays in pour plating on bacterial counts. J. Dairy Sci. 52:1456–1457; 1969.
4. BISHOP, J.R.; WHITE, G.H.; FIRSTENBERG-EDEN, R.A. A rapid impedimetric method for determining the potential shelf-life of pasteurized whole milk, J. Food Prot. (In press).
5. BUTTERFIELD, C.T. Experimental studies of natural purification in polluted waters. VII. The selection of a dilution water for bacteriological examinations. J. Bacteriol. 23:355–368; 1932. Pub. Health Rep. 23:681–691; 1933.
6. COURTNEY, J.L. The relationship of average standard plate count ratios to employee proficiency in plating dairy products. J. Milk Food Technol. 19:336–344; 1956.
7. DRUCE, R.G.; THOMAS, S.B. Preliminary incubation of milk and cream prior to bacteriological examination: A review. Dairy Sci. Abstr. 30:291–307; 1968.
8. ELIKER, P.R.; SING, E.L.; CHRISTENSEN, L.J.; SANDINE, W.E. Psychrophilic bacteria and keeping quality of pasteurized dairy products. J. Milk Food Technol.27:69–75; 1964.
9. FOWLER, J.L.; CLARK, JR., W.S.; FOSTER, J.F.; HOPKINS, A. Analyst variation in doing the standard plate count as described in *Standard Methods for the Examination of Dairy Products*. J. Food Prot. 41:4–7; 1978.
10. FRUIN, J.T.; CLARK, JR., W.S. Plate count accuracy; analysts and automatic colony counter versus a true count. J. Food Prot. 40:552–554; 1977.
11. HARTMAN, P.A.; WEBER, J.A. Holding times of raw milk dilutions: a reassessment. J. Milk Food Technol. 35:713–714; 1972.
12. HUHTANEN, C.N.; BRAZIS, A.R., ARLEDGE, W.L.; COOK, E.W.; DONNELLY, C.B.; FERRELL, S.E.; GINN, R.E.; LEMBKE, L.; PUSCH, D.J.; BREDVOLD, E.; RANDOLPH, H.E.; SING, E.L.; THOMPSON, D.I. A comparison of horizontal versus vertical mixing procedures and plastic versus glass petri dishes for enumerating bacteria in raw milk. J. MIlk Food Technol. 33:400–404; 1970.
13. HUHTANEN, C.N.; BRAZIS, A.R.; ARLEDGE, W.L.; COOK, E.W.; DONNELLY, C.B.; GINN, R.E.; JEZESKI, J.J.; PUSCH, D,; RANDOLPH, H.E.; SING, E.L. Effects of time of holding dilutions of counts of bacteria from raw milk. J. Milk Food Technol. 35:126–130; 1972.
14. HUHTANEN, C.N.; BRAZIS, A.R.; ARLEDGE, W.L.; COOK, E.W.; DONNELLY, C.B.; GINN, R.E.; MURPHY, J.N.; RANDOLPH, H.E.; SING, E.L.; THOMPSON, D.I. Effect of dilution bottle mixing methods on plate counts of raw-milk bacteria. J. Milk Food Technol. 33:269–273; 1970.
15. HUHTANEN, C.N.; BRAZIS, A.R.; ARLEDGE, W.L.; DONNELLY, C.B.; GINN, R.E.; RANDOLPH, H.E.; KOCH, E.J. Temperature equilibration times of plate count agar and a comparison of 50 versus 45C for recovery of raw-milk bacteria. J. Milk Food Technol. 38:319–322; 1975.
16. JOHNS, C.K. Preliminary incubation for raw milk samples. J. Milk Food Technol. 23:137–141; 1960.
17. JOHNS, C.K. Use of counts after preliminary incubation to improve raw milk quality for a Denver plant. J. Milk Food Technol. 38:481–482; 1975.
18. JOHNS, C.K.; CLEGG, L.F.L.; LEGGATT, A.G.; NESBITT, J.M. Relation between milk production conditions and results of bacteriological tests with and without preliminary incubation of samples. J. Milk Food Technol. 27:326–332; 1964.
19. PEELER, J.T.; LESLIE, J.E.; DANIELSON, J.W.; MESSER, J.W. Replicate counting errors by analysis and bacterial colony counters. J. Food Prot. 45:238–240; 1982.
20. READ, JR., R.B.; BRADSHAW, J.G.; FRANCIS, D.W. Effect of freezing raw milk on standard plate count. J.Dairy Sci. 52:1720–1723; 1969.
21. REINBOLD, G.W.; JOHNS, C.K.; CLARK, JR., W.S. Modification of the preliminary incubation treatment for raw milk samples. J. Milk Food Technol. 32:42–43; 1969.
22. KOBURGER, J.A. Stack-pouring of petri plates: a potential source of error. J. Food Prot. 43:561–562; 1980.
23. TOMASIEWICZ, D.M.; HOTCHKISS, D.K.; REINBOLD, G.W.; READ, JR., R.B.; HARTMAN, P.A.

The most suitable number of colonies on plates for counting. J. Food Prot. **43**:282–286; 1980.
24. U.S. DEPT. OF HEALTH AND HUMAN SERV. Grade A pasteurized milk ordinance. No. 017-001-00419-7. Washington, DC: U.S. Gov. Printing Office; 1980.
25. YALE, M.W.; PEDERSON, C.S. Optimum temperature of incubatioin for standard methods of milk analysis as influenced by the medium. Amer. J. Pub Health and the Nation's Health. **26**:344–349; 1936.

CHAPTER 7

ALTERNATIVE MICROBIOLOGICAL METHODS

G.A. Houghtby, L.J. Maturin, and W.R. Kelley
(J.W. Messer,Tech. Comm)

7.1 Introduction

The Standard Plate Count (SPC) is relatively accurate for determining viable bacteria in dairy products, but it is expensive and time-consuming. Direct microscopic counts and reduction tests of raw milk are not sufficiently accurate. The following methods have merit as alternative tests for determining viable bacterial counts.

The oval tube count (OTC) and plate loop count (PLC) procedures are specifically for examining raw milk. Although both are single loop measurements (0.001 mL), the limited growth area of medium in each oval tube limits precision when more than 100 colonies develop, while plate loop counts are satisfactorily precise within the range of 25 to 250 colonies. Limitations of the plate loop method have been reported with manufacturing grade raw milk, when counts exceed 200,000/mL.[12]

The spiral plate loop count method (SPLC) is applicable to raw milk, pasteurized milk and milk products. The SPLC has been collaboratively tested with milk[10] and food products[6] and found equivalent to the SPC Method. The SPLC is a mechanical method that requires less skilled manpower, time, equipment, and space than does the SPC procedure.

The impedimetric method is applicable to estimating the initial SPC in raw milk. This instrumental method measures impedance changes associated with the metabolic activity of bacteria growing on an agar surface. Such changes have been correlated with the original SPC in both comparative and collaborative studies.[3,4]

7.2 Collection and Storage of Samples Before Testing

Collect and store samples in accordance with the procedures specified in 4.2C, 4.2D, and 6.5.

7.3 Statistical Protocol for an Analyst

Before any alternative viable count method is used, the individual analyst must test his/her ability to perform that method against his/her results with the SPC. The alternative viable count method and the SPC must be done on the same samples of milk. The following protocol is recommended when an analyst needs to evaluate his/her performance:

Samples are plated in duplicate by both the alternative method and the SPC until 25 samples give plate counts of from 25 to 250 colonies by the SPC and the PLC methods, or 25 to 100 colonies by the OTC method. Depending on the bacterial count range per mL, a count of 20 to 75 colonies in a segment or wedge, or greater than 20 colonies on the total plate surface, can be used for the SPLC method. All samples do not have to be plated on the same day; however, all preparations (both alternative and SPC methods) on one sample must be done at the same time. Duplicate counts from the alternative technique are converted to $\log_{10}$ counts; the difference between $\log_{10}$ counts for duplicate plates is squared and the squared values are added for all samples. In addition, the $\log_{10}$ counts are added together and divided by 50 to obtain the mean $\log_{10}$ count for each alternative count method. The SPC results are tabulated and analyzed in the same manner.

For each alternative method to be rated satisfactory, the mean $\log_{10}$ count for the 50 observations for the alternative method must not differ from the same value for the SPC by more than 0.036. For satisfactory duplicate data, the sum of the squared differences (log variances) between duplicate samples for both alternative and SPC methods, divided by 50, should not be greater than 0.005. Examples of data included in Table 7.1 show how to evaluate results to determine compliance with the two criteria.

7.4 Plate Loop Method[11] (Class O)

A. Equipment and supplies:
1. Measuring and transfer instrument: The total equipment assembly cannot be purchased as a unit, but is assembled from the following parts:
    a. Loop, 0.001 mL: Preferably welded, calibrated to contain 0.001 mL, made of B & S gauge No. 26 platinum-rhodium (3.5%) or platinum-iridium (15%) wire, true-circle loop, ID 1.45 mm ± 0.06 mm, attached to a 75 mm length of wire (American Scientific Products 1430 Waukegam Rd. McGaw Park, IL 60085 N-2075-2, or equivalent). (Loops may be reshaped by being fitted over the top of an awl, then checked over a No. 54 twist drill. Loops of proper size should not fit over a No. 53 twist drill.) Make an approximately 30° bend about 3 to 4 mm from the loop, with the loop opening toward the hub. Kink the opposite end of the wire in several places.

7.4 Plate Loop Method (Class O)

Table 7.1 Example of Statistical Comparison of Plate Loop Count (Alternative Viable Count) with Standard Plate Count

Sample	SPC/mL	Log SPC/mL	SPC (log a-log b)	SPC (log a-log b)	PLC/mL	Log PLC/mL	PLC (log a-log b)	PLC (log a-log b)
1 a	83,000	4.9191	−0.1223	0.0150	110,000	5.0414	−0.0725	0.0053
b	110,000	5.0414			130,000	5.1139		
2 a	170,000	5.2304	−0.0249	0.0006	180,000	5.2553	−0.0235	0.0006
b	180,000	5.2553			190,000	5.2788		
3 a	100,000	5.0000	0.0177	0.0003	130,000	5.1139	0.1139	0.0130
b	96,000	4.9823			100,000	5.0000		
24 a	270,000	5.4314	0.0164	0.0003	270,000	5.4314	0.0335	0.0011
b	260,000	5.4150			250,000	5.3979		
25 a	220,000	5.3424	0.0202	0.0004	210,000	5.3222	−0.0202	0.0004
b	210,000	5.3222			220,000	5.3424		
Sum		250.5432		0.0312		250.3631		0.0737
Sum ÷ 50 SPC results (by means)		5.0109[y]		0.0006[z]		5.0073[y]		0.0015[z]

Note: [z]Mean Log Variance for Duplicates:
SPC = 0.0006
PLC = 0.0015
Acceptable, as both results are less than 0.005
[y]Difference between Log Mean SPC and Log Mean PLC
5.0109 − 5.0073 = 0.0036
Acceptable, as difference is less than 0.036

b. Luer-Lok hypodermic needle: 13-gauge, sawed-off 24 to 36 mm from the point where the barrel enters the hub.
c. Cornwall continuous pipetting outfit: Becton, Dickinson & Co. (Available through scientific suppliers such as American Scientific Products Op. Cit.) No. 1251 (consisting of a metal pipetting holder, a Cornwall Luer-Lok syringe, and a filling outfit), 2 mL capacity. Adjust to deliver 1.0 mL. Rubber tubing, attached to the syringe and extending into a capped dilution blank, should be of sufficient length to facilitate the sampling process (Fig. 7.1). Because repeated autoclaving will cause the rubber tubing to deteriorate, have additional tubing on hand.
d. Cover: Protective glass test tube for 0.001-mL loop during sterilization and handling (optional).

The above parts are assembled in the following manner:
e. Insert the kinked end of the wire shank into the sawed-off needle to point where the bend is about 12 to 14 mm from the end of the barrel. Attach the needle to the pipetting outfit. Place glass tube, if used, over the loop and needle.

Sterilize the assembled apparatus by autoclaving at 121°C for 15 min. Alternatively, sanitize the part before assembly by submerging the parts in boiling water for 10 min., cool and aseptically assemble.

2. Other equipment and supplies [6.2]:
   a. Materials:
   [6.3.]
   b. Procedure:
   1. After sterilization of the assembled pipetting outfit, allow equipment to cool. Depress syringe plunger rapidly several times to pump dilution water into glass syringe (which has previously been adjusted to deliver 1 mL with each depression of the plunger).
   2. Before initial transfer is made in examining a series of samples, briefly flame the loop (in a clean, high-temperature gas flame) and allow it to cool at least 15 sec. Discharge several 1 mL portions of water to waste; then discharge a 1 mL portion of water into first petri dish. Label this dish "sterile instrument control". Shake raw milk samples [6.6B], carefully dip the loop into the sample (avoiding foam) as far as the bend in the shank (the bend serves as a graduation mark and permits vertical removal of the loop). To measure the sample, insert the loop vertically into the sample three times, moving the loop with a uniform up-down movement over a distance of about 2.54 cm. Avoid dilution-water droplets that may rinse off the loop. Each downward movement should be at a rate

7.4 Plate Loop Method (Class O)   155

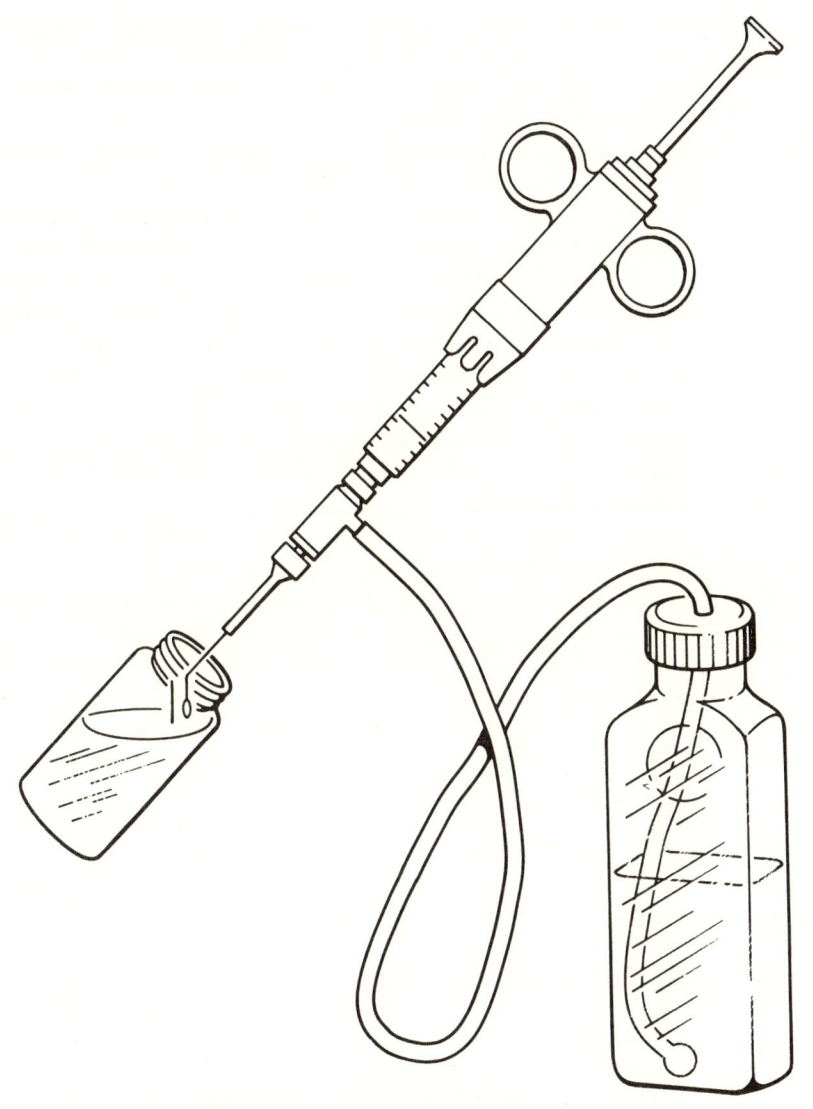

Figure 7.1. Plate loop outfit, showing vertical removal of loop.

of about 55 to 60 beats per minute. A metronome or watch may be used to establish uniform timing.

CAUTION: The speed of removal of the loop from the surface of the milk sample affects accuracy of the measurement. Removing the loop slowly causes less than 0.001 mL to adhere; jerking the loop out rapidly causes more than 0.001 mL to adhere.

3. Raise the cover of a sterile petri dish, insert loop and depress the plunger, causing 1 mL of sterile dilution water to flow across the charged loop and thus wash a measured 0.001 mL of sample into the dish.

   CAUTION: Do not depress the plunger so rapidly that water fails to follow the shank and flow across the loop.
4. Normally, the residue remaining on the loop after discharging the sample is not significant. However, small imperfections in the welding of the loop or in smoothness of the metal surface may lead to incomplete rinsing. This possibility of significant residue retention should be determined for each loop by doing a series of control plates, a minimum of at least one for every 20 samples plated. If the loop is determined to be free-rinsing, no flaming between samples is necessary.
5. Pour plates with 12 to 15 mL of agar and incubate 48 ± 3 h at 32 ± 1°C.

c. Counting colonies and recording results: Count plates and record the number of colonies on each plate counted or estimated [6.2K].

d. Reporting counts: Counts from plates having 25 to 250 colonies are multiplied by 1,000 and reported as Plate Loop Count per milliliter (PLC/mL). If plates have fewer than 25 colonies each, multiply count by 1,000 and report count as Estimated Plate Loop Count per milliliter (EPLC/mL). If plates have no colonies, and inhibitory substances have not been detected, report count as <1,000 EPLC/mL. If plates have greater than 250 colonies each, multiply estimated count by 1,000 and report count as EPLC/mL.

Where plates prepared from samples 1) have excessive spreader growth, see [6.2K1a6]; or 2) are known to be contaminated or mislabeled or are otherwise unsatisfactory, record as Laboratory Accident (L.A.). If inhibitory substances have been detected [Chapter 14], record the PLC as Growth inhibitors (G.I.). Do not record the count/milliliter on such samples.

7.5 Spiral Plate Count Method[1,2,5,6,8,10] (Class A2)

A. *Equipment and supplies:*
1. Spiral plater. (Spiral Systems Marketing, Inc., 4853 Cordell Ave., Suite A-10, Bethesda, Maryland 20014)
2. Spiral colony counter with special grid for relating deposited sample volumes to specific portions of petri dishes.
3. Vacuum trap for disposal of liquids (a vacuum bottle of 2 to 4 L capacity to act as a vacuum reservoir and a vacuum source of 50 to 60 cm of Hg).
4. Disposable micro beakers, 5 mL.

7.5 Spiral Plate Count Method (Class A2)

5. Petri dishes, plastic or glass, 150 × 15 mm, or 100 × 15 mm.
6. Standard methods agar [5.9E].
7. Polyethylene bags, about 30 cm × 20 cm × 40 cm.
8. Commercial sodium hypochlorite solution, approximately 5% NaOCl (commercial bleach solution).
9. Sterile dilution water [5.3B].
10. Syringe, with Luer-Lok tip, capacity not critical.
11. Other equipment and supplies [6.2B].

Preparation of agar plates:

An automatic dispenser with a sterile delivery system is recommended to prepare agar plates. The agar volume dispensed into plates is reproducible and the contamination rate is low compared to hand pouring of agar in an open laboratory. When possible, a laminar air flow hood should be used in conjunction with an automated dispenser.

The same quantity of agar should be poured into all plates so that the same height of agar will be presented to the Spiral Plate stylus tip to maintain contact angle. The agar plates should be level during cooling. The following is a suggested method for prepouring agar plates:

1. Use an automatic dispenser or pour a constant amount (about 15 mL/100 mm plate; 50 mL/150 mm plate) of sterile agar, at 60 to 70°C, into each petri dish.
2. Allow agar to solidify on a level surface with poured plates stacked no higher than ten dishes.
3. Place the solidified agar plates in polyethylene bags, close with ties or heat-sealer, and store inverted at 0 to 4.4°C.
4. Bring prepared plates to room temperature before inoculation.

C. Preparation of samples:
Shake samples as described in 6.2F.

D. Description of spiral plater:

The Spiral Plate Count (SPLC) plater [Fig. 7.2] inoculates the surface of a prepared agar plate to permit measurement of numbers of bacteria in solutions containing between 500 and 500,000 bacteria per mL. An operator with a minimum of training can inoculate 50 plates per hour. Within the range stated, dilution bottles or pipettes and other auxiliary equipment are not required. Required bench space is minimal, and time to check instrument alignment is less than 2 min.

The plater deposits a decreasing amount of sample in the form of an Archimedean spiral on the surface of a prepoured agar plate. The volume of sample on any portion of the plate is known. After incubation, colonies appear along the line of the spiral [Fig. 7.3]. If colonies on a portion of the plate are sufficiently spaced from each other, they are counted on a special grid [Fig. 7.4] which associates a calibrated volume with each area. The number of bacteria in the sample is estimated by dividing the number of colonies in a defined area by the volume contained in the same area. Studies

Figure 7.2. Spiral plater.

7.5 Spiral Plate Count Method (Class A2)

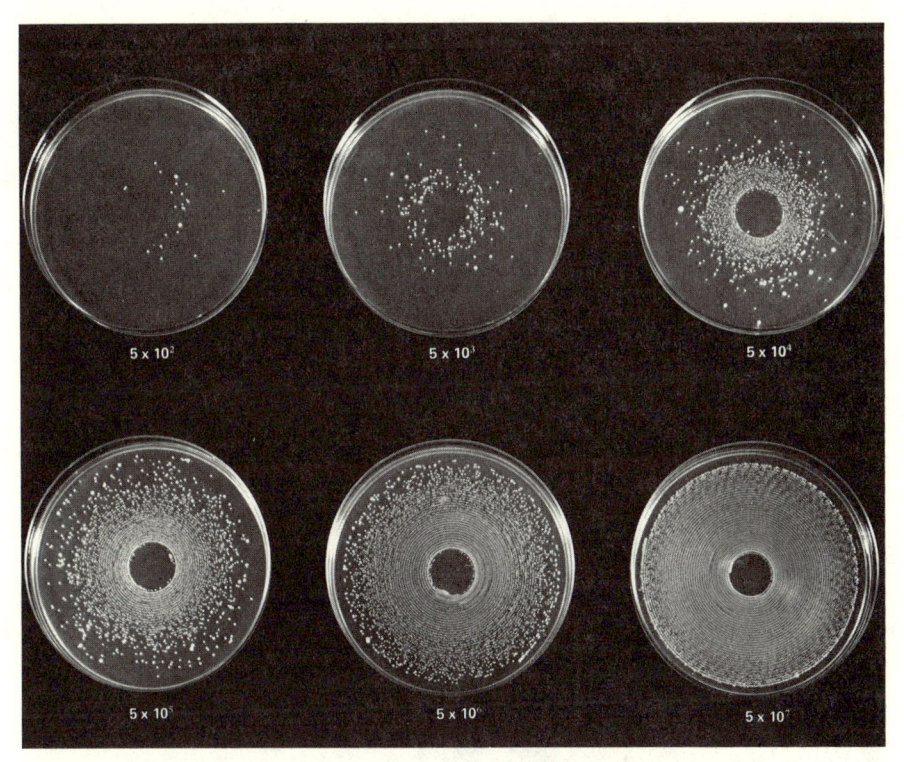

Figure 7.3. Spiral plates representing 500 to 500,000 *Escherichia coli/mL* **and estimate counts of 5 and 50 million.**

have shown the method is proficient not only with milk analysis but also with other foods[1,6,8] and in yeast and mold determinations.[13]

E. Plating procedure:

Check the stylus tip angle daily and adjust, if necessary. (Use vacuum to hold a microscope cover slip agaqinst the face of the stylus tip; if the cover slip plane is parallel at about 1 mm from the surface of the platform, the tip is properly oriented.) Liquids are moved through the system by vacuum. Cleanliness of the stylus tip is assured by a rinse for 1 sec using sodium hypochlorite solution followed by sterile dilution water for 1 sec before sample introduction. The rinse procedure between processing of each sample minimizes cross contamination.

After rinsing, the sample is drawn into the tip of the Teflon tubing by the vacuum applied to the two-way valve. When tubing and syringe are filled with sample, the valve attached to the syringe is closed. An agar plate is placed on the platform, the stylus tip placed on the agar surface, and the motor started. During inoculation the petri plate lid may be labeled. After

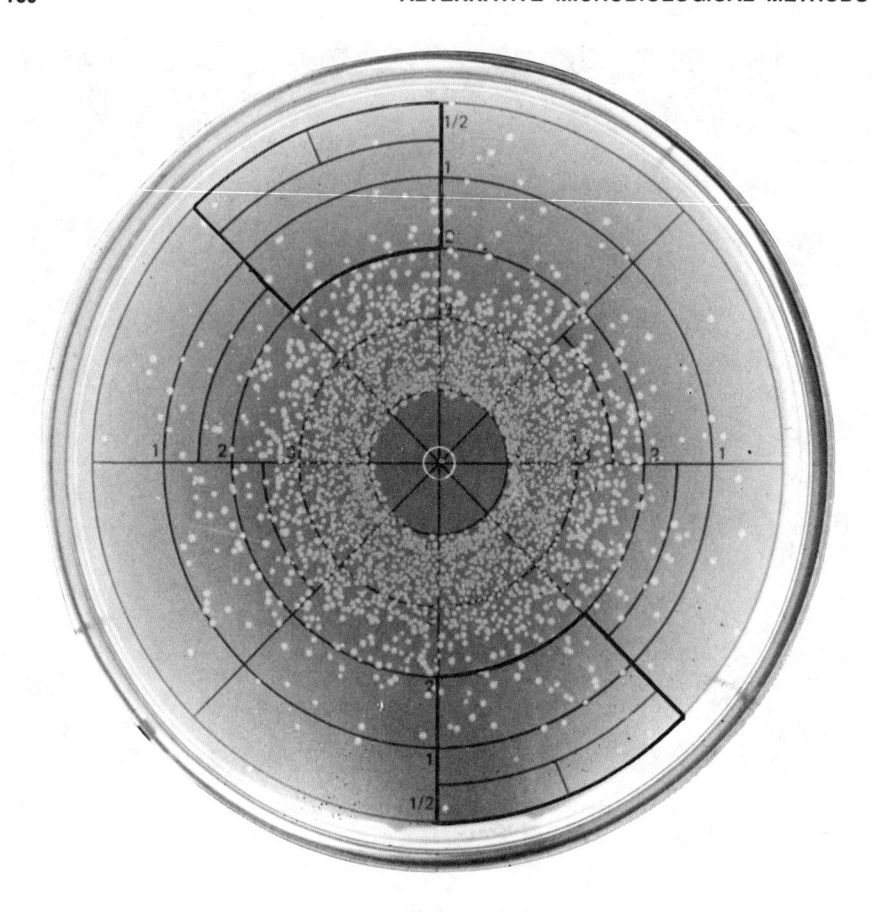

Figure 7.4. Spiral plate inoculated with *Bacillus subtilis* 77,000/mL. Two segments of opposite wedges are outlined.

the agar has been inoculated, the stylus lifts from the agar surface, and the spiral plater automatically stops. The inoculated plate is removed from the platform and covered. The stylus is moved back to the starting position. The system is vacuum-rinsed with hypochlorite, followed by water, then a new sample is introduced. Invert plates and promptly place in an incubator for 48 ± 3 h at 32 ± 1°C.

F. Sterility controls:

For each series of samples, check sterility of the spiral plater by plating sterile dilution water.

G. Cautions:

Prepoured plates should not:

   1. Be contaminated by a surface colony.

7.5 Spiral Plate Count Method (Class A2)

2. Be below room temperature: condensed water welling up from agar can occur.
3. Be excessively dry, as indicated by large wrinkles or excessive glazed appearance.
4. Have water droplets on surface of agar.
5. Have differences greater than 2 mm in depth of agar.
6. Be stored at 0 to 4.4°C for longer than one month.

Reduced flow rate through tubing indicates obstructions or material in system. Obstructions may be cleared by removing valve from syringe, inserting a hand-held syringe with a Luer-Lok fitting containing water and applying pressure. An alcohol rinse may remove residual material adhering to walls of system, or acid detergent may be used to dissolve accumulated residue. Rinse well after cleaning.

H. Counting grid:

1. Description: The same counting grid is used for either 100 mm or 150 mm petri dishes, with a mask being supplied for use with the 100 mm dishes. The counting grid is divided into eight equal wedges; each wedge is divided by four arcs labeled 1, 2, 3, and 4 from the outside grid edge. Other lines within these arcs are added for ease of counting. A segment is the area between two arc lines within a wedge. Areas counted, e.g., three, refers to the number of segments within a wedge that were counted.

    The spiral plater reproducibly deposits on an agar plate a volume of sample in the same way each time. The grid is used to relate colonies on a spiral plate to the volume in which they were contained. When counting colonies with the grid [7.5I], the sample volume becomes greater as the counting starts at the outside edge of the plate and proceeds toward the center of the plate. In Table 7.2, the

Table 7.2 Determination of Volume Constants for 150-mm Spiral Plates

Colonies/mL by SPC	Segments of opposite wedges counted	Colony count for opposite wedges	Grid area constant (mL)
500	Total	20	0.040
3,800	Total	150	0.040
8,000	4	45 + 35	0.0100
12,000	3	23 + 20	0.0035
120,000	2	56 + 62	0.0010
690,000	1	65 + 72	0.00020
2,000,000	½	58 + 64	0.00006

Example area 4, $\dfrac{45 + 35 \text{ colonies}}{8,000 \text{ SPC/mL}} = 0.010 \text{ mL}$

fourth column (labeled "grid area constant") shows how the sample volume increases from the amount contained in one segment of a wedge to the amount contained in four segments of a wedge.
2. Calibration: The volume of sample represented by various parts of the counting grid are shown in the operator's manual accompanying the Spiral Plater. For precise work, the grid area constants should be checked by the responsible analyst. This may be done by preparing 11 bacterial concentrations in the range of 10^6 to 10^3 cells/mL, by making 1:1 dilutions of a bacterial suspension (use a nonspreader). Plate all dilutions in duplicate by both the SPC [Chapter 6] and the SPLC using the same medium and incubator. Incubate both sets of plates for 48 ± 3 h at 32 ± 1°C. Calculate the SPC/mL for each of the dilutions. Count the spiral plates over the grid surface; use the counting rule of 20 [7.5I] to record the number of colonies counted and the grid area over which they were counted. Each of the SPLC colony counts for a particular grid area, divided by the SPC/mL for corresponding spirally plated bacterial concentrations, will indicate how much volume was deposited on that particular grid area. The following formula should be used:

$$\text{Volume (mL) for grid area} = \frac{\text{Spiral colonies counted in area}}{\text{Bacterial count/mL (SPC)}}$$

If the SPC is not used, a count of between 50 and 200 colonies on the whole spiral plate may be used to determine the concentration of the plated bacterial suspension; this can be done by dividing the colonies counted by 0.035 (mL). The concentration of the other ten dilutions of plated suspension may be calculated by using the appropriate dilution factor.

One example of the calculation required to determine the area volume is given in Table 7.2.

I. *Examination and reporting spiral plate counts:*
1. Counting rule of 20: After incubation, the spiral plate is centered over the grid by adjusting the holding arms on the viewer. Choose any wedge and begin counting colonies from the outer edge of the first segment toward the center until 20 colonies have been counted. The count is completed by counting the remainder of the colonies observed in the segment in which the 20th colony occurs. This number is recorded along with the number of the segment (marked on Fig. 7.4 as 1, 2, 3, and 4), that included the 20th colony. In this counting procedure, the preceding numbers refer to the total area in all the segments from the outer edge of a wedge to the designated arc lines as illustrated in Figure 7.4. Any count irregularities in sample composition are controlled by counting the same segments in the opposite wedge and recording results. If there is not a total

7.5 Spiral Plate Count Method (Class A2)

of 20 colonies in the four segments of the wedge, then all of the colonies on the whole plate should be counted. The number of bacteria is estimated by dividing the count obtained by the volume contained in all the segments counted.

If the number of colonies exceed 75 in the 2nd, 3rd, or 4th segment, which also contains the 20th colony, the estimate of number of bacteria will generally be low because of coincidence error associated with crowding of colonies. It is recommended that the count then be made by counting each of the circumferentially adjacent segments in all eight wedges, counting at least 50 colonies. As an example, if the first two segments of a wedge contain 19 colonies and the 3rd segment contains the 20th and 76th (or more) colony, count the colonies in all circumferentially adjacent 1st and 2nd segments in all eight wedges. Calculate the contained volume in the counted segments of wedges and divide into the number of colonies as shown below.

Figure 7.4 shows an example of a spirally inoculated plate that demonstrates the method for determining a bacterial count. Two segments of each wedge were counted on opposite sides of the plate with 21 and 33 colonies, respectively.

From Table 7.2, the grid area volume suggested is 0.0010 mL.

$$\frac{21 + 33}{0.0010} = 54,000 \text{ SPLC/mL}$$

2. When fewer than 20 colonies are counted on the total plate, report results as "less than 500 Estimated (E) SPLC per mL". If the colony count exceeds 75 in the first segment of a wedge, report results as "greater than 500,000 Estimated (E) SPLC per mL".
3. Do not count spiral plates with irregular distributions of colonies caused by dispensing errors. Report results of such plates as laboratory accident (LA).
4. If a spreader covers the entire plate, the plate should be discarded. When a spreader covers one-half of the plate area, count colonies only when well distributed in spreader-free areas [6.2K1a6].
5. Compute SPLC, unless restricted by detection of inhibitory substances in the sample, excessive spreader growth, or laboratory accidents [6.2K4]. Record only the first two left-hand digits; raise the digit to the next highest number only when the third digit from the left is five or greater and use zeroes for each successive digit toward the right from the second digit.

Report counts as the Spiral Plate Count (SPLC), or the Estimated (E) SPLC/mL, or as applicable.

7.6 Oval Tube Method[9] (Class D)

A. *Equipment and supplies:*
   1. Oval tube: Round neck, plugged with cotton or preferably capped with stainless-steel closures (Bellco B-16); bulb dimension, 100 × 17 × 27 mm, modified form, with 19.1 mm neck (Fisher Scientific 711 Forbes Ave, Pittsburgh, PA 15219 No. 14-928).
   2. Water bath: [6.2L].
   3. Loop 0.001 mL mounted in a suitable holder [7.4A1a].
   4. Racks: Noncorrosive wire, for holding oval tubes during sterilization, cooling, inoculation, and incubation. Size for oval tubes is to be, 40 or 50 spaces, 2.43 × 3.175 cm mesh; and two shelves, the top shelf is to have a 1.25 cm support on one side (long dimension) to tilt the rack slightly when placed on its side during incubation. Or, use a tube cabinet containing six shelves, each holding 12 oval tubes, shelf angle adjustable (Curtin Matheson Scientific Inc., PO Box 1546, Houston, TX 77251, No. 376-848 or equivalent).
   5. Containers: [4.A3].
   6. Others: Colony counter, incubator, incubator room, work area, storage space, refrigerator, thermometer, tally, autoclave, hot-air sterilizing oven [6.2B].

B. *Materials:*
[5.9E, 6.2 A and B].

C. *Procedure:*
   1. Shake each raw milk sample thoroughly 25 times through a 30-cm arc in 7 sec. Flame sterilize the 0.001 mL standard loop and allow it to cool at least 15 sec. Transfer a loopful of milk to an oval tube containing 4 mL of sterile melted agar, tempered to 44 to 46°C, being careful to dip the standard loop only 2 to 3 mm below the surface of the milk, in an area as free from foam as possible [7.4A262]. Hold the plane of the loop in a vertical position when withdrawing it from milk and move the loop back and forth several times through the agar to insure removal of all milk from the loop.

      CAUTION: The speed of removal of the loop from the milk sample affects the accuracy of the measurement. Removing the loop slowly causes less than 0.001 mL to adhere; jerking the loop out rapidly causes more than 0.001 mL to adhere.
   2. Replace the closure, mix agar and milk thoroughly by swinging the tube back and forth rapidly through a small arc for 5 sec (about 25 complete arcs), and lay the oval tube flat on the table or, preferably, slant it slightly. After the agar has solidified, incubate oval tubes horizontally in a wire rack or tube cabinet (with agar adhering to upper side of the tube) for 48 ± 3 h at 32 ± 1°C.

7.7 Electrical Impedance Method (Class A2)

D. Counting colonies and recording results:
Using a colony counter, count colonies from tubes having 100 or fewer colonies [6.2K] and record the number.

E. Reporting counts:
Counts from tubes having 25 to 104 colonies are multiplied by 1,000 and reported as "Oval Tube Count per milliliter" (OTC/mL). If tubes have fewer than 25 colonies each, multiply count by 1,000 and report count as "Estimated Oval Tube Counter per milliliter" (EOTC/mL). If tubes have no colonies, and inhibitory substances have not been detected, report count as <1,000 EOTC/mL. If a tube has a colony count in excess of 100 colonies, report count as >100,000 EOTC/mL. Where tubes prepared from samples: 1) have excessive spreader growth, see 6.2K1a6; or 2) are known to be contaminated, mislabeled, or other otherwise unsatisfactory, record as "Laboratory Accident" (L.A.). If inhibitory substances have been detected [Chapter 14], record the OTC as "Growth Inhibitors" (GI). Do not record the count/milliliter on such samples.

7.7 Electrical Impedance Method[3,4,7] (Class A2)

A. Principle:
Bacterial concentrations in milk are determined by measuring impedance changes associated with the metabolic activity of the reproducing microflora present in each sample when inoculated onto an agar surface. The microbial activity results in the conversion of nonelectrolytes (such as lactose) to ionic compounds (such as pyruvic acid or lactic acid).

Impedance (defined as the resistance to the flow of an alternating electrical current through the medium) remains relatively constant until the number of microorganisms present in the incubated sample reaches a threshold of 10^6 to 10^7 per mL. When this threshold level is reached, marked impedance changes occur due to changes in the ionic constituency of the medium. Fatty acids, amino acids and organic acids produced through breakdown of fats, proteins and carbohydrates, respectively, contribute toward this change in impedance. The time required for the initial inoculum to reach the threshold level (10^6 to 10^7 cells/mL) is designated as the impedance detection time (IDT). The instrument automatically registers an IDT.

The time required for the bacteria in the sample to reach the instrument's threshold level correlates with the original bacterial number present as determined by the SPC Method. An IDT can be achieved with a single organism, providing it can grow to the threshold level. After a calibration curve relating IDT to SPC has been established, samples are impedimetrically tested and their IDT levels are used together with the calibration curve to determine the original microbial concentration of the sample. IDT "cut off" levels are established to designate whether or not a specific sample contains microbial concentrations above or below specifications. Impedance measurements can provide earlier estimates of initial numbers than can standard methods. For example, 10^4/mL of mesophiles or psychrotrophs in milk can be detected in 4.0 and 24.0 h, respectively.[3]

B. *Equipment and supplies:*
  1. Bactometer Microbial Monitoring System M123-SC and disposable modules. (Bactomatic Inc., A Division of Medical Technology Corporation, PO Box 3103, Princeton, NJ 08540)
  2. Appropriate medium, see D, below.
  3. Pipettes—0.1 mL and 1.0 mL.

C. *Description of the bactometer:*

The Bactometer Microbial Monitoring System M123 includes an LS1-11/23 microcomputer, color video terminal and keyboard, printer and M120SC Bactometer Processing Unit and incubator (Fig. 7.5). Samples are loaded into sterile modules that are then plugged into the M120SC incubator system. Each module contains 16 sample wells (Fig. 7.6). Sterile plastic caps are provided to cover the sample wells.

The system automatically monitors impedance changes in up to 128 samples per Bactometer Processing Unit (BPU). Four BPUs may be linked to one computer. The computer uses an algorithm that determines the onset of acceleration in the impedance curve. The impedance detection times (IDT) are displayed on the video screen as soon as they are obtained. A color graphics software package enables the operator to display impedance change curves for any sample on the video screen.

D. *Medium for the impedance method:*

The accuracy and precision of the results by the impedimetric test of dairy products depend on the culture medium used. The same basic considerations given to standard culture media [Chapter 5] should also be given to the medium prepared for impedance measurements. The following medium is required:
  1. Modified Standard Methods Agar (MSMA).

Yeast extract	20.0 g
Pancreatic digest of caserin	20.0 g
Dextrose	4.0 g
Agar	10.0 g
MS water to make	1 L

Suspend the ingredients in water, heat, and boil 1 min to dissolve completely. Dispense 100-mL portions of medium into screw-capped containers. Sterilize at 120°C for 20 min. Final pH should be 7.0 ± 0.2 @ 25°C after sterilization. Store medium in the refrigerator until ready for use. In order to prepare each module, place a bottle of MSMA in a beaker of water and heat to completely melt the agar. Aseptically remove the caps from the module wells and dispense 0.5 mL of melted MSMA into each well. Once the medium has been added to all the wells, gently agitate the

7.7 Electrical Impedance Method (Class A2)

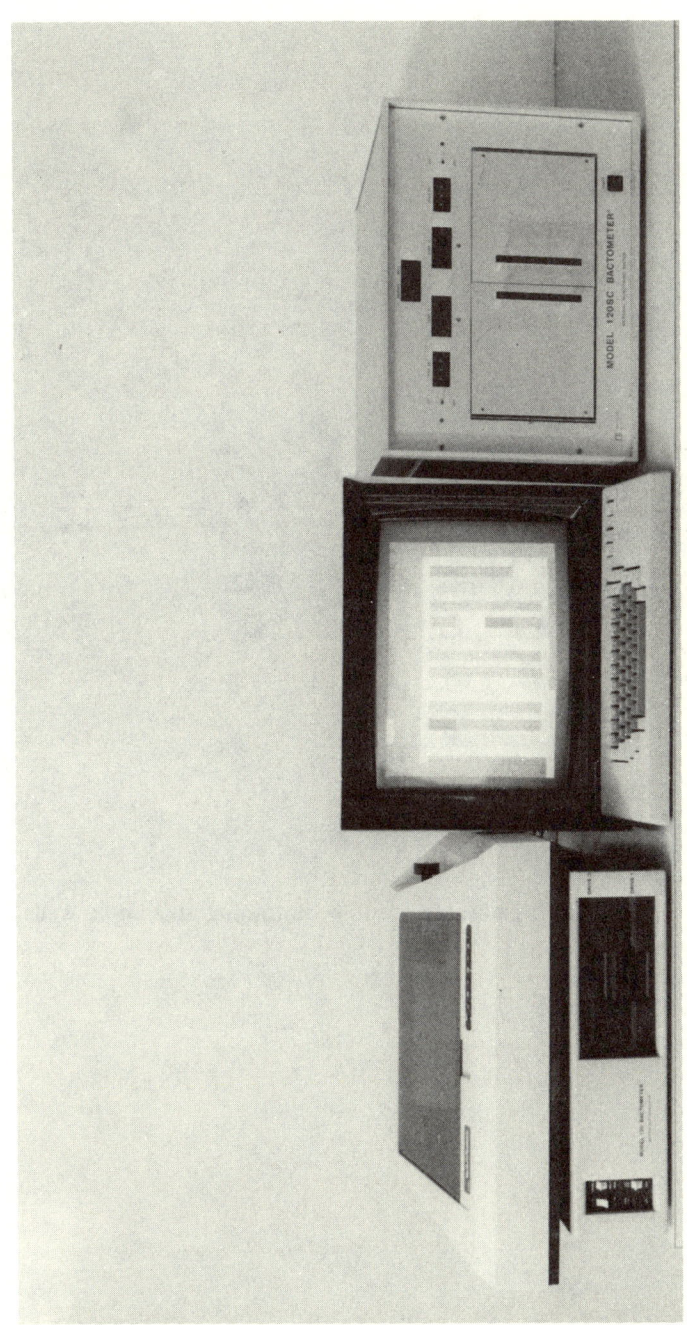

Figure 7.5. The Bactometer microbial monitoring system M123.

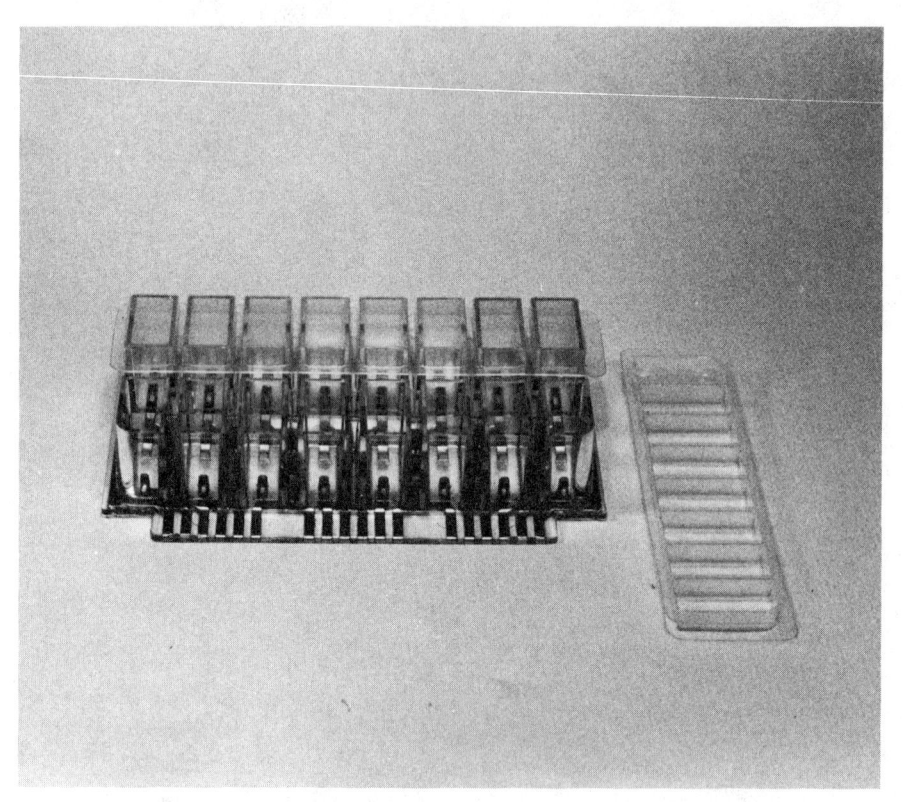

Figure 7.6. Disposable module containing 16 sample wells.

7.7 Electrical Impedance Method (Class A2)

module to ensure even distribution of agar at the bottom of the wells. Replace caps over the top of the wells. Use the modules after the agar has solidified or place the modules in sealed, moisture-impermeable plastic bags and store (inverted) at 0 to 4.4°C. Modules must be used before the agar begins to dry (approx. 2 mo.).

E. *Preparation of system:*

Power up all system components. Insert computer system disk into drive 0 and data disk into drive 1 and follow appropriate commands as described in the user's manual for the M123 system.

F. *Preparation of samples for total aerobic count:*
1. Shake the sample vial 25 times in a 30-cm arc for 7 sec.
2. Remove caps from module wells.
3. Pipette 0.1 mL of each sample into two agar (MSMA) filled module wells.
4. Aseptically place the caps back on all wells.

G. *Operation of system:*
1. Set the incubator temperature at 18°C.
2. Operate the system for 48 h.

Since mesophiles outnumber psychrotrophs in many raw milk samples, it is frequently desirable to determine bacterial counts by incubating Bactometer modules at 35°C. This results in earlier IDTs (4 h vs 16 h for 10^5 cells/mL). The user should check the milk supply to determine if a significant correlation exists between SPC at 32°C and IDTs at 35°C.

Immediately after the appropriate medium and samples have been introduced into the modules, plug the inoculated modules into their designated incubator position. Close the door and initiate the test. Time indicated is maximum running time to assure attaining IDTs with samples well within product specification limits. When products to be tested exceed specification limits, a more rapid IDT and shorter system run time will be possible.

H. *Preparation of calibration curve:*

A calibration curve is calculated using the IDT and the SPC data.

Simultaneously test milk samples in parallel using the instrument and the appropriate SPC method. Use the calibration system/data disk to correlate IDT with the standard method. Enter data for the IDT and SPC values of at least 70 samples, with the SPCs evenly distributed over 4 to 5 log cycles (i.e., 100 to 1,000,000). At least 10% of the data points should be at least a log cycle above the specified level of contamination. Execute the command for Regression Analysis and input a cut-off level (e.g., 10^5 SPC per mL) to determine the caution and cut off impedance detection times for this microbial concentration.

The calibration curve is used to obtain Bactometer Predicted Counts (BPCs). The computer software uses the IDT result together with the calibration curve to calculate the BPC, which can be printed after the tests are terminated. Bactometer Predicted Counts are obtained with bacterial concentrations in ranges of from 5×10^3 to 5×10^7 SPC/mL. Bactometer Plate Counts can be obtained above and below these values but with reduced accuracy.

I. Determination and interpretation of results:

Once the calibration line has been completed, modify the product code by changing the cut-off and caution times to those values that were determined by the regression analysis. Test samples in the instrument as previously described [7.7F and G]. Samples that detect before the cut-off time contain bacterial concentrations in excess of the specified level. Their IDTs will be displayed on the video screen in red. Samples that detect after the caution time contain fewer bacteria than the specified level and their IDT values will be displayed in green. Samples that detect within the cut-off and caution times are marginally close to the specified level and will be displayed in yellow.

J. Verification of the instrument results:

Upon test completion, the operator should obtain a hard-copy printout of the data.

Evaluate duplicate samples for discrepancies in detection times. Examine the module well if discrepancies of 0.5 h for IDTs shorter than 10 h or of 1 h for IDTs longer than 10 h exist. If any well shows signs of drying agar, make a note next to the appropriate IDT on the hard-copy printout and select the data from the normal well.

The operator could also run the "Verification of Detection Time" function. This function should alert the user to any gross error in the IDT by placing a question mark next to its detection time on the printout.

If a discrepancy in detection time larger than mentioned above is observed in duplicate samples where no abnormality is observed in wells, the IDT curves should be compared simultaneously. The curves of the duplicate wells are displayed on the color monitor. The operator can then determine whether the IDT for one of the two curves is incorrect. Curves that are either discontinuous or associated with noise can result in false detections. Any curve that shows excessive drift in either direction should be considered as "questionable." If such abnormalities are observed, describe their appearance on the hard-copy printout adjacent to the model well in question and select the data from the normal curve.

K. Sterility control:

Check sterility of the medium by dedicating a few wells in each run to the sterile medium. Assign these wells [7.7E] and run the system [7.7G]. No IDTs should be obtained.

7.8 References

1. CAMPBELL, J.E; GILCHRIST, J.E. Spiral plating technique for counting bacteria in milk and other foods. Dev. Ind. Microbiol. **14**:95–102; 1973.
2. DONNELLY, C.B.; GILCHRIST, J.E.; PEELER, J.T.; CAMPBELL, J.E. Spiral plate count method

7.8 References

for the examination of raw and pasteurized milk. Appl. Environ. Microbiol. **32**:21–27; 1976.
3. FIRSTENBERG-EDEN, R.; TRICARICO, M.K. Impedimetric determination of total, mesophilic and psychrotrophic counts in raw milk. J. Food Sci. **48**:1750–1754; 1983.
4. FIRSTENBERG-EDEN, R.A. Collaborative study of the impedence method for examining raw milk samples. J. Food Prot. **47**:707–712; 1984.
5. GILCHRIST, J.E.; CAMPBELL, J.E.; DONNELLY, C.B.; PEELER, J.T.; DELANEY, J.M. Spiral plate method for bacterial determination. Appl. Microbiol. **25**:244–252; 1973.
6. GILCHRIST, J.E.; DONNELLY, C.B.; PEELER, J.T.; CAMPBELL, J.E. Collaborative study comparing the spiral plate and aerobic plate count methods. J. Assoc. Office. Anal. Chem. **60**:807–812; 1977.
7. GNAN, S.; LUEDECKE, L.O. Impedance measurements in raw milk as an alternative to the standard plate count. J. Food Prot. **45**:4–7; 1982.
8. JARVIS, B.; LACH, V.H.; WOOD, J.M. Evaluation of the spinal plate maker for the enumeration of micro-organisms in foods. J. Appl. Bacteriol. **43**:149–157; 1977.
9. MYERS, R.P.; PENCE, J.A. A simplified procedure for the laboratory examination of raw milk supplies. J. Milk Technol. **4**:18–25; 1941.
10. PEELER, J.T.; GILCHRIST, J.E.; DONNELLY, C.B.; CAMPBELL, J.E. A collaborative study of the spiral plate method for examining milk samples. J. Food Prot. **40**:462–464; 1977.
11. THOMPSON, D.I.; DONNELLY, C.B.; BLACK, L.A. A plate loop method for determining viable counts of raw milk. J. Milk Food Technol. **23**:167–171; 1960.
12. WRIGHT, E.O.; REINBOLD, G.W.; BURMEISTER, L.; MELLON, J. Prediction of standard plate count of manufacturing-grade raw milk from the plate loop count. J. Milk Food Technol. **33**:168–170; 1970.
13. ZIPKES, M.R.; GILCHRIST, J.E.; PEELER, J.T. Comparison of yeast and mold counts by spiral, pour, and streak plate methods. J. Assoc. Offic. Anal. Chem. **64**:1465–1469; 1981.

CHAPTER 8
COLIFORM BACTERIA

P. A. Hartman and W. S. LaGrange
(J.C. Bruhn, Tech. Comm.)

8.1 Introduction

The coliform group of bacteria comprises all aerobic and anaerobic, gram-negative, nonspore-forming rods able to ferment lactose with production of acid and gas at 32°C within 48 h. One source of these organisms is the intestinal tract of warm-blooded animals; certain bacteria of nonfecal origin also are members of this group. Typically, these organisms are classified in the genera *Escherichia, Enterobacter,* and *Klebsiella.* A few lactose-fermenting species of other genera are included in the coliform group. In proportion to the numbers present, existence of any of these types in dairy products is suggestive of unsanitary conditions or practices during production, processing, or storage.

Application of the test for coliforms is intended neither to detect fecal pollution specifically nor to identify *Escherichia coli* in dairy products, but rather to measure the quality of the practices used to minimize bacterial contamination of processed dairy products.[4,19] Such tests are conducted following pasteurization primarily to detect significant bacterial recontamination of milk, cream and other processed dairy products.[28]

Results of tests on raw samples are to be interpreted differently from those obtained by testing pasteurized milk.[22,27] Coliforms in small numbers may enter raw milk and cream under normal conditions of production and handling.[2]

With increased use of farm bulk tanks and pipeline milkers, some laboratories are now making coliform counts on raw milk to determine the degree of contamination during milk production. With both raw and pasteurized samples, in the absence of storage-interval and temperature records for milk or cream before testing, it is impossible to conclude from positive tests alone to what degree bacterial growth following initial contamination may have contributed to test results. With cultured products, some limitations on age of sample before testing seem necessary because of the rapid and marked reductions encountered in coliform counts within 24 h of processing.[9]

For information concerning coliforms in other dairy products, refer to discussions of concentrated and dry milk products, butter, margarine and related products, cheese and other cultured products, and ice cream and related products [Chapter 10].

8.2 Definitions

A positive *presumptive test* for members of the coliform group in milk and cream products is indicated when: a) typical dark red colonies (normally measuring at least 0.5 mm in diameter on uncrowded plates) appear within 24 ± 2 h of incubation at 32 ± 1°C on violet red bile agar [8.8, 8.10],[1,8,15,31] or b) fermentation tubes containing 2% brilliant green lactose bile broth[8,15,16] show gas formation within 48 ± 3 h incubation at 32 ± 1°C [8.9, 8.11].

A *confirmed test* should be made of doubtful colonies from violet red bile agar plates by transferring each of five colonies (if available) to tubes of 2% brilliant green lactose bile broth.

A *completed test* is made by streaking material, from typical colonies on solid media [8.10] or from 2% brilliant green lactose bile broth tubes showing gas production [8.11], onto plates of eosin methylene blue agar. Coliform colonies on eosin methylene blue agar are dark or have dark centers and colorless peripheries; coliform colonies on violet red bile agar are red. Pure cultures so isolated should produce gas in fermentation tubes of lactose broth at 32 ± 1°C within 48 h and consist only of gram-negative, nonspore-forming rods.

8.3 General Interpretations

Since results of determinations of total numbers of bacteria in dairy products by the agar plate and direct microscopic methods are reported per milliliter or per gram, the generally accepted practice when solid media are used is to report results of coliform determinations as "Coliform colony count per milliliter" (or per gram). Workers in some areas prefer to use the fermentation tube technique, which is one standard procedure used in water and wastewater control laboratories, and to report results similarly—for example, as "Coliforms, Most Probable Number/100 mL" (or /100 g), or "Coliforms, MPN/100 mL" (or /100 g).

Because small test quantities often will fail to yield appreciable numbers of coliforms, negative results should be interpreted by plant operators and by control officials with due consideration to the volume of sample tested. With the quantities ordinarily tested, any postitive result on a pasteurized product suggests recontamination and indicates the presence of defective equipment or a need for improved plant sanitation. Regardless of the volume basis for reporting results, proper emphasis should be placed on all positive findings, and the cause should be ascertained and eliminated.

When coliforms are enumerated in manufactured products, the question always remains whether the counts actually reflect the total number of viable coliforms present. In products that contain bacterial cells that may have been stressed either during processing or in storage, and where debilitation or sublethal injury of coliforms may occur, both solid and liquid selective media can be inhibitory. Thus, counts on some manufactured milk products may be low and may not indicate the true population of viable coliforms present. Environmental stresses such as sublethal heat or chemicals, freezing, or dehydration may interfere with quantitative recovery of coliform organisms on selective media.[24] This has been found true for fluid milk, cheese, and acidified or frozen dairy products.[10,14,26] In such instances, modified procedures are available that increase the recovery of injured coliforms.[6,11,12,25,26]

8.4 Sampling

Sampling is done in accordance with procedures given in 4.1C.

8.5 Equipment, Supplies, and Media

Refer to Chapter 2 for standard equipment and supplies, except for special items as follow:
    a. Inoculation loop (3-mm) and needle.
    b. Hardwood applicator sticks (1.27-cm dia., 15.2-cm long). Disposable applicator sticks must be dry-heat sterilized.
    c. Fermentation tubes.

A. Media:

The preparation of appropriate media is described in Chapter 5. Preferably use dehydrated media. When media are prepared from component ingredients, the analyst must assume full responsibility for correct formulation.
    1. Violet red bile agar.
    2. Brilliant green lactose bile broth, 2%.
    3. Eosin methylene blue agar, Levine.
    4. Lactose broth.
    5. Nutrient agar.

B. Gram stain:
    1. Gram stain reagents or commercial kit.
    2. Glass slides.
    3. Microscope with oil-immersion lens.

8.6 General Procedure

Examine samples of fluid milk and cream by using either solid or liquid media [8.8, 8.9]. Since organisms giving positive coliform tests are practically nonexistent in properly pasteurized milk collected directly from pasteurizers, presence of such organisms is usually interpreted as the result of contamination introduced by equipment and/or processing downstream from the pasteurizer. Therefore, a completed test is generally not needed in routine and official control of pasteurized fluid milk products or manufactured products made from milk given a heat treatment equivalent to or greater than that of pasteurization. However, questionable colonies should be confirmed by transfer to 2% Brilliant Green Lactose Bile Broth.

When fermentable substances other than lactose are present in samples, confirm the presence of lactose fermenters by appropriate transfers of dark red colonies on solid medium, or of broth from tubes showing gas, to 2% Brilliant Green Lactose Bile Broth [8.10]; or proceed with the completed test [8.11].

8.7 Relative Values of Plate and Tube Methods

Choice of method will depend largely upon limits for the number of coliforms in which the analyst is interested. Other considerations are materials required, laboratory facilities, and labor available to conveniently process the volume of work involved.

When five or more coliform colonies appear on single plates, counts are usually more precise than are counts obtained by using five ferementation tubes.[16] For routine examination of milk or cream with numbers of coliforms usually greater than 5/mL, or in preliminary surveys of pasteurizing plants, one or two plates, each containing 1 mL of sample, are sufficient. Well operated pasteurization plants normally achieve coliform counts less than 5/mL on processed samples. For increased sensitivity on samples routinely less than 5/mL, use a larger sample: 4-mL portions in plates poured with 15-20 mL of medium are satisfactory. When counts are still less than five coliform colonies per plate, and more accurate results are desired, a multiple-tube procedure must be used so that larger samples can be examined.

The multiple-tube method results in a most probable number that is an estimate of the number of bacteria within limitations of the procedure.[23] For example, the 95% confidence limits, based on coliforms per 100 mL, could vary as follows [Table 8.1]: between less than 1 to 11 bacteria with one positive tube; between 1 and 20 bacteria with two positive tubes; and between 4 and 38 bacteria with four positive tubes. Because of the irregular distribution of bacteria, even when coliforms average 1/mL, the probabilities are that some 37% of 1-mL portions of a sample will contain no coliforms; but if each test is made with five portions of 1 mL each, completely negative results may be expected less than 1% of the time.

Caution should be exercised before drawing conclusions from the results of single examinations by either tube or plate method.[18,19] The precision of results obtainable is low when: a) few positive tubes appear among several

8.8 Coliform Test with a Solid Medium

inoculated with one or more sample volume(s), or b) a very low colony count is obtained from several plates.

8.8 Coliform Test with a Solid Medium

A. Violet Red Bile Agar procedure (Class O):

Transfer 1 mL of milk sample or decimal dilution thereof into a sterile plate. Add to each plate 10 to 15 mL of Violet Red Bile Agar tempered to 44 to 46°C. Pour an agar control for each flask of medium used.

To facilitate mixing of the sample and the agar medium when plating high-solids and viscous products such as cream, ice cream, and buttermilk, prepare initial dilutions of 1:2 or 1:10 in warmed (about 40°C) dilution blanks and distribute 10 mL into three dishes, using 15 to 20 mL of medium per dish. If high counts are consistently encountered, distribute 1 mL of a 1:10 dilution into one plate.

Mix contents of plates thoroughly by tilting and rotating each dish. Allow the mixture to solidify (5 to 10 min) on a level surface, then distribute an additional 3 to 4 mL of the plating medium as an overlay, completely covering the surface of the solidified medium to inhibit surface colony formation. Invert and incubate plates for 24 ± 2 h at 32 ± 1°C. Dark red colonies measuring 0.5 mm or more in diameter on uncrowded plates are counted, and results are recorded as directed in 8.12.

On crowded plates (greater than 150 colonies per plate), coliform colonies may show atypical characteristics and may measure less than 0.5 mm in diameter. Furthermore, on crowded plates strains of some noncoliforms and yeasts may produce colonies comparable in color and size to colonies of coliforms.[20] In such instances, pick representative colonies and confirm the presence of lactose fermenters [8.6, 8.10], or subject the chosen colonies to the completed test [8.11]. Results from plates having excessive numbers of colonies may be unreliable, particularly for samples containing substantial numbers of organisms able to grow at refrigeration temperatures. Avoid crowded plates, if possible, by plating a sufficient number of dilutions with selected volumes of sample.

B. Modified violet red bile agar procedures (Class C):

As stated in section 8.3, some dairy foods may contain injured cells that will not grow on Violet Red Bile Agar when the procedures described in 8.8A are used. In such instances, the coliform test may be modified when using solid media in an effort to obtain more realistic counts. One modification is to incubate the dairy food sample at 32 ± 1°C for 3 to 4 h prior to plating with Violet Red Bile Agar. Preincubation of the test sample may allow the repair of damaged coliform cells before they are exposed to the growth-restricting chemicals included in Violet Red Bile Agar.

Modified plating procedures also are effective.[6,11,12,25,26] The most convenient method is to prepare plates as in 8.8A, with the following modifications: a) The base layer should consist of 8 to 10 mL of Standard Methods Agar

Table 8.1 Selected MPN Estimates and 95% Confidence Limits of Estimates for Fermentation Tube Tests When Five Tubes With 10-, 1.0-, and 0.1-mL Volumes Are Used.[5]

No. of Positive Tubes Out of —			MPN/100 mL*	95% Confidence Limits	
5–10 mL	5–1.0 mL	5–0.1 mL		Lower	Upper
0	0	0	<2	<1	—
0	0	1	2†	<1	10
0	1	0	2	<1	10
1	0	0	2	<1	11
1	0	1	4†	1	15
1	1	0	4	1	15
1	2	0	6†	2	18
2	0	0	4	1	17
2	0	1	7†	2	20
2	1	0	7	2	21
2	1	1	9†	4	25
2	2	0	9	4	25
3	0	0	8	3	24
3	0	1	11	4	29
3	1	0	11	4	30
3	1	1	14†	6	35
3	2	0	14	6	35
3	2	1	17†	8	40
3	3	0	17†	8	41
4	0	0	13	5	38
4	0	1	17	7	46
4	1	0	17	7	47
4	1	1	21	9	55
4	2	0	22	10	57
4	2	1	26†	12	65
4	3	0	27	13	67
4	3	1	33†	16	77
4	4	0	34†	16	80
5	0	0	23	10	87
5	0	1	31	13	110
5	1	0	33	14	120
5	1	1	46	20	150
5	1	2	63†	27	190
5	2	0	49	21	170
5	2	1	70	30	210
5	2	2	94†	40	250
5	3	0	79	30	250
5	3	1	110	40	300
5	3	2	140	60	360
5	4	0	130	50	390
5	4	1	170	70	480
5	4	2	220	100	580

8.10 Confirmed Test from a Solid Medium (Class O)

Table 8.1 (*Continued*)

No. of Positive Tubes Out of —			MPN/100 mL*	95% Confidence Limits	
5–10 mL	5–1.0 mL	5–0.1 mL		Lower	Upper
5	4	3	280†	130	700
5	4	4	350†	160	830
5	5	0	240	100	940
5	5	1	350	150	1400
5	5	2	540	220	2000
5	5	3	920	400	3000
5	5	4	1600	600	5300
5	5	5	>1600	800	∞

* Normal results, obtained in 95% of tests, *are not* followed by a dagger. Less likely results, obtained in only 4% of tests, *are* followed by a dagger. Combinations of positive tubes not shown in the table occur in less than 1% of tests, and their frequent occurrence is an indication of faulty technique or that assumptions underlying the MPN estimate are not being fulfilled. MPN estimates for combinations that are *not* shown in the table may be obtained by extrapolation to the next highest combination that is shown in the table; for example, a result of 4/0/2 would have an MPN of approximately 21, which is the MPN for a more likely result of 4/1/1.

Note: All figures under "MPN/100 mL" in this table may be divided by 100 for reporting "MPN/milliliter" (or "MPN/gram").

or Tryptic Soy Agar. b) Plates are incubated at room temperature for 2 ± 0.5 h before the overlay is added. c) The overlay consists of 8 to 10 mL of Violet Red Bile Agar. By using this method, the 2-h resuscitation period in a nonselective medium permits repair of injured cells before the highly selective Violet Red Bile Agar is added.

The use of Standard Method Agar as a base layer may yield higher counts than when Tryptic Soy Agar is used. The glucose in Standard Methods Agar facilitates cell repair.[6] In some instances, however, higher counts may be caused by the recovery of glucose fermenters as well as lactose fermenters. Thus, when Standard Methods Agar is used as the base layer, the method becomes an "*Enterobacteriaceae* count" rather than a "coliform count."[21]

8.9 Coliform Test with a Liquid Medium (Class O)

Inoculate fermentation tubes of 2% Brilliant Green Lactose Bile Broth with appropriate volumes (such as 10.0, 1.0, 0.1, 0.01 mL, etc.) of sample, using five portions of each test volume in separate tubes. Prepare decimal dilutions to obtain at least one positive and one negative tube in each series inoculated for use in calculating the most probable number present. Incubate the tubes at 32 ± 1°C for 24 and 48 ± 3 h [8.12B].

8.10 Confirmed Test from a Solid Medium (Class O)

If additional information is needed to confirm a positive presumptive test from Violet Red Bile Agar, promptly transfer typical and/or atypical colonies to 2% Brilliant Green Lactose Bile Broth fermentation tubes and incubate for 48 h at 32 ± 1°C. Gas production indicates a positive confirmed test. Failure to show gas production within 48 h excludes members of the coliform group. If growth on a solid medium is without clearly isolated typical colonies, streak sample material on the surface of solidified Violet Red Bile Agar plates for isolation. Incubate for 24 h at 32 ± 1°C and transfer the isolated growth to 2% Brilliant Green Lactose Bile Broth fermentation tubes. Incubate tubes until gas appears, but no longer than 48 h; proceed with the completed test of positive tubes as in 8.11.

8.11 Completed Test from a Liquid Medium (Class O)

When additional information is needed following a positive test from a liquid medium [8.9, 8.10], promptly streak an Eosin Methylene Blue Agar plate from the liquid culture to produce discrete colonies after the plate has been incubated for 24 h at 32 ± 1°C. From these, inoculate a Nutrient Agar slant and a lactose fermentation tube from one typical colony or, if only atypical colonies are present, from two or more colonies that most closely conform to coliform-like colonies. Incubate the slant(s) and the Lactose Broth tube(s) for 24 to 48 h at 32 ± 1°C. Gas formation in Lactose Broth and the presence of gram-negative, nonspore-forming, rod-shaped bacteria on the slant constitute a positive completed test.

8.12 Reporting Results

A. Tests with a solid medium (procedure 8.8A):

If *no* coliforms appear on plate(s), report "less than n per milliliter" (or per gram), substituting for "n" the number that would be reported if only one coliform colony had been secured from the total volume plated. For example, if 2 mL of undiluted milk was plated, report results as "Coliform count less than 0.5 per milliliter."

Preferably record results from only those plates that contain *between 15 and 150* coliform colonies per plate. If plates from two consecutive decimal dilutions yield colony counts of 15 to 150, compute the counts for each dilution by multiplying the number of colonies per plate by the dilution used and report the arithmetical average as coliform count per milliliter or per gram, unless the higher count is greater than twice the lower count, in which case report the lower computed count. Use only two significant figures followed by zeros. Keep in mind, however, that precision decreases as the number of colonies per plate decreases.[3] When fewer than 15 coliform colonies routinely develop on the plate(s) from diluted sample(s), lower dilutions should be plated. If fewer than five coliform colonies routinely develop on the plate(s) from undiluted sample(s),[18] use of multiple-tube procedures should be considered. [See 8.7.]

8.12 Reporting Results

If *more than 150* colonies develop on the highest-dilution plate, record count as >150 times the reciprocal of the dilution.

Where multiple plates are used to distribute several milliliters of the same dilution, sum the counts for the plates and multiply by the appropriate factor.

B. *Tests with a modified solid medium (procedure 8.8B):*

The procedures described in 8.8B are not normally used for control purposes. They do, however, provide information that may be useful to the dairy food processor. The specific modification(s) that were used should be included in the report of results.

C. *Tests with a liquid medium:*[16,17,23]

Record results, indicating number of tubes used, volume introduced into each, and number of positive tubes obtained; for example, if three of five tubes with 10-mL portions in each have gas and none of five tubes with 1-mL portions in each have gas, record as follows:

$$\frac{10 \text{ mL}}{3/5} \quad \frac{1 \text{ mL}}{0/5}$$

The numerator indicates the number of positive tubes and the denominator the number of tubes inoculated. Interpret the laboratory record from Table 8:1, showing most probable numbers [8.13] and report "Coliforms, MPN per milliliter" (or per gram). The coliform count given in the example would be 8/100 mL or 0.08/mL.

Table 8.1 lists the most probable numbers of coliforms per 100 mL. These values correspond to the frequency of positive fermentation tests that were secured from a variety of multiple-portion inoculation systems, beginning with 10-mL test portions. Since more than one organism may be responsible for each positive test, the most probable number of organisms is a logarithmic function of results when test portions are decimally related. Therefore, if the greatest volume inoculated is 1 mL rather than 10 mL, multiply the most probable number listed in Table 8:1 by 10. Thus for results of

$$\frac{1 \text{ mL}}{3/5} \quad \frac{0.1 \text{ mL}}{1/5}$$

the corresponding most probable number, 0.11 per mL (11/100 mL), is multiplied by 10 to arrive at 1.1 per mL (110/100 mL) as the final most probable number of coliforms in the sample. Similarly, if the greatest volume inoculated is 0.01 mL, multiply the most probable number listed in the table by 1,000.

When more than three different volumes are tested, use results from only three consecutive volumes. Use results of the smallest volume having a reading of 5/5 and results obtained with the next two smaller volumes.

Significant results in the following examples do not have asterisks: values with asterisks are not to be used in the computations:

Example	1.0 mL	0.1 mL	0.01 mL	0.001 mL
(a)	*5/5	5/5	2/5	0/5
(b)	5/5	4/5	2/5	*0/5
(c)	0/5	1/5	0/5	*0/5
(d)	5/5	3/5	1/5	1/5
(e)	5/5	3/5	2/5	*0/5
(f)	*5/5	5/5	5/5	5/5

In Example (a), only the three smallest volumes are used. In Example (b), the three largest volumes are used. In Example (c), where none of the volumes tested were 5/5, the first three volumes should be taken so as to place positive results in the middle volume. If a positive test occurs in a volume smaller than the three chosen [Example (d)], add the value appearing in the smallest positive volume to the result obtained for the next smallest volume, to produce a result that might read as in Example (e). When all tubes are positive, as in Example (f), the smallest three volumes tested are chosen and a "greater than" sign (>) is used to indicate that the most probable number is greater than the one given.

8.13 Table of MPN Coliforms

Table 8:1 lists commonly occurring MPN results and computed values for coliform organisms, the most probable number per 100 mL (Coliforms, MPN/100 mL) of sample for the multiple portion inoculating system employed,[5,13,29,30] and confidence limits of estimates for fermentation tube tests.

8.14 Impedimetric Coliform Detection Method: Raw Milk (Class B)

A. Principle:

Instrumentation to estimate initial microbial flora based upon impedimetric measurements [7.7] can be used for estimation of coliforms (presumptive). A selective medium is used to evaluate coliforms in raw and pasteurized milk, cream, and ice cream mix.[7] The following procedure applies to raw milk only.

B. Equipment and supplies:
  1. The basic instrumentation is described in 7.7.
  2. Sample and instrument incubators set at 35°C.
  3. Broth medium
     Proteose peptone .... 10 g
     Yeast extract 6
     Lactose 20

Sodium desoxycholate 0.1
Sodium laurel sulfate 1.0
Brom cresol purple 0.035
Bile salts 1.0
MS Water .... make to 1 L

The dry ingredients should be reconsituted, the medium pH should be adjusted to pH 7.0 with 0.1 N NaOH, dispensed in 99-mL amounts in dilution bottles, and sterilized at 121°C for 15 min.

C. Preparation of samples and analysis:
1. Mix the sample container as specified [6.2F2].
2. Add 11 mL of mixed sample into 99-mL coliform medium broth. Replace the bottle cap and shake bottles using 25 complete up and down cycles through an arc of about 30 cm in 7 sec.
3. Preincubate the diluted samples for 3 h at 35°C.
4. Shake the dilutions again, remove caps from module wells, and pipette 1.5 mL into each of two module wells. Replace the caps.
5. Insert modules with samples into instrument and initiate data collection using the appropriate product code and coliform software. Monitor the impedence for 24 h at 35°C.
6. Report the results as impedence coliform count per milliliter or gram as appropriate.

8.15 Hydrophobic Grid Membrane Filter Method: (Class A2) (Presumptive Coliform Count in Cheddar Cheese)

A. Principle

Hydrophobic grid membrane filter (HGMF) uses membrane filter imprinted with hydrophobic material in grid pattern. Hydrophobic lines act as barriers to spread of colonies, thereby dividing membrane filter surface into sep. compartments of equal and known size. Number of squares occupied by colonies is enumerated and converted to most probable number value of organisms by using formula given below.

B. Equipment and supplies
1. *Hydrophobic grid membrane filter (HGMF).*—Membrane filter has pore size of 0.45 μm and is imprinted with nontoxic hydrophobic material in grid pattern. (QA Laboratories Ltd, 135 The West Mall, Toronto, Ontario, Canada M9C 1C2) or equiv. meets these specifications.
2. *Filtration units for HGMF.*—Equipped with 5 μm mesh prefilter to remove food particles during filtration. One unit is required for each sample. (QA Laboratories Ltd) or equiv. meets these specifications.
3. *Pipets.*—1.0 mL serological with 0.1 mL graduations; 1.1 mL or 2.2 mL milk pipets are satisfactory. 5.0 mL serological with 0.1 mL graduations.

4. *Blender.*—Waring Blender, or equiv., multispeed model, with low-speed operation at 10,000–12,000 rpm, and 250 mL glass or metal blender jars with covers. One jar is required for each sample.
5. *Vacuum pump.*—Water aspirator vac. source is satisfactory.
6. *Manifold or vacuum flask.—Filter paper.—*
7. Whatman No. 1 or No. 4, or equiv.
8. *Peptone/Tween 80 diluent.*—Dissolve 1.0 g peptone (Difco 0118) and 10.0 g Tween 80 in 1 L H_2O. Dispense enough vol. into diln bottles to give 90 ± 1 mL after autoclaving 15 min at 121°.
9. *M-FC agar.*—10.0 g tryptose, 5.0 g proteose peptone No. 3, 3.0 g yeast ext, 5.0 g NaCl, 12.5 g lactose, 1.5 g bile salts No. 3, 0.1 g aniline blue, and 15.0 g agar dild to 1 L with H_2O (M-FC Agar, Difco 0677, is satisfactory). Heat to boiling. Autoclave 15 min at 121°. Temper to 50–55°. Adjust pH to 7.4 ± 0.1. Dispense ca 18 mL portions into 100 × 15 mm petri dishes. Surface-dry plated medium before use.
10. *Tryptone bile agar (TBA).*—20.0 g tryptone, 1.5 g bile salts No. 3, and 15.0 g agar dild to 1 L with H_2O (Tryptone bile agar, Oxoid CM595, is satisfactory). Heat to boiling. Temper to 50–55°. Adjust pH to 7.2 ± 0.1. Dispense ca 18 mL portions into 100 × 15 mm petri dishes. Surface-dry plated medium before use. (k) *Tryptic soy-magnesium sulfate agar (TSAM).*—15.0 g tryptone, 5.0 g phytone (or soytone), 5.0 g NaCl, 1.5 g $MgSO_4.7H_2O$, and 15.0 g agar dild to 1 L with H_2O. Heat to boiling. Autoclave 15 min at 121°. Temper to 50–55°. Adjust pH to 7.3 ± 0.1. Dispense ca 18 mL portions into 100 × 15 mm petri dishes. Surface-dry plated medium before use.
11. *Indole reagent.—(1) Soln A:* Dissolve 2.5 g *p*-dimethylamino-benzaldehyde and 10 mL HCl in 90 mL alcohol. *(2) Soln B:* Dissolve 2.0 g potassium persulfate in 200 mL H_2O. Mix equal vols of Soln A and Soln B just before use.
12. *Tris buffer.*—1.0M, pH 7.6. Dissolve 121.1 g tris(hydroxymethyl-amino)methane and dil. to 1L with H_2O. Adjust pH to 7.6 with 1N HCl.
13. *Trypsin stock soln.*—Dil. 10 g trypsin to 100 mL with tris buffer. Warm to 35° if necessary to aid soln. Filter thru Whatman No. 1 paper (or equiv.) to remove insoluble material, then filter-sterilize using 0.45 μm membrane filter.

C. *Sample preparation and analysis:*

Aseptically weigh 10 g sample of cheese into a sterile blender jar. Add 90 ml sterile tryptone/tween diluent and blend 2 min. at low speed (10,000–12,400 rpm). Aseptically combine in 16 × 150 mm tube 5.0 ml of this 1:10 homogenate and 1.0 ml trypsin solution. Incubate 20–30 min at 35 ± 0.5°C in H_2O bath. Vortex to remix suspension just before doing filtrations.

8.15 Hydrophobic Grid Membrane Filter Method (Class A2)

Follow instructions of the manufacturer for use of the HGMF and filtration units.

Aseptically add 15–20 ml sterile H_2O to funnel. Pipet 2 ml of homogenate trypsin mixture into the funnel. Apply suction to draw liquid through the prefilter only. Add 10–15 ml sterile H_2O to the funnel and draw through the prefilter as before. Close the clamp to draw the liquid through the HGMF. Release the vacuum and aspetically transfer the HGMF to the surface of a pre-dried M-FC agar plate. Avoid trapping bubbles between the filter and the agar surface.

(a) *Total coliform count.*—Incubate the agar plate 24 ± 2 h at 35°C. Count all squares containing one or more blue colonies. Include any shade of blue. Score each square as pos(blue) or neg. Convert pos square count to MPN with the formula

$$MPN = [N \log_e (N/(N - X))]$$

where N = total number of squares and X = the number of pos. squares. Multiply by reciprocal of dilution factor (filtering dilution 10^{-1}, multiplication factor 10) and report as MPN of total coliform bacteria/g.

(b) *Fecal coliform count.*—Place HGMF on surface of pre-dried TSAM (k.) Incubate 4–5 h at 25° for dry foods and 4–5 h at 35° for all other foods. Transfer HGMF to surface of pre-dried M-FC agar (1) and incubate 24 ± 2 h at 44.5 ± 0.5° in closed container. Proceed as in (a), and report as MPN of fecal coliform bacteria/g.

(c) *E. coli count.*—Place HGMF on surface of pre-dried TSAM (k). Incubate 4–5 h at 25° for dry foods and 4–5 h at 35° for all other foods. Transfer HGMF to surface of pre-dried TBA (j) and incubate 24 ± 2 h at 44.5 ± 0.5° in closed container. Prepare indole reagent (1) by combining equal vol. of Soln A and Soln B. Place 9 cm filter paper disk in petri dish lid and flood with 1–2 mL indole reagent (l). Transfer HGMF to filter paper, ensuring that no air bubbles are trapped between HGMF and paper. Let stand 10–15 min, then transfer HGMF back to surface of TBA. Count all squares contg one or more pink (indole pos.) colonies. Score each square as either pos. (pink) or neg. Convert pos. square count to MPN with formula above. Multiply by reciprocal of diln factor and report as MPN of *E. coli* (biotype 1)/g.

8.16 References

1. BARTRAM, M.T.; BLACK, L.A. Detection and significance of the coliform group in milk. I. A comparison of media for use in isolation. Food Res. 1:551–563; 1936.
2. BARTRAM, M.T.; BLACK, L.A. The detection and significance of Escherichia-Aerobacter in milk. III. Correlation of total bacterial count and presence of the coli-aerogenes group. J. Dairy Sci. 20:105–112; 1937.

3. COWELL, N.D.; MORISETTI, M.D. Microbiological techniques—some statistical aspects. J. Sci. Food Agr. **20**:573–579; 1969.
4. DELAY, P.D. The significance of the coliform test in pasteurized milk. J. Milk Food Technol. **10**:297–299; 1947.
5. DE MAN, J.C. The probability of most probable numbers. European J. Appl. Microbiol. **1**:67–78; 1975; and personal communication.
6. DRAUGHON, F.A.; NELSON, P.J. Comparison of modified direct-plating procedures for recovery of injured *Escherichia coli.* J. Food Sci. **46**:118–1191; 1981.
6a. ENTIS, P. Enumeration of total coliforms, fecal coliforms, and *Echerichia coli* in foods by hydrophobic grid membrane filter: Collaborative study. J. Assoc. Offic. Anal. Chem. **67**:709–712; 1984.
7. FIRSTENBERG-EDEN, R.; VAN SISE, M.L.; ZINDULIS, J.; KAHN, P. Impedimetric coliform estimation in dairy products. J. Food Sci. (In press).
8. FOURNELLE, H.J.; MACY, H. A comparative study of presumptive media for the coliform group. Amer. J. Pub. Health and the Nation's Health **40**:934–942; 1950.
9. GOEL, M.C.; KULSHRESTHA, D.C.; MARTH, E.H.; FRANCIS, D.W.; BRADSHAW, J.G.; READ, R.B. JR. Fate of coliforms in yogurt, buttermilk, sour cream, and cottage cheese during refrigerated storage. J. Milk Food Technol. **34**:54–58; 1971.
10. HALL, H.E.; BROWN, D.F.; LEWIS, K.H. Examinaiton of market foods for coliform organisms. Appl. Microbiol. **15**:1062–1069; 1967.
11. HARTMAN, P.A. Modification of conventional methods for recovery of injured coliforms and salmonellae. J. Food Prot. **42**:356–361; 1979.
12. HARTMAN, P.A.; HARTMAN, P.S.; LANZ, W.W. Violet Red Bile-2 agar for stressed coliforms. Appl. Microbiol. **29**:537–539, 865; 1975.
13. HOSKINS, J.K. Most probable numbers for evaluation of coli-aerogenes tests by fermentation tube method. Pub. Health Rep. **49**:393–405; 1934.
14. MAXCY, R.B. Non-lethal injury and limitations of recovery of coliform organisms on selective media. J. Milk Food Technol. **33**:445–448; 1970.
15. MCCARTHY, J.A.; THOMAS, H.A. JR.; DELANEY, J.E. Evaluation of the reliability of coliform density tests. Amer. J. Pub. Health and the Nation's Health **48**:1628–1635; 1958.
16. MCCRADY, M.H. The numerical interpretation of fermentation tube results. J. Infect. Dis. **17**:183–212; 1915.
17. MCCRADY, M.H. Tables for rapid interpretation of fermentation tube results. Pub. Health J. (Toronto) **9**:201–220; 1918.
18. MCCRADY, M.H.; ARCHAMBAULT, J. Examining dairy products for members of the *Escherichia-Aerobacter* group. Amer. J. Pub. Health **24**:122–128; 1934.
19. MCCRADY, M.H.; LANGEVIN, E.M. The coli-aerogenes determination in pasteurization control. J. Dairy Sci. **15**:321–329; 1932.
20. MORRIS, R.L.; CERNY, J. Significant abnormalities in the violet red bile technique for coliforms in milk. J. Milk Food Technol. **17**:185–187; 1954.
21. MOSSEL, D.A.A.; MENGERINK, W.H.J.; SCHOLTS, H.H. Use of a modified MacConkey agar medium for the selective growth and enumeration of *Enterobacteriaceae.* J. Bacteriol. **84**:381; 1962.
22. NELSON, F.E.; BAKER, M.P. The influence of time and temperature of plate incubation upon bacterial counts of market milk and related products, particularly after holding under refrigeration. J. Milk Food. Technol. **17**:95–100; 1953.
23. OBLINGER, J.L.; KOBURGER, J.A. Understanding and teaching the most probable number technique. J. Milk Food Technol. **38**:540–545; 1974 and **39**:78; 1975.
24. RAY, B. Methods to detect stressed microorganisms. J. Food Prot. **42**:346–355; 1979.
25. RAY, B.; SPECK, M.L. Plating procedure for the enumeration of coliforms from dairy products. Appl. Environ. Microbiol. **35**:820–822; 1978.

8.15 References

26. REBER, C.L.; MARSHALL, R.T. Comparison of VRB and VRB-2 agars for recovery of stressed coliforms from stored acidified half-and-half. J. Food Prot. **45**:584–585; 1982.
27. REINBOLD, G.W. Bacteriological testing of milk for regulatory purposes—usefulness of current procedures and recommendations for change. III. Raw milk quality—where do we go from here? J. Milk Food Technol. **34**:260–263; 1971.
28. U.S. DEPARTMENT OF HEALTH, EDUCATION AND WELFARE. Grade A pasteurized milk ordinance—Recommendations of the Public Health Service. Food and Drug Administration. Washington, D.C.; 1978.
29. WOODWARD, R.L. How probable is the most probable number? J. Amer. Water Works Assoc. **49**:1060–1068; 1957.
30. YALE, M.W. Escherichia-Aerobacter group in ice cream. (b) Its significance and importance. **2**:17-30. Proc. 36 Ann. Conv. Intern. Assoc. Ice Cream Manufacturers; 1936.
31. YALE, M.W. Comparisons of solid with liquid media as a means of determining the presence of lactose fermenting bacteria in pasteurized milk. Amer. J. Pub. Health and the Nation's Health **27**:564–569; 1937.

CHAPTER 9

TESTS FOR GROUPS OF MICROORGANISMS

J.F. Frank, L. Hankin, J.A. Koburger, and E.H. Marth
(E.H. Marth, Tech. Comm.)

9.1 Thermoduric Bacteria

A. Introduction:

Thermoduric bacteria can survive exposure to temperatures considerably higher than the maximum temperatures for their growth. In the dairy industry, the term refers to microorganisms that survive pasteurization but do not grow at pasteurization temperatures. Thermoduric bacteria isolated from milk usually include species of *Micrococcus, Microbacterium, Streptococcus, Lactobacillus*, the coryneform group of bacteria, *Bacillus, Clostridium,* and occasionally, gram-negative rods.[64] Primary sources of contamination are poorly cleaned and sanitized udders, utensils, and equipment. High thermoduric counts are consistently associated with unhygienic production practices. Hence, the thermoduric or laboratory pasteurization count is used as an indicator of the thoroughness of equipment sanitation and to detect sources of organisms responsible for high-count pasteurized milk products. Thermoduric bacteria can contribute appreciably to high Standard Plate Counts in pasteurized milk and milk products. Since milk is presently handled and stored at low temperatures, thermoduric psychrotrophic contamination (especially with *Bacillus* species) can hinder efforts to increase the shelf-life of pasteurized milk.[44,49]

B. Collection of samples:

Normally, samples of individual producers' milk are obtained as described in Chapter 4. Samples of finished products may also be analyzed.

C. Equipment:
1. Test tubes: Sterile, screw-capped, 20 x 125 mm; caps with rubber or plastic liners.
2. Pipettes: Sterile, graduated, 5-mL or 11-mL delivery (TD).
3. Thermometers: Thermometers used in the water bath and in the pilot tube with milk should cover the critical range of temperature during pasteurization. They should have divisions of 0.1°C, and

should be checked at least biennially with a certified (NBS) thermometer.
4. Water bath: Electrically heated, thermostatically controlled at 62.8 ± 0.5°C, equipped with stirrer and thermometer. The volume of water should be sufficient to absorb the cooling effect of tubes placed in the bath without a drop in temperature of more than 0.5°C.
5. Metal, plastic, or wire rack: For holding test tubes.

D. *Procedure (Class O):*

Aseptically transfer 5 mL of a thoroughly mixed milk sample to a test tube. Be careful to avoid contaminating the lip and upper portions of the tube with milk. During this step, all samples and filled tubes should be kept in ice water at 0 to 4.4°C. A pilot tube containing a thermometer and 5 mL of milk should be placed in the test tube rack with the other samples so that the temperature can be easily monitored. Place the rack of tubes in the pasteurizing bath at 62.8°C. Either immerse the closed, water-tight tubes completely, or immerse them to approximately 4 cm above the level of milk in the tubes. The temperature of the milk should reach at least 62.8°C *within 5 min*. If bubbles come from immersed tubes during heating, the heat treatment is suspect, and contaminants may be drawn into tubes when they are cooled. Do not attempt to simulate high temperature-short time (HTST) pasteurization.

Start timing the 30-min holding period when the temperature in the pilot tube reaches 62.8°C. At the end of the holding period, immediately place the rack of tubes in an ice water bath and cool to 10°C or lower. Determine the Laboratory Pasteurization Count (LPC) by the Standard Plate Count procedure [Chapter 6]. Alternative methods to obtain counts may be used [Chapter 7].

E. *Interpretation:*

Presence of appreciable numbers of thermoduric bacteria in an individual producer's samples indicates contamination from poorly cleaned or sanitized farm equipment. Poorly sanitized udders or processing equipment in milk plants can also be a source of contamination. For a more detailed discussion, see the paper by Thomas et al.[64]

F. *Reporting:*

Report results as "Laboratory Pasteurization Count per milliliter" (LPC/mL). Where an alternative aerobic count method is used, add the appropriate designation to the reported count to identify the method.

G. *Limitations:*

The thermoduric count simulates low-temperature long-time (LTLT) pasteurization. Consequently, estimates of numbers of bacteria obtained with this procedure might not correspond precisely to bacterial survival when the HTST process is used. Suitable laboratory techniques that simulate currently used pasteurization conditions have been developed, but their applications within the dairy industry have not yet been evaluated.[14,59]

The nutrient medium and culturing conditions used in this test may not recover all heat-injured viable cells, and they may not reflect the recovery pattern in fluid milk or other dairy products. This method is adequate, however, for current applications of the LPC as an indicator of adequacy of sanitation.

9.2 Thermophilic Bacteria

A. Introduction:

In the dairy industry, the term "thermophilic bacteria" refers to organisms that grow in milk or milk products held at elevated temperatures (55°C or higher), which include LTLT pasteurization conditions.

Thermophilic bacteria are usually species of *Bacillus*, which enter milk from various sources on the farm or from poorly cleaned equipment in the processing plant. These bacteria rapidly increase in numbers when present in milk or dairy products that are held at high temperatures for long periods.

B. Procedures (Class O):

Prepare dilutions and plates as described for the Standard Plate Count [Chapter 6], but use 15 to 18 mL of agar per plate. Incubate plates for 48 h at 55 ± 1°C. Caution: maintain humidity during incubation to prevent drying of the agar medium.

C. Reporting:

Report the colony count as "Thermophilic Bacterial Count per milliliter" (or per gram), i.e., TBC/mL or TBC/g.

9.3 Proteolytic Microorganisms

A. Introduction:

Many bacteria responsible for spoilage of refrigerated dairy products are highly proteolytic and can cause flavor defects. Proteolytic enzymes produced by psychrotrophic bacteria during growth in milk often remain active after HTST and ultra-high-temperature heat treatments, and they may damage the quality of stored, heat-treated products. A test for proteolytic bacteria may be useful in identifying practices during handling and storage of raw or processed milk that can cause contamination of products.

B. Skim milk agar method (Class C):

Proceed as for an SPC, but prepare the medium immediately before pouring plates by mixing 100 mL of sterile skim milk (10% solids) with 1.0 L of melted Standard Methods Agar (SMA) adjusted to 50°C. Incubate poured plate for 48 to 72 h at 32 ± 1°C or, to increase the recovery of psychrotrophic bacteria, incubate for 72 h at 21 ± 1°C. Flood incubated plates for 1 min with a solution of 1% HCl or 10% acetic acid. Pour off the excess acid solution and count the colonies surrounded by clear zones.

C. Standard methods caseinate agar method[41,43] *(Class B):*

Prepare Standard Methods Caseinate Agar (SMCA) using dehydrated SMA [Chapter 6] as a base. Prepare 0.015 M trisodium citrate by dissolving

4.41 g of hydrated trisodium citrate in enough MS water to make 1 L. Dissolve 23.5 g of dehydrated SMA in 500 mL of the citrate solution. Dissolve 10.0 g of sodium caseinate in the other 500 mL of citrate solution, using a blender to aid in dispersion. Mix the two solutions and autoclave at 121°C for 15 min. Also prepare a calcium chloride solution by adding 218.99 g of calcium chloride hexahydrate/L and autoclave at 121°C for 15 min. Before pouring plates, add 20 mL of a sterile 1.0 M calcium chloride solution to the molten agar and mix.

Pour SMCA into plates so that the medium is 2 mm thick (e.g., 12 mL for an 8.5-cm plate). After the medium has hardened, plate 0.1-mL quantities of the sample or diluted sample on the agar surface and spread evenly with a sterile, bent-glass rod. To insure absorption of the sample, allow inoculated plates to dry for 15 min and incubate plates for 48 to 72 h at 32 ± 1°C or for 72 h at 21 ± 1°C. Colonies surrounded by a white or off-white zone of casein precipitate are proteolytic. Highly proteolytic bacteria will also produce a clear inner zone.

D. Reporting:

Report results as Proteolytic Count (PC)/mL or PC/g. Designate the medium used.

E. Interpretation:

Large numbers of proteolytic bacteria in raw milk indicate potential quality problems.[67] Caseinate Agar is preferred over Milk Agar for enumerating this group of bacteria, because Caseinate Agar is more sensitive for detecting weakly proteolytic organisms and rarely produces false-positive reactions.[43] Acid-producing organisms can cause false-positive reactions on Milk Agar. Aerobic plate counts can be estimated using Caseinate Agar. However, to measure the proportion of proteolytic organisms in the total microflora, only uncrowded plates (< 80 colonies/plate) can be read accurately. For best results, plates may have to be read twice—once at 48 h to estimate the total count and again after 72 h to insure that slower-to-develop proteolytic colonies are included.

9.4 Lipolytic Microorganisms (Class C)

A. Introduction:

Growth of lipase-producing microorganisms can contribute to flavor defects in milk and high-fat dairy products. Many psychrotrophic bacteria as well as molds and some yeasts are lipolytic.[39]

B. Materials:

1. Spirit blue agar. (available from media manufacturers).
2. Lipase reagent (available from media manufacturers).

C. Procedure:

Prepare Spirit Blue agar by hydrating and sterilizing (121°C, 15 min) it as instructed by the manufacturer. Cool melted Spirit Blue Agar to 50 to 55°C, and add 3% Lipase Reagent with thorough mixing. Optionally use 1 to 2% of the Lipase Reagent and sonify the mixture to uniformly emulsify the reagent in the agar. Pour 10 to 12 mL of the medium into a sterile petri plate and allow the medium to solidify.

Prepare 1:10 or other suitable dilutions of product to be tested. Spread 0.1 mL of the desired dilution over the surface of the solidified medium. Invert plate and incubate at 32 ± 1°C for 48 ± 3 h. The complete medium is pale lavender, and colonies of lipolytic microorganisms develop a clear zone and/or a deep-blue color around and under each colony. Count colonies of lipolytic microorganisms using procedures detailed in Chapter 6.

D. Reporting:

Report as Lipolytic Count per g or mL.

E. Interpretation:

Presence of appreciable numbers of lipolytic microorganisms suggests contamination or mishandling of the product. Occurrence of rancid-type (lipolyzed) flavors also can be traced to activity of this group of microorganisms.

9.5 Enterococci (Class B)

A. Introduction:

The enterococcus count is more reliable than the coliform count as an index of the sanitary quality of churned butter.[4,55] This is because enterococci are better able to survive in the unfavorable microenvironment of salted, churned butter.

B. Procedure:

Prepare samples of butter for plating as described in [Chapter 4]. Prepare samples of cream, milk, water, etc., as outlined in the appropriate Chapters 4, 10, and 15.

After petri dishes are inoculated, add 10 to 12 mL of melted and tempered (45°C) Citrate Azide Agar per plate. The medium must be allowed to solidify while plates are on a level surface. Add 3 to 4 mL of the plating medium as an overlay to completely cover the surface of the solidified agar. When the overlay has solidified, incubate plates at 37 ± 1°C for 48 to 72 h. To facilitate counting of colonies, a thin sheet of white tissue paper may be placed under the petri dish on the illuminated colony counter. This enhances the contrast between colonies and the background. Count only blue colonies.

C. Reporting:

Report as Enterococcus Count per g or mL.

D. Interpretation:

This method is useful for evaluating sanitation in butter manufacturing plants. An Enterococcus Count of less than 10/g of butter is not considered unduly stringent for a well-managed butter manufacturing plant.

9.6 Aerobic Bacterial Spores (Class B)

A. Introduction:

Spores of the genus *Bacillus* are commonly found in raw milk. They survive pasteurization, and may survive in products given higher heat treatments. Product defects associated with growth of aerobic sporeforming bacteria include the sweet-curdling defect in pasteurized milk,[49] which is caused by growth of psychrotrophic *Bacillus cereus*, and coagulation of canned evaporated milk caused by growth of *Bacillus subtilis* and *Bacillus coagulans*.[23] The initial mesophilic spore count of raw milk is not a good indicator of the keeping quality of the heated product.[44]

B. Equipment:
1. Erlenmeyer flasks: Sterile, screw-capped, 500-mL size; cap with rubber or plastic liner.
2. Water or ethylene glycol bath: With thermostatic control for 80 and 82°C, equipped with shaker for constant agitation of flasks.
3. Thermometers: For use in ethylene glycol bath and pilot flask; [9.1C].
4. Standard Methods Agar with 0.1% soluble starch added.
5. Pipettes: Sterile; [9.1C].

C. Procedure[44]*:*

Place 200 mL of each thoroughly mixed milk sample into a sterile Erlenmeyer flask and screw on its cap. Seal the outside of the cap with masking tape to prevent airborne contamination. Prepare a pilot unit by filling a flask with 200 mL of milk and sealing it with a rubber stopper equipped with a thermometer. Place the prepared flasks in a water bath at 82°C and agitate them throughout the heat treatment. When contents of the pilot flask reach 79°C, lower the temperature of the bath to 80°C, or move flasks to a second bath in which the water is at 80°C. After contents of the pilot flask reach 80°C, keep samples in the bath for an additional 12 min. Cool samples immediately in an ice bath. Plate samples on SMA with starch and incubate the plates at 32 ± 1°C for 48 h for mesophilic counts or at 7°C for 10 d for psychrotrophic spore counts. To detect low levels of psychrotrophic contamination, the heated milk can be incubated at 7 ± 1°C for 5 to 7 d before plating. If samples are to be taken during preincubation, a separate flask can be used for each sampling, to reduce the possibility of contamination.

D. Reporting:

Results should be reported as Mesophilic or Psychrotrophic Aerobic Spore Count/mL. The preincubation time should be noted if applicable.

9.7 Psychrotrophic Bacteria

A. Introduction:

From the viewpoint of both consumers and processors, nothing is more detrimental to acceptability of milk than off-flavors and poor keeping quality.

9.7 Psychotrophic Bacteria

The overall length of time both raw and pasteurized milks are held under refrigeration has increased through less frequent pick-ups of raw milk at the farm, extended storage of raw milk at processing plants, increased length of time between purchases by consumers, and prolonged storage times in home refrigerators. Such extended storage provides adequate opportunity for growth of psychrotrophic bacteria, which can cause off-flavors and reduce keeping quality.

Psychrotrophic organisms can grow at temperatures below 7°C, although their optimal growth temperature may be in the range of 20 to 30°C.[15,65] The term psychrophile (cold-loving) has been used in the past to describe this group of organisms, but "psychrotrophic" is the more appropriate term. Psychrotrophic, at least in the dairy industry, implies relatively rapid growth at from 2 to 7°C without reference to an optimal temperature for growth.[9,29,45,60,61,68] Psychrotrophs found in milk are mainly gram-negative, rod-shaped bacteria in the genera *Pseudomonas*, *Archromobacter*, *Alcaligenes*, *Enterobacter*, *Acinetobacter*, and *Flavobacterium*, but gram-positive psychrotrophs in the genera *Bacillus* and *Clostridium* have also been isolated.[9,22,56,62,63] Some streptococci belonging to groups III and IV, as listed in the 8th edition of *Bergey's Manual*,[5] and some yeasts and molds found in dairy products are also psychrotrophic.

Psychrotrophic bacteria are generally non-pathogenic, but when present in dairy products, they can cause a variety of off-flavors that include fruity, stale, bitter, putrid, and rancid, as well as physical defects.[9,16,18,23,46,60,61] They are also involved in losses of quality by cultured dairy products.[50] In pooled raw milk intended for manufacturing, psychrotrophic bacteria can influence yield, processing efficiency, and final quality (body, texture, flavor) of the end product.[11,12,17,38] Many of the detrimental effects perceived in the finished product occur in raw milk during refrigerated storage before processing.

Psychrotrophic bacteria are rarely present in the udder. The numbers of psychrotrophic bacteria in raw milk depend on sanitary conditions during production and on duration and temperature of storage before the milk is processed. The influence of psychrotrophic bacteria on the shelf life of pasteurized milk will depend mainly on how many are present after packaging, their rates of growth and levels of biochemical activity, and the length of storage. Even if no detectable changes occur in the product, an increase in numbers of psychrotrophs may cause problems in meeting bacterial standards.

Most psychrotrophs are destroyed by pasteurization[68] except for some *Bacillus* spp., which cause spoilage during extended storage of products.[22,47,49] Thus, the presence of psychrotrophs in pasteurized products can be attributed largely to post-pasteurization contamination.

The greatest problem caused by psychrotrophs is associated with the extracellular thermostable enzymes that they produce in raw milk.[7,21,37] The most commercially important of these enzymes are the proteases and lipases, both of which can withstand HTST (high temperature-short time) and UHT (ultra-high-temperature) treatment.[7,40] Even at low concentrations, these thermostable enzymes can degrade casein and milkfat in milk and milk products kept refrigerated for long times.[6,10,35,36,51] Studies of the heat-stability of enzymes from psychrotrophs have been done mainly with those from *Pseudomonas* spp.,[1,51,53] although there are some data on stability of enzymes from other psychrotrophs.[7,31,37,58]

Although it does not provide results rapidly, the Standard Plate Count method (with low-temperature incubation of plates) is still used to enumerate psychrotrophic organisms. For uniformity, $7 \pm 1°C$ and incubation for 10 d have been stipulated. Unfortunately, extended refrigerated storage favors growth of psychrotrophs. It is desireable, therefore, to determine their presence as quickly as possible to decide if the product is heavily contaminated. Then the source of the contamination can be ascertained and the problem remedied so that subsequent batches of product will not become contaminated. Toward that end, numerous techniques have been proposed as enumerating psychrotrophs in less time than is required by the standard method. None, however, has been subjected to collaborative study.

Tests that may be effective with raw milk may not be applicable to pasteurized milk and vice versa. Tests for milk may not be applicable to other products such as cottage cheese. Many of the proposed tests are widely used in industry, as they have proved useful for specific products and for monitoring production. The more promising of these techniques involve selective media for gram-negative bacteria, preliminary incubation of samples or plates, use of SPC in conjunction with the oxidase test and enumeration of microcolonies by optical or electronic counting procedures.[24,25,27,42,61] Impedance detection time instruments have proven useful [Chapter 6]. To determine psychrotrophs in pasteurized cream, incubation of plates at 21°C for 25 h appears to yield results comparable to those from incubation at 6°C for 14 d.[20] Spread or spiral plating gave better results than pour plates. None of these tests, however, distinguish spoilage organisms from inert types.

Certain non-standard methods are useful for specific needs. Only trials will determine the best one for a given purpose. It is claimed that extending preliminary incubation may give more reliable psychrotrophic counts [Chapter 6], but this method does not shorten the time needed for obtaining results.[13,30,52] It does provide information, however, on the possible relationship between psychrotrophs and keeping quality of products. Use of selective media could help to signal potential psychrotrophic-related problems before they become explosive.[19,26] Use of selective media also allows manipulation of incubation temperature so results can become available in less time than is required by the standard method. Selective media, however, may inhibit organisms that are causing the problem under investigation.[57]

Use of SPC agar in conjunction with the oxidase test does not involve any change in laboratory procedures, and both the SPC and potential psychrotrophic count can be obtained on the same plate.[24] The keeping-quality test of Moseley[16] [Chapter 6] would be improved in usefulness if the recommended 5- or 7-d incubation at 7 ± 1°C could be shortened by combining the test with selective inhibitors and a higher incubation temperature.

None of these methods has been accorded Standard Methods status. As of now there is no single, all-inclusive method for determining psychrotrophs under the widely varying conditions found in the dairy industry. Quality control specialists should examine all options and use whichever applies to their conditions.

B. Procedure (Class O):

Prepare dilutions of product and then plate as described for the Standard Plate Count [Chapter 6], except incubate plates at 7 ± 1°C for 10 d. Take *special care* that melted medium is cooled to 45 ± 1°C before pouring plates to prevent destruction of some psychrotrophic bacteria.[66] At temperatures above 7°C, other organisms may grow, possibly yielding misleading results.

C. Reporting:

After determining the colony count, report as "Psychrotrophic Bacteria Count per milliliter" or per gram; i.e., PBC/mL or PBC/g.

9.8 Yeast and Mold Counts

A. Introduction:

Although the acidified agar method has traditionally been used for enumerating yeasts and molds in dairy products, the antibiotic agar method is now the method of choice. The literature[3,8,32,34,48] clearly indicates that use of antibiotics rather than acid for suppression of bacteria results in improved recovery of injured (acid-sensitive) fungal cells, better control of bacteria, and less interference during counting from precipitated food particles. In addition, the antibiotic-supplemented medium is more amenable to certain modifications, thus allowing for increased versatility.[33] A recent collaborative study[28] showed, however, that similar results were obtained with acidified and antibiotic-fortified media when dairy products contained few or no debilitated cells. Both methods are given. Before the analyst employs the acidified agar method, however, its disadvantages should be carefully considered. Additional information on the role of yeasts and molds in foods is available.[2,54]

B. Antibiotic plate count method (Class A2):
  1. Equipment and supplies:
     a. Glassware and other equipment: The same as for the Standard Plate Count [Chapter 6].
     b. Medium: Standard Methods Agar.
     c. Antibiotic solution: Suspend 500 mg each of chlortetracycline HCl and chloramphenicol (U.S. Biochemical Corp., Cleveland,

OH 44128; Sigma, St. Louis, MO 63178) in 100 mL of sterile phosphate-buffered diluent [Chapter 6]. Filter-sterilization is generally not required. The suspension can be stored for 2 mo at 5°C without loss of inhibitory properties.
2. Procedure:
   a. Preparation of medium: Prepare Standard Methods Agar according to manufacturer's directions, sterilize and cool to 45 ± 1°C. Add 2 mL of antibiotic solution per 100 mL of medium and mix.
   b. Plating and incubation: Prepare and plate samples as for the SPC. Incubate at 20 to 25°C for 5 to 7 d and count as for the SPC. Report as yeast and mold per gram of sample.

C. *Acidified potato dextrose (glucose) agar (Class O):*
1. Equipment and supplies:
   a. Glassware and other equipment: Same as for the Standard Plate Count.
   b. Medium: Potato Dextrose (Glucose) Agar.
   c. Acidifying agent: Sterile 10% tartaric acid. A 10% solution of tartaric acid may be sterilized at 121°C for 10 min.
2. Procedure:
   a. Preparation of medium: Prepare Potato Dextrose (Glucose) Agar according to manufacturer's directions, sterilize and cool to 46°C. Add sufficient sterile, 10% tartaric acid to adjust pH of medium to 3.5 ± 0.1.
   b. Plating and incubation: Prepare and plate samples as for the Standard Plate Count. Incubate at 20 to 25°C for 5 to 7 days and count as for the SPC. Report as yeast and mold per gram of sample.

9.9 References

1. ADAMS, D.M.; BARACH, J.T.; SPECK, M.L. Heat resistant proteases produced in milk by psychrotrophic bacteria of dairy origin. J. Dairy Sci. **58**:828–834; 1975.
2. BEUCHAT, L.R., ED. Food and beverage mycology. Westport, CT: AVI Publishing Co. Inc.; 1978.
3. BEUCHAT, L.R. Comparison of acidified and antibiotic-supplemented potato dextrose agar from three manufacturers for its capacity to recover fungi from foods. J. Food Prot. **42**:427–428; 1979.
4. BLANKENAGEL, G.; GIBSON, D.L.; SHIH, C.N. The enterococcus count of butter for evaluating creamery sanitation. Can. Dairy Ice Cream J. **46**:17–19; 1967.
5. BUCHANAN, R.E.; GIBBONS, N.E. Bergey's manual of determinative bacteriology, 8th ed. Baltimore, MD: Williams & Wilkins Co.; 1974.
6. CHRISTEN, G.L.; MARSHALL, R.T. Thermostability of selected lipases produced by psychrotrophic bacteria. (Abstr.) J. Dairy Sci. **63**:46; 1980.
7. COGAN, T.M. A review of heat resistant lipases and proteinases, and the quality of dairy products. Irish J. Food Sci. Technol. **1**:95–105; 1977.
8. COOKE, W.B.; BRAZIS, A.R. Occurrence of molds and yeasts in dairy products. Mycopathol. Mycol. Appl. **35**:281–289; 1968.

9.9 References

9. COUSIN, M.A. Presence and activity of psychrotrophic microorganisms in milk and dairy products: A review. J. Food Prot. **45**:172–207; 1982.
10. COUSIN, M.A.; MARTH, E.H. Changes in milk proteins caused by psychrotrophic bacteria. Milchwissenschaft **32**:337–341; 1977.
11. COUSIN, M.A.; MARTH, E.H. Cheddar cheese made from milk that was precultured with psychrotrophic bacteria. J. Dairy Sci. **60**:1048–1056; 1977.
12. COUSIN, M.A.; MARTH, E.H. Psychrotrophic bacteria cause changes in stability of milk to coagulation by rennet or heat. J. Dairy Sci. **60**:1042–1047; 1977.
13. DESAI, M.N.; CLAYDON, T.J. Preliminary incubation of raw milk samples as an aid in evaluating bacteriological quality. J. Milk Food Technol. **27**:333–338; 1964.
14. DICKERSON, R.W., JR.; READ, R.B., JR. Instrument for study of microbial thermal inactivation. Appl. Microbiol. **16**:991–997; 1968.
15. EDDY, B.P. The use and meaning of the term psychrophilic. J. Appl. Bacteriol. **23**:189–190; 1960.
16. ELLIKER, P.R.; SING, E.L.; CHRISTENSEN, L.J.; SANDINE, W.E. Psychrophilic bacteria and keeping quality of pasteurized dairy products. J. Milk Food Technol. **27**:69–75; 1964.
17. ELLIOTT, J.A.; EMMONS, D.B.; YATES, A.R. The influence of the bacterial quality of milk on the properties of dairy products. A review. Can. Inst. Food Sci. Technol. J. **7**:32–39; 1974.
18. FOSTER, E.M.; NELSON, F.E.; SPECK, M.L.; DOETSCH, R.N.; OLSON, J.C., JR. Dairy microbiology. Englewood Cliffs, NJ: Prentice-Hall, Inc.; 1957.
19. FREEMAN, T.R.; NANAVATI, N.V.; GLENN, W.E. Faster enumeration of psychrophilic bacteria in pasteurized milk—A preliminary report. J. Milk Food Technol **27**:304–307, 1964.
20. GRIFFITHS, M.W.; PHILLIPS, J.D.; MUIR, D.D. Rapid plate counting techniques for enumeration of psychrotrophic bacteria in pasteurized double cream. J. Soc. Dairy Technol. **33**:8–10; 1980.
21. GRIFFITHS, M.W.; PHILLIPS, J.D.; MUIR, D.D. Thermostability of proteases and lipases from a number of species of psychrotrophic bacteria of dairy origin. J. Appl. Bacteriol. **50**:289–303; 1981.
22. GROSSKOPF, J.C.; HARPER, W.J. Role of psychrophilic sporeformers in long life milk. (Abstr.) J. Dairy Sci. **52**:897; 1969.
23. HAMMER, B.W.; BABEL, F.J. Dairy bacteriology, 4th ed. New York, NY: John Wiley & Sons, Inc.; 1957.
24. HANKIN, L.; DILLMAN, W.F. A rapid test to find "potentially" psychrophilic organisms in pasteurized dairy products. J. Milk Food Technol. **31**:141–145; 1968.
25. HANKIN, L.; PERNICE, A.R.; DILLMAN, W.F. Quality control of milk production by means of the cytochrome oxidase test. J. Milk Food Technol. **31**:165–170; 1968.
26. HARTLEY, J.C.; REINBOLD, G.W.; VEDAMUTHU, E.R. Effects of medium and incubation temperature on recovery of microorganisms from manufacturing-grade, Grade-A and pasteurized milk. J. Milk Food Technol. **32**:90–93; 1969.
27. HEESCHEN, W.; REICHMUTH, J.; TOLLE, A; ZEIDLER, H. Zur Bestimmung der psychrotrophen Keimflora in der Anlieferungsmilch. Milchwissenschaft **24**:721–726; 1969.
28. HENSON, O.E.; HALL, P.A.; ARENDS, R.E.; ARNOLD, E.A., JR.; KNECHT, R.M.; JOHNSON, C.A.; PUSCH, D.J.; JOHNSON, M.G. Comparison of four media for the enumeration of fungi in dairy products—A collaborative study. J. Food Sci. **47**:930–932; 1982.
29. INTERNATIONAL DAIRY FEDERATION. Recommendations adopted at the IDF seminar on psychrotrophic organisms held at the University of Sussex, Falmer (United Kingdom), 8–10 April 1968. Conclusions. Intern. Dairy Fed. Ann. Bull., Part I: 46–50, 1969.
30. JOHNS, C.K. Preliminary incubation for raw milk samples. J. Milk Food Technol. **23**:137–141; 1960.
31. KISHONTI, E. Influence of heat resistant lipases and proteases in psychrotrophic bacteria on product quality. Intern. Dairy Fed. Ann. Bull. Document No. 86, 121–124; 1975.

32. KOBURGER, J.A. Fungi in foods. 1. Effect of inhibitor and incubation temperature on enumeration. J. Milk Food Technol. 33:433–434; 1970.
33. KOBURGER, J.A. Fungi in foods. 3. The enumeration of lipolytic and proteolytic organisms. J. Milk Food Technol. 35:117–118; 1972.
34. KOBURGER, J.A. Fungi in foods. 5. Response of natural populations to incubation temperatures between 12 and 32C. J. Milk Food Technol. 36:434–435; 1973.
35. LAW, B.A.; SHARPE, M.E.; CHAPMAN, H.R. The effect of lipolytic gram-negative psychrotrophs in stored milk on the development of rancidity in Cheddar cheese. J. Dairy Res. 43:459–468; 1976.
36. LAW, B.A.; ANDREWS, A.T.; SHARPE, M.E. Gelation of ultra-high-temperature-sterilized milk by proteases from a strain of *Pseudomonas fluorescens* isolated from raw milk. J. Dairy Res. 44:145–148; 1977.
37. LAW, B.A. Review of the progress of dairy science: enzymes of psychrotrophic bacteria and their effect on milk and milk products. J. Dairy Res. 46:573–578; 1979.
38. LAW, B.A.; ANDREWS, A.T.; CLIFFE, A.J.; SHARPE, M.E.; CHAPMAN, H.R. Effect of proteolytic raw milk psychrotrophs on Cheddar cheese-making with stored milk. J. Dairy Res. 46:497–509; 1979.
39. LAWRENCE, R.C. Microbial lipases and related esterases. Part 1 Detection, distribution and production of microbial lipases. Dairy Sci. Abstr. 29:1–8, 59–70; 1967.
40. LAWRENCE, R.C. Microbial lipases and related esterases. Part II. Estimation of lipase activity. Characterization of lipases. Recent work concerning their effect on dairy products. Dairy Sci. Abstr. 29:59–70, 1967.
41. LEE, J.S. Proteolytic microorganisms. In M.L. Speck Ed., Compendium of methods for the microbiological examination of foods. Washington, D.C.: Amer. Pub. Health Assoc.; pp. 190–193; 1976.
42. LÜCK, H. Bacteriological quality tests for bulk-cooled milk. Dairy Sci. Abstr. 34:101–102; 1972.
43. MARTLEY, F.G.; JAYASHANKAR, S.R.; LAWRENCE, R.C. An improved agar medium for the detection of proteolytic organisms in total bacterial counts. J. Appl. Bacteriol. 33:363–370; 1970.
44. MIKOLAJCIK, E.M.; SIMON, N.T. Heat resistant psychrotrophic bacteria in raw milk and their growth at 7C. J. Food Prot. 41:93–95; 1978.
45. MOSSEL, D.A.A.; ZWART, H. The rapid tentative recognition of psychrotrophic types among Enterobacteriaceae isolated from food. J. Appl. Bacteriol. 23:185–188; 1960.
46. MURRAY, J.G.; STEWART, D.B. Advances in the microbiology of milk and dairy products. J. Soc. Dairy Technol. 31:28–35; 1978.
47. OLSON, H.C. Spoilage of milk by thermoduric psychrophiles. J. Dairy Sci. Abstr. 46:362; 1963.
48. OVERCAST, W.W.; WEAKLEY, D.J. An aureomycin-rose bengal agar for enumeration of yeast and mold in cottage cheese. J. Milk Food Technol 32:442–445; 1969.
49. OVERCAST, W.W.; ATMARAM, K. The role of *Bacillus cereus* in sweet curdling of fluid milk. J. Milk Food Technol. 37:233–236; 1974.
50. PARKER, R.B.; ELLIKER, P.R. Effect of spoilage bacteria on biacetyl content and flavor of cottage cheese. J. Dairy Sci. 36:843–849; 1953.
51. PINHEIRO, A.J.R.; LISKA, B.J.; PARMALEE, C.E. Heat stability of lipases of selected psychrotrophic bacteria in milk and Purdue Swiss-type cheese. J. Dairy Sci. 43:459–468; 1965.
52. REINBOLD, G.W.; JOHNS, C.K.; CLARK, W.S., JR. Modification of the preliminary incubation treatment for raw milk samples. J. Milk Food Technol. 32:42–43; 1969.
53. RICHARDSON, B.C.; TE WHAITI, I.E. Partial characterization of heat stable extracellular proteases of some psychrotrophic bacteria from raw milk. N.Z. J. Dairy Sci. Technol. 13:172–176; 1978.
54. RHODES, M.E., ED. Food mycology. Boston, MA: G.K. Hall & Co.; 1979.

9.9 References

55. SARASWAT, D.S., REINBOLD, G.W.; CLARK, W.S. The relationship between enterococcus, coliform and yeast and mold counts in butter. J. Milk Food Technol **28**:245–249; 1965.
56. SHEHATA, T.E.; COLLINS, E.B. Isolation and identification of psychrophilic species of *Bacillus* from milk. Appl. Microbiol. **21**:466–469; 1971.
57. SMITH, T.L.; WITTER, L.D. Evaluation of inhibitors for rapid enumeration of psychrotrophic bacteria. J. Food Prot. **42**:158–160; 1979.
58. STADHOUDERS, J.; MÜLDER, H. Fat hydrolysis and cheese flavour, IV. Fat hydrolysis in cheese from pasteurised milk. Neth. Milk Dairy J. **14**:141–147; 1960.
59. STROUP, W.H.; DICKERSON, R.W., JR.; READ, R.B., JR. Two-phase slug flow heat exchanger for microbial thermal inactivation research. Appl. Microbiol. **18**:889–892; 1969.
60. THOMAS, S.B. Psychrophilic micro-organisms in milk and dairy products. Part I and Part II Dairy Sci. Abstr. **20**:355–370, 447–468; 1958.
61. THOMAS, S.B. Methods of assessing the psychrotrophic bacterial content of milk. J. Appl. Bacteriol. **32**:269–296; 1969.
62. THOMAS, S.B. The microflora of bulk collected milk.—Part 2. Dairy Ind. **39**:279–282; 1974.
63. THOMAS, S.B.; DRUCE, R.G. Types of psychrophilic bacteria in milk. Dairy Eng. **80**:378–381; 1963.
64. THOMAS, S.B.; DRUCE, R.G.; PETERS, G.J.; GRIFFITHS, D.G. Incidence and significance of thermoduric bacteria in farm milk supplies: A reappraisal and review. J. Appl. Bacteriol. **30**:265–298; 1967.
65. THOMAS, S.B.; DRUCE, R.G. Psychrotrophic bacteria in refrigerated pasteurised milk. A review. Dairy Ind. **34**:351–355, 430–433, 501–505; 1969.
66. VANDERZANT, C.; MATTHYS, A.W. Effect of temperature of the plating medium on the viable count of psychrophilic bacteria. J. Milk Food Technol. **28**:383–388; 1965.
67. WHITE, C.H.; GILLIS, W.T.; SIMMLER, D.L.; GALAL, M.K.; WALSH, J.R.; ADAMS, J.T. Evaluation of raw milk quality tests. J. Food Prot. **41**:356–360; 1978.
68. WITTER, L.D. Psychrophilic bacteria.—A review. J. Dairy Sci. **44**:983–1015; 1961.

CHAPTER 10
MICROBIOLOGICAL METHODS FOR DAIRY PRODUCTS

R. T. Marshall, D. M. Adams, D. R. Morgan, N. F. Olsen and C. H. White
(R. T. Marshall, Tech. Comm.)

10.1 Introduction

Determining the microbiological quality of manufactured milk products commonly involves performing various plate counts. These include the Standard Plate Count (SPC) [Chapter 6], coliform count, yeast and mold count, and psychrotrophic count. Techniques used for plating are the same for the respective tests except for sampling and preparation of samples. Additional testing for some products is necessary. The SPC has limited value for cultured products and cheeses (except indirectly acidified and hot-packed products) because numbers of viable microorganisms are large.

Because the procedures for performing the various plate counts are similar, this chapter considers in detail only those characteristics of products that make their analyses unique. Methods for sampling and sample treatment are also discussed. Table 10.1 lists suggested dilutions for the six most used tests.

10.2 Evaporated, Concentrated, and Sweetened Condensed Milks and Liquid Infant Formulas

Evaporated milk is obtained by removing a portion of the water from milk. Following homogenization, evaporated milk is further processed by heat (either before or after sealing in a can) to prevent spoilage. The resulting product is commercially sterile or microbiologically stable. If the bacteria content is a pure culture, highly heat-resistant spores usually are found during a microbiological examination of canned, evaporated milk. Mixed cultures indicate gross under-processing, contamination after heat processing, or contamination during microbiological examination.

Liquid infant formulas are preserved similarly to evaporated milk and should be microbiologically examined using the same methods.

Concentrated milk is obtained by removing a portion of the water from

Table 10.1: Suggested Dilutions for Plating Dairy Products

Product	Standard Plate	Coliform	Yeast & Mold	Proteolytic	Psychrotrophic	Thermoduric
Evaporated milk	1:2	—	—	—	—	—
Concentrated milk	1:10–1:100	—	1:1–1:10	—	1:100	1:10
Sweetened condensed milk	1:10–1:100	—	1:1–1:10	—	—	1:10
Dry milk products	1:10–1:100	1:10	1:1–1:10	—	1:10	1:10
Butter and margarine	1:100	1:2–1:10	1:2–1:10	1:2–1:10	1:2–1:10	—
Cheese	—	1:10	—	—	—	—
Cottage cheese cultured	—	1:1	1:10	1:10–1:100	1:10–1:100	—
Cottage cheese acidified	1:10–1:100	1:1	1:10	1:10–1:100	1:10–1:100	—
Soft or semi-fluid products						
cultured	—	1:1	1:10	1:10–1:100	1:10–1:100	—
acidified	1:10–1:100	1:1	1:10	1:10–1:100	1:10–1:100	—
Frozen desserts	1:10–1:100	1:1	1:10	—	—	—
Unfrozen dessert mixes	1:10–1:100	1:1–1:100	1:10	—	1:10–1:100	—
Coloring						
dry form	1:100	1:100	1:100	—	—	1:100
solublized	1:10–1:100	1:1	1:1	—	—	1:1
Extracts, flavors	1:1–1:10	1:1	1:1	—	—	—
Fruits, nuts	1:10–1:100	1:10	1:10	—	—	—
Pastry, cake, confection	1:10–1:100	1:10	1:10	—	—	—
Stabilizers and emulsifiers	1:100	1:100	1:100	—	—	—

10.2 Milks and Liquid Infant Formulas

milk. It is pasteurized, but is not processed by heat to prevent spoilage. Liquid infant formulas and concentrated milk may be homogenized.

Although sweetened condensed milk (whole or skim), is commonly retailed in hermetically sealed cans, large quantities of sweetened and unsweetened products are stored temporarily in bulk containers such as 38-L cans, barrels, tank trucks, or cars, to be incorporated into other foods.

Proper heat processing destroys all coliform bacteria in condensed products. Since bulk, concentrated milk usually is exposed to contamination more often than canned milk, and since canned milk, once opened, is subject to recontamination, the coliform test may be used to indicate possible recontamination after processing or after exposure of contents of opened cans.

Yeasts and molds are the most common cause of spoilage of sweetened condensed milk. Presence of yeast or mold indicates lack of sanitary care.

A. Collection and preparation of samples:

The work area preferred for examining samples in hermetically sealed containers, e.g., evaporated milk, is a class 100 laminar-flow work station.[5] This equipment can provide an environment free of particles 0.3 μm in diameter or larger at an efficiency of 99.99%. The interior of the unit should be disinfected, before use, with iodophor or quaternary ammonium compound at the concentration recommended by the manufacturer. After chemical disinfection, operate the blower for at least 1 h before doing analyses within the unit. Monitor air quality within the unit by exposing sterile media in control plates during the transfer period. Place one plate each at the left, center, and right of the work area.

Alternatively, transfer of material from hermetically sealed containers may be made on a counter surface that has been thoroughly scrubbed with detergent solution and disinfected with iodophor or quaternary ammonium compound.

Open hermetically sealed cans using a "Bacti-Disc Cutter" can opener (Laboratory Manual for Food Canners and Processors Vol. 1, Microbiology and Processing, 1968. National Canners Association, AVI, Westport, CT, p. 45) [Figure 10.1].[22]

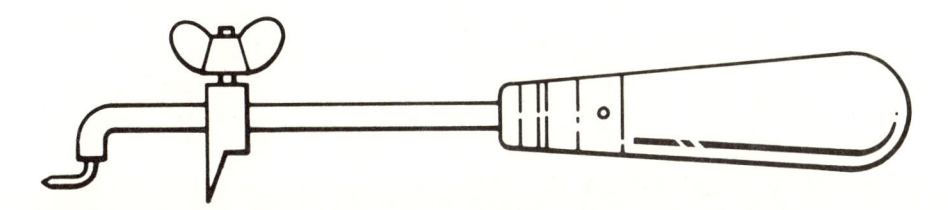

Figure 10.1 Bacti-Disc Cutter for opening hermetically sealed cans prior to microbiological examination.

Personnel making tests for commercial sterility should scrub their hands thoroughly with germicidal hand soap before doing the analyses. They should wear clean laboratory clothing. Long hair and beards should be restrained.

Collect samples using apparatus and procedures as given in [4.2A] except as follows: When necessary to insure uniformity of contents immediately before plating, as with viscous, sweetened condensed milk, heat in water bath to a temperature not above 45°C, mix contents thoroughly, and transfer promptly to plates or dilution blanks.

Weigh 11 g of sample into a wide-mouth bottle and add 99 mL of MS water [6.2 C, D]. Optionally, with evaporated milk and products no more viscous than evaporated milk, transfer to dilution blank volumetrically. For direct plating, transfer 0.5-mL portions of sample or transfer 5 mL of a 1:10 dilution to each of two petri dishes. If the sample is coagulated, use MS water containing 2% sodium citrate as the first dilution blank to dissolve the sample. Use MS water to prepare additional dilutions of any product. Suggested dilutions are given in Table 10.1

B. Microbiological analyses:
1. Plate counts: Use SPC method [Chapter 6] except as follows.

    For samples in hermetically sealed containers, prepare four sets of plates in duplicate for each dilution. Incubate two sets at 32°C for 48 ± 3 h – one aerobically and one anaerobically.[23] Incubate two sets similarly at 55°C for 48 ± 3 h. Use 15 to 18 mL of agar in plates incubated at 55°C and provide a high humidity in the incubator by placing a shallow container of water therein if high humidity is not otherwise provided.[1] For samples in unsealed containers make only the counts of aerobic microorganisms at 32 and 55°C. Count plates as per [6.11]. Record number of colonies on control plates.

    Report counts according to incubation method as follows:
    SPC—32°C, aerobic
    Anaerobic Plate Count—32°C, anaerobic
    Thermophilic Plate Count—55°C, aerobic
    Thermophilic Anaerobic Plate Count—55°C, anaerobic

    Plates having few colonies should be considered possibly contaminated, and additional tests should be performed on the same lot of product.
2. Coliform count: Divide 10 mL of a 1:10 dilution approximately equally among three petri dishes and proceed as directed in Chapter [8.8]. Apply test to heat-treated but not to sterilized, concentrated products.
3. Yeast and mold count: Commonly make 1:1 and/or 1:10 dilutions. For 1:1 dilution divide 10 mL of a 1:10 dilution approximately equally among three petri dishes.

    Prepare plates as per [9.8] with 15 to 18 mL of Antibiotic-

Supplemented Agar. Invert and incubate plates at 23°C ± 2°C. Normally, count plates after a 5-d incubation. Count colonies as per [9.11] after 3 d if colonies are numerous.
4. Thermoduric count: Dilute as per Table 10.1 and proceed as per Chapter [9].
5. Determining sterility of sterilized product: Commercial sterility of product can be determined by incubation of unopened cans at 32°C and 55°C for one week followed by plating the SPC as described in Chapter 6. If the contents have an abnormal appearance after incubation, large numbers of organisms usually are present. Presence of nonsporeforming organisms usually indicates post-sterilization contamination, frequently caused by imperfect cans. Cans containing milk that appear normal after incubation and yield no colonies on plates may be assumed to be commercially sterile. Negative controls should be required on agar, plates, and pipettes used in plating.

Alternatively, test for presence of viable anaerobes by aseptically transferring 1 mL of sample to each of four tubes containing 9 mL of Cooked Meat medium. Incubate anaerobically at 32 and 55°C for two days. Prepare smears, stain, and examine under a light microscope for bacilli. Presence of spore-formers is presumptive evidence of survivors of the heating process.

Where it is desirable to establish the significance of microorganisms from hermetically sealed cans, consult the chapters on canned foods in the *Compendium of Methods for the Microbiological Examination of Foods*.[27]

10.3 Ultra-high Temperature Processed, Aseptically Packaged, Low-acid Products (Class B)

This type of product is relatively new in the Western hemisphere. Control is exercised through regulations published in Part 113 of Title 21 of the Code of Federal Regulations.

The primary concern related to microbiology is whether the contents of packages of such products are sterile. Methods for examining canned foods are provided in Chapter XXI of the *Bacteriological Analytical Manual* of the U.S. Food and Drug Administration.[31]

A. *Collection and preparation of samples:*
1. Select packaged products representative of the lot and transfer them unopened, in clean containers, to the laboratory: keep the product temperatures 0 to 40°C.
2. Swab thoroughly a flat surface of the container with 70% alcohol.
3. Insert the needle of a sterile, single-service hypodermic syringe through the package wall, stay within the disinfected area and away from any seam. Operate plunger to obtain the required amount of sample.

4. Because the usual purpose of sampling is to determine whether contents are sterile, it is normally unnecessary to make precise measurements of each sample. When accuracy of volume dispensed is important, it may be necessary to expell entrained air from the syringe. This must be done carefully to avoid contaminating the sample.

B. *Examining samples:*

Examine contents of packages using applicable methods published in the chapter entitled "Examination of Canned Foods" in the *Bacteriological Analytical Manual.*[31]

10.4 Dry Dairy Products

Dry dairy products in tote bins or in 22.5 to 45 kg bags frequently are incorporated into ice cream mix to provide the required milk solids. They also are used in cottage cheese, cheese-food preparations and for reinforcing low-fat milk and yogurt. These products also may be reconstituted in the home for beverage or culinary use. They may be made from grade A or manufacturing-grade raw milk.

The coliform group, when present in dry dairy products, has the same significance as when present in pasteurized milk [8.1]. Food poisoning outbreaks caused by coagulase-positive staphylococci have been associated with dry milk and special foods prepared thereform.[4,5,22] Although not all coagulase-positive cocci in foods are toxigenic, their presence in large numbers in milk suggests a need for additional tests to determine if enterotoxin is present [3.2]. Occurrence of *Salmonella* in dry milk and related products has been the object of intensive study.[18] In one study, about 15% of over 500 isolates of *Salmonella* from dried milk fermented lactose.[8] *Salmonellae* entering dry milk during the final stages of processing are quite heat resistant.[20] Mold content of dry dairy products is influenced by exposure and handling during and after drying.

Standards for dry milks include SPC, Coliform Count and Direct Microscopic Count.[3,32] These methods generally are applicable to all types of dry-milk products, including those sold for infant feeding and for special diets. Limits in numbers of coagulase-positive *Staphylococci, Salmonellae,* and *Enterobacteriaciae* have been suggested.[10]

Instant-type dry milk is processed by special methods, resulting in a product that is easily rehydrated. Nonfat dry milk may be classified by heat treatment based on use of the whey-protein nitrogen test.[3] This is of practical importance in indicating the suitability of spray-process, nonfat, dry milk for specific purposes.

A. *Collection and preparation of samples:*

Collect samples, using apparatus and procedures as noted in [4.2B].

Preferably examine samples soon after manufacture. Whenever possible,

10.4 Dry Dairy Products

record the date or code of manufacture. Before opening a container for transferring a test portion, make the contents homogeneous by shaking or by rolling and inverting the sample containers; or mix the contents carefully by using a sterile spatula or spoon.

Weigh 11-g portions in a sterile weighing scoop or onto hard-surface paper or aluminum foil cut into squares and sterilized. The scoop may be sterilized between weighings by brushing clean, dipping in alcohol and flaming.

For initial dilutions temper blanks to 40 to 45°C. Aseptically weigh 11-g portions into sterile, dry, wide-mouth, screw-capped dilution bottles and add diluent to each. Optionally, transfer 11 g of dry product directly into 99 mL of water dilution water as per [5.3]. Disperse sample completely in initial dilution blank before making further dilutions. For samples of low solubility, dissolve in sterile, buffered, MS water containing 2% sodium citrate and having a pH below 8.0. Allow incompletely dispersed sample to soak 2 min; shake diluted sample 25 times in up-and-down or back-and-forth movements of 30 cm in 7 sec.

B. Microbiological analyses:

Using appropriate dilutions (Table 10.1), determine counts by one or more of the following methods, reporting results as the count per gram of the specific dry dairy product.

1. Standard Plate Count: [Chapter 6].

    Optionally pour 3 to 5 mL of agar overlay on surfaces of solidified plates before incubation. This reduces spreading of colonies.

    If colonies cannot be discerned from undissolved particles of dry product, verify identity of objects in doubt under low power microscopic or magnifying lens.

    Since determinations by the agar plate method do not reveal all the sanitary conditions of production, processing, and storage, the direct microscopic method[14] and one or more of the following microbiological tests may be of value.

2. Coliform Count: [Chapter 8].
3. Yeast and Mold Count: [9.8].
4. Thermoduric, thermophilic, and psychrotrophic bacteria: [Chapter 9].

C. Direct microscopic count [Chapter 11]:

Because of the progressively lethal effect of processing and storage on microorganisms in dry milk, determination of the number of viable bacteria may not reliably indicate the previous sanitary quality or the sanitary quality of handling of raw milk before drying. Direct microscopic examination of stained preparations of dry milk may give valuable additional information. [See 11.4]. Count differences may be large, however, when single sample comparisons are made between laboratories.[25]

1. Preparing and staining films: Dissolve 11 g of each readily soluble sample in 99 mL of dilution water. For samples that dissolve less

readily, use 2% sodium citrate solution to make 1:10 dilutions. Make films [11.11] and stain them [11.12 to 11.13].
2. Examining films: Using a binocular microscope with a microscopic factor of 500,000 to 600,000 [11.9], count clumps of microorganisms including poorly stained cells [11.15 to 11.18].
3. Reporting results: Multiply count by 10 (to compensate for the 1:10 dilution) and by the microscopic factor. Report results as Direct Microscopic Count per gram (DMC/g) of dry milk.

10.5 Butter, Margarine and Related Products

Methods to measure the microbiological quality of butter or similar foods must take into account their composition and physical structure. The water content is relatively low (16 to 17%) and, in well-worked butter or margarine, is distributed in numerous microscopic droplets throughout the product. If salt is added, all of it is dissolved in water, thus producing a relatively high sodium chloride concentration in the dispersed aqueous phase. These conditions provide significant protection against microbial spoilage of butter or margarine.

On the other hand, if butter has been contaminated in the manufacturing process, and if conditions such as poor dispersion of water and high temperature favor microbial growth, spoilage may occur. Psychrotrophic bacteria are prominent in this type of deterioration. Yeasts and molds[17] and coliforms[9] do not survive pasteurization and are rarely found in butter made under sanitary conditions. Estimates of numbers of coliforms, yeasts and molds, or enterococci in samples of cream or butter taken at various stages of processing are useful in detecting sources of contamination. Enterococci are more salt-tolerant and survive cold storage better than do coliform bacteria.[26] The SPC can be used as a measure of processing sanitation since viable cultures of lactic acid bacteria are seldom used during processing of commercial butter or margarine. Plating methods to detect proteolytic and psychrotrophic bacteria are useful in evaluating the keeping quality of butter.

Microbiological methods described here are also applicable to butter-margarine combinations and low-fat spreads.

A. *Collection and preparation of samples:*
Follow instructions for sampling given in [4.23].
Before plating, warm dilution blanks [5.3] and pipettes to 40 to 45°C. Rotate containers, filled to not more than 50% of capacity with butter or similar product, in a water bath at not over 40°C (usually less than 15 min) until product is fluid enough to pipette. Avoid separation of fat and serum fractions. Wet the pipette by drawing into it 11 mL of sterile, warm water from the dilution blank and discharging the contents back into the blank. With a minimum of product adhering to the exterior of the pipette, immediately transfer 11 g of thoroughly mixed, melted sample into 99 g of

warm (40°C) dilution water. Shake dilution bottle [6.2F]. Pipette immediately to avoid uneven sampling of serum and fat.

To prepare 1:2 dilution divide 5 mL of 1:10 dilution equally into two petri dishes. For 1:10, 1:100 and higher dilutions, follow the procedure for milk described in [Figure 6.1]. Plating of undiluted butter or similar product is unsatisfactory because it prevents breakup of clumps of cells and uniform dispersal of fat in the plates.

B. *Microbiological analyses:*

Using appropriate dilutions (Table 10.1) determine counts by one or more of the following methods, reporting results as the count per gram of butter, margarine or similar product.

1. Standard Plate Count: [Chapter 6].
2. Coliform Count: [Chapter 8].
3. Proteolytic Count: [9.3].
4. Psychrotrophic Count: [9.7].
5. Lipolytic Count: [9.4].
6. Enterococcus Count: [9.5].
7. Yeast and Mold Count: [9.8].

10.6 Cheese and Other Cultured and Culture-added Products

Cheese making generally depends on the natural presence in milk, or the addition to milk or curd, of specific types of microorganisms. The sequence of changes that usually occurs during manufacturing can be predicted within biological limitations; however, growth and activity of microorganisms are influenced by time, temperature, salt, pH, nutrient requirements and other factors during the ripening process. Different groups of bacteria may develop sequentially. One group often enhances growth of others, but the metabolic by-products of most microbial species eventually may inhibit their growth. Some of these changes are essential for development of characteristic flavor in certain types of cheese. The microbial flora varies among different types of cheese and even among different cheeses of the same type. Numbers of bacteria in natural or processed cheeses are not generally related to public health or safety. For example, the absence of coliforms should not be used as an index of sanitation, because growth and death rates of coliforms are unpredictable in cheese during ripening.[24] Although comparatively few cheese-borne disease outbreaks have been reported, some have occurred in recent years.[2,15,21,29]

The pattern of microbial growth in other cultured products is less complex than in ripened cheese. In cottage cheese, cultured milks, yogurt, sour cream and similar products, the specific organisms producing lactic acid initially are present in large numbers and may provide, to some degree, an index of culture development. Organisms such as coliform bacteria and yeasts and molds are present as contaminants and may be used as indices of sanitary conditions.

Certain dairy foods are produced by adding food-grade acids directly to pasteurized milk or milk products. Lactic-acid-producing cultures are normally not added to these foods, and the expected microflora of the freshly prepared food is that of the pasteurized milk product from which the food was made. Viability of acid-sensitive bacteria, however, decreases with time after acidification.

A. *Collection and preparation of samples:*
   1. Collect samples using apparatus and procedures as given in [4.2D].
   2. For preparation of samples provide the following equipment and supplies:
      a. Spatula: sterile.
      b. Blender: electro-mechanical.
      c. Container for blender, sterile: Stainless steel or other nontoxic metal or glass, with leakproof blending-blade assembly. The unit may be autoclaved (preferred) or sanitized with a 200-ppm chlorine solution for 5 min. Rinse chlorine solution from container with an ample, single rinse of sterile 0.1% aqueous sodium thiosulfate. Select blender to produce rapid and complete emulsification of cheese.
      d. Balance: Select balance with capacity and pan size to accommodate sample, container, and container cover, and with a precision of ± 0.1 g.
   3. Procedure for cheese other than cottage: Using aseptic technique, thoroughly comminute or mix each sample until representative portions can be removed. Heat 99 mL of sterile, freshly prepared, aqueous 2% sodium citrate to 40 to 45°C. Aseptically transfer 11 g of cheese to a sterile blender container previously warmed at 40 to 45°C and add the warmed dilution blank. Mix for 2 min at a speed sufficient to thoroughly emulsify the sample, invert the container to rinse particles from the interior walls, and remix for approximately 10 sec. Determine extent of temperature rise during 2 min of blending and when temperatures exceed 40°C, take corrective action such as chilling blender containers. If heating is unavoidable with equipment available, mixing periods of less than 2 min may be used, provided that complete emulsification is obtained. Plate or further dilute the 1:10 dilution immediately, taking care to exclude foam from the pipette.

   Alternatively, rapidly weigh 1 ± .01 g into a presterilized 177-mL Whirl-Pak bag (NASCO, Inc., Fort Atkinson, WI 58538.) or its equivalent.[30] Close the bag, place it on a flat surface, and macerate the contents into a fine paste by rolling a 15 × 125 mm test tube or similar cylindrical object over the bag. Do not force the sample into the corners of the sealed area of the bag. Open the bag and add 9 mL of 2% sodium citrate at 40°C. Reclose the bag and roll the

10.6 Cheese and Other Cultured and Culture-added Products

contents, as described above, to form a fine emulsion, and proceed with plating immediately. Because of limited agitation, breakup of chains and clumps of bacteria may not be complete. Thus, enumeration of bacterial species such as lactic streptococci, may not be feasible with this method.[19] This problem was not evident, however, in a nine-laboratory, collaborative study involving analysis of Cheddar and Romano cheeses.

A second alternative method[28] of preparing cheeses for plating is to place 11 g of sample and 99 mL diluent in an 18 × 30 cm polyethylene bag. Lock the bag into the jaws of a Stomacher 400 (Dynatech Laboratories Inc., 900 Slaters Lane, Alexandria, VA 22314) and blend for 2 min.

4. Procedure for cottage cheese: Place the sterile, blender-container on the balance and tare. With a sterile spatula, mix contents of the cottage cheese container; or if in a tightly closed, plastic, sample pouch, gently knead and mix the enclosed curd. Aseptically remove the cover of the blender and place it upside down on the balance beside the container. Weigh 11 g of cottage cheese into the sterile container. Add 99 mL of warmed (40 to 45°C), sterile, 2% sodium citrate solution and proceed as described in [10.6A3]. The alternative bag method or Stomacher method described in [10.6A3] may also be used.

5. Procedure for cultured and acidified milk and cream products: After thoroughly mixing the sample, weigh 11 g of product into a sterile, wide-mouth container, add 99 mL of water (40 to 45°C), shake until a homogeneous dispersion is obtained, and withdraw appropriate amounts of this 1:10 dilution for plating or further dilution. Alternatively, the same may be dispersed in 2% sodium citrate with a mechanical blender or Stomacher as per [10.6A3].

B. *Microbiological analyses:*

1. SPC: (applicable only to cultured products that are heat treated after culturing and to acidified products). Make suitable dilution (Table 10.1) and proceed as per [Figure 6.1].

2. Coliform Count: Make suitable dilutions (Table 10.1) and proceed as per 8.8.

    Samples of cultured or acidified cottage cheese, sour cream, yogurt and milk products must be plated within 24 h after manufacture to obtain meaningful results.[12] Coliform-like colonies in Violet Red Bile Agar should be confirmed for yogurt and other products containing fruit or added sweetener.[7]

    Alternatively, inoculate a series of tubes of Brilliant Green Bile (2%) Broth with appropriate volumes of diluted sample and proceed as per [8.9]. Report as Coliform MPN/gram or other appropriate quantity.

3. Psychrotrophic Count: Make suitable dilutions (Table 10.1) and proceed as per [9.7].
4. Yeast and Mold Count: Make appropriate dilution (Table 10.1) and proceed as per [10.2B3]. Where counts are expected to be low, distribute 10 mL of the 1:10 dilution about equally among three plates. Include the total number of colonies appearing on the three plates in the count.
5. Lactobacillus Count[11] (Class B): Since the usual purpose of determining counts of lactobacilli is to ensure that numbers are high in products to which they are added, commonly plate at 10^{-4} through 10^{-6} dilutions. Prepare sterile dilution blanks containing 0.1% nonfat, dry milk (0.1 g/100 mL buffered MS water) [5.3]. Plate samples according to [6.8], except pour plates with Lactobacillus Selective (LBS) agar as described in [5.9B] and overlay with 3 to 5 mL LBS agar.

Place plates in CO_2-incubator or place in a plastic bag and flush with CO_2 for 1 min then seal. Incubate at 37°C for 72 ± 3 h.

Count plates [6.2K] and compute counts as per [6.2L].

To enumerate bile-resistant lactobacilli, primarily *Lactobacillus acidophilus,* add 0.15% oxgal to LBS agar and proceed as described above.

Record results as Lactobacillus Count or Bile-resistant Lactobacillus Count/g (mL).

10.7 Ice Cream and Related Food Products

Methods to determine the sanitary quality of frozen dairy foods are similar in application and limitations to those used for fluid milk and cream. Directions to determine the bacteriological quality of certain product ingredients are given elsewhere as follows: evaporated, concentrated, and sweetened milk [10.2], dry milk [10.3], butter [10.4] and fluid milk and cream [Chapter 6].

Microbiological methods described here are also applicable to frozen dairy foods in which vegetable fats have been substituted for milkfat (e.g., Mellorine and Parevine) and to water ices, pops, and non-dairy creamers.

SPCs of bacteria in frozen dairy foods and ingredients, as in other dairy products, are indices of plant sanitation and of processing and handling conditions. The method may be applied to samples taken at successive stages of manufacture to detect bacterial contamination of growth or both.

Ingredients added to frozen dessert mixes before pasteurization normally contribute little to the microbial flora of the pasteurized mix except as these ingredients contain bacterial spores; however, microbiological analysis may be important in establishing the quality of ingredients such as chocolate, cocoa, eggs, milk, cream, other dairy ingredients, and stabilizers, emulsifiers, and other food additives.[27] The Standard Plate Count, Coliform Count,

10.7 Ice Cream and Related Food Products

Yeast and Mold Count, and Thermoduric Count are useful tests to determine the quality of such ingredients. With coliforms, the problem of false counts in fruit ice cream should be recognized.[7] There can be organisms other than coliforms present in fruit ice cream that give positive results with the coliform test. Therefore, confirmed tests and caution are required when interpreting positive-presumptive coliform tests.

Coloring materials, extracts and flavors, fruits, nuts,[13] cakes, pastries, and confections (ingredients frequently added to pasteurized ice cream mix or to the partially frozen product) may be important sources of bacterial contamination of frozen dairy foods.

A. *Collection and preparation of samples:*
  1. Collect samples using apparatus and procedures as given in [4.2E].
  2. Prepare samples using equipment and supplies described in [Chapter 6] except as specified in the following. Whereas the following calls for adding diluent to sample weight in a sample container, optionally weigh samples on sterile foil and transfer quantitatively and aseptically to dilution blank.
  3. Mixes and frozen desserts: Aseptically weigh 11-g portions of frozen or unfrozen product into a sterile, wide-mouth container and add 99 mL of MS dilution water. Thaw partially frozen samples at room temperature for not more than 15 min; stir to make homogeneous before weighing.
  4. Coloring, flavoring, or other added materials: When necessary to ensure homogeneity, warm samples in a water bath at 40°C for up to 15 min. To improve solubility of some samples, use dilution blanks tempered to 40 to 45°C for the initial dilution.

     Weigh 1 g of well-mixed dry colors into a sterile, wide-mouth container then add 99 mL of dilution water.

     Mix liquid samples as described in [6.2F2] before pipetting to plates or dilution blanks. Owing to the inhibitory properties of some components of extracts and flavors, more colonies may appear on plates of higher than of lower dilutions.

     Weigh representative 11-g portions of nuts, cake, pastry, confection, or fruits directly into wide-mouth containers before adding 99 mL of dilution water. Allow the mixture to soak for 5 min, shake vigorously then plate.
  5. Stabilizers and emulsifiers: Mix contents of sample container thoroughly by stirring or shaking. Where materials are coarse, e.g., 10-mesh, aseptically grind or pulverize sufficiently that sample will pass through a sterile sieve of 40-mesh or smaller. Weigh 1 g into a wide-mouth container and add 99 mL of MS dilution water. Optionally, weigh sample onto sterile scoop or square of foil or hard-surfaced paper and transfer sample to dilution blank. Shake dilution vigorously for 15 sec, then allow hydration at 20 to 40°C for up to 20 min.

Optionally hold dilution(s) at 0 to 4.4°C for up to 4 h to allow hydration.

B. *Microbiological analyses:*
1. SPC: Proceed with general instructions for diluting and plating samples and for incubating and counting plates and described in [Chapter 6].

    Because of the relatively long time required to prepare initial dilutions of frozen dairy foods, ordinarily limit the number of samples in a single series of plating 6 to 8 so that exposure of mixtures to temperatures that might permit bacterial growth will not exceed 20 min.
2. Coliform count: Proceed with general instructions given in [8.8] for diluting and plating samples and for incubating and counting plates. Confirm representative colonies as per [8.10], because false-positive results may arise when certain microorganisms ferment the sucrose or glucose in frozen desserts and grow rapidly at 32°C.
3. Psychrotrophic count: Evaluate ice cream mixes, milk shake mixes, and other mixes intended for freezing or semifreezing, and sold through refrigerated distribution channels using procedures in [9.3B].
4. Yeast and mold count: Using appropriate dilutions (Table 10.1) proceed as per [9.8].

Where molds may be present in products such as frozen, crushed and pureed fruit, prepare a homogeneous mixture of pulp from a representative portion of sample and examine microscopically for mold filaments according to methods outlined in the AOAC's *Official Methods of Analysis.*[16]

10.8 References

1. ALEXANDER, R.N.; MARSHALL, R.T. Moisture loss from agar plates during incubation. J. Food Prot. **45**:162–163; 1982.
2. ALLEN, V.D.; STOVALL, W.D. Laboratory aspects of staphylococcal food poisoning from Colby cheese. J. Milk Food Technol. **23**:271–274; 1960.
3. AMERICAN DRY MILK INSTITUTE. Standards for grades of dry milks, including methods of analysis. ADMI Bull. 916. Chicago, IL: Amer. Dry Milk Inst.; 1971.
4. ANDERSON, P.H.R.; STONE, D.M. Staphylococcal food poisoning associated with spray-dried milk. J. Hyg. **53**:387–397; 1955.
5. ANONYMOUS. Federal Standard 209a. No. 252-522/290. Washington, DC: U.S. Gov. Printing Office; 1967.
6. ARMIJO, R.; HENDERSON, D.A.; TIMOTHÉE, R.; ROBINSON, H.B. Food poisoning outbreaks associated with spray-dried milk - an epidemiologic study. Amer. J. Pub. Health and the Nation's Health. **47**:1093–1100; 1957.
7. BARBER, F.W.; FRAM, H. The problem of false coliform counts on fruit ice cream. J. Milk Food Technol. **18**:88–90; 1955.
8. BLACKBURN, B.O.; ELLIS, E.M. Lactose-fermenting *salmonella* from dried milk and milk-drying plants. Appl. Microbiol. **26**:672–674; 1973.
9. BUCHBINER, L.; ALFF, E.C. Studies of coliform organisms in dairy products. Mycopathol. Mycol. Appl. **35**:281–289; 1947.

10.8 References

10. GALESLOOT, TH. E.; STADHOUDERS, J. The microbiology of spray-dried milk products with special references to Staphylococcus aureus and salmonellae. Neth. Milk and Dairy J. 22:158–172; 1968.
11. GILLILAND, S.E.; SPECK, M.L. Enumeration and identity of lactobacilli in dietary products. J. Food Prot. 40:760–762; 1977.
12. GOEL, M.C.; KULSHRESTHA, D.C.; MARTH, E.H.; FRANCIS, D.W.; BRADSHAW, J.G.; READ, JR., R.B. Fate of coliforms in yogurt, buttermilk, sour cream, and cottage cheese during refrigerated storage. J. Milk Food Technol. 34:54–58; 1971.
13. HALL, H.E. The significance of *Escherichia coli* associated with nut meats. Food Technol. 25:230–232; 1971.
14. HEINEMANN, B. Factors affecting the direct microscopic clump count of nonfat dry milk. J. Dairy Sci. 43:317–328; 1960.
15. HENDRICKS, S.L.; BELKNAP, R.A.; HAUSLER, JR., W.J. Staphylococcal food intoxication due to Cheddar cheese. I. Epidemiology. J. Milk Food Technol. 22:313–317; 1959.
16. HORWITZ, W., ED. Official methods of analysis, 13th ed. Washington, DC: Assoc. Offic. Anal. Chem.; 1980.
17. MACY, H.; COULTER, S.T.; COMBS, W.B. Observations on the quantitative changes in the microflora during the manufacture and storage of butter. St. Paul, M.N.: University of Minnesota: Minn. Agr. Exper. Sta. Tech. Bull. 82; 1932.
18. MARTH, E.H. Salmonellae and salmonellosis associated with milk and milk products. A review. J. Dairy Sci. 52:283–315; 1969.
19. MARTLEY, F.G. The effect of cell numbers in streptococcal chains on plate-counting. New Zealand J. Dairy Sci. Technol. 7:7–11, 1972.
20. McDONOUGH, F.E.; HARGROVE, R.E. Heat resistance of salmonella in dried milk. J. Dairy Sci. 51:1587–1591; 1968.
21. MINOR, T.E.; MARTH, E.H. *Staphylococcus aureus* and staphylococcal food intoxications. A review. III. Staphylococci in dairy foods. J. Milk Food Technol. 35:77–82; 1972.
22. NATIONAL CANNERS ASSOC. RES. LAB. Canned foods, principles of thermal process control and container closure evaluation. pp 45–58. Berkeley, CA: Food Processors Inst.; 1975.
23. NORRIS, J.R.; RIBBONS, D.W., EDS. Methods in microbiology. New York, NY: Academic Press; 1969: p. 79–115. (Willis, A.T. Techniques for the study of anaerobic spore-forming bacteria; vol. 3B.)
24. PARK, H.S.; MARTH, E.H.; OLSON, N.F. Fate of enteropathogenic strains of *Escherichia coli* during the manufacture and ripening of Camembert cheese. J. Milk Food Technol. 36:543–546; 1973.
25. PEDRAJA, R.; CHOI, R.; MENGELIS, A.; SMALL, E. Joint collaborative study of the direct microscopic clump count method in dry milks and its statistical considerations. J. Dairy Sci. 46:810–818; 1963.
26. SARASWAT, D.S.; REINBOLD, G.W.; CLARK, JR., W.S. The relationship between enterococcus, coliform and yeast and mold counts in butter. J. Milk Food Technol. 28:245–249; 1965.
27. SPECK, M.L., ED. Compendium of methods for the microbiological examination of foods, 2nd ed. Washington, D.C.: Amer. Public Health Assoc.; 1985.
28. THRASHER, S.; RICHARDSON, G.H. Comparative study of the stomacher and the waring blender for homogenization or high-fat dairy foods. J. Food Prot. 43:763–764; 1980.
29. TODD, E.; SZABO, R.; ROBERN, H.; GLEESON, T.; PARK, C.; CLARK, D.S. Variation in counts, enterotoxin levels and TNase in Swiss-type cheese contaminated with *Staphylococcus aureus*. J. Food Prot. 44:839–848; 1981.
30. WILKINSON, R.F.; RICHARDSON, G.H. Comparative and collaborative studies on the preparation of hard cheese samples for microbiological assay using a plastic pouch technique. J. Milk Food Technol. 38:583–585; 1975.

31. U.S. DEPT. OF HEALTH AND HUMAN SERVICES. U.S. Food and Drug Admin. Bacteriological analytical manual. 5th ed. Washington, DC; 1978.
32. U.S. DEPT. OF HEALTH AND HUMAN SERVICES. Grade A condensed and dry milk products and condensed and dried whey. Suppl. 1 to the Grade A pasteurized milk ordinance. Washington, DC: U.S. Gov. Printing Office; 1978.

CHAPTER 11

DIRECT MICROSCOPIC METHODS FOR BACTERIA OR SOMATIC CELLS

V. S. Packard, Jr. and R. E. Ginn
(R. E. Ginn, Tech. Comm.)

11.1 Introduction

The Direct Microscopic Clump (DMC) method provides for the examination of milk or certain milk products for numbers of clumps of bacteria or for numbers of somatic cells in milk. In the latter instance, the method is one of several that has official recognition for confirming somatic cell counts previously estimated by one of several screening tests [12.3A, B and C]. In the former application, bacterial clumps may be counted and, at the same time, an evaluation made of the distinctive morphology and arrangement of bacteria.[4] To a limited extent, cell morphology and configuration of clumps allow the analyst to assess the cause(s) of quality problems. Large numbers of somatic cells indicate mastitis or other abnormal conditions of the udder.

Over the years, the term "somatic cells" has come to replace the word leucocytes as a more meaningful expression of those body cells usually associated with inflammation of the udder. Accordingly, the method is now referred to as the Direct Microscopic Somatic Cell Count (DMSCC). As such, it reflects the research findings of several investigators[13,22,25,26,27,28] and is applied as one of several confirmatory procedures in evaluating milk previously screened for presence of somatic cells. The extent of variations in counts made by different analysts and the same analyst have been reported for certain screening and confirmatory methods.[23]

The accuracy and reproducibility of the microscopic method depends largely upon the training and skill of the technician. Test results are reported in actual counts of clumps of bacteria or individual somatic cells.

11.2 Application to Raw Milk (Class O)

The microscopic method offers a rapid technique for determining the extent of bacterial contamination of samples of raw milk or cream taken directly from the producer or from pooled supplies in tankers or storage tanks. Individual samples may be examined in 10 to 15 min. Causes of high counts may be suggested by evaluation of morphology of cells, configuration of clumps, or presence of somatic cells. Such interpretations should always be made cautiously, even by well-trained analysts. Precise estimates on milk of very low count are generally impractical because of the time required to count the large number of fields necessary to achieve precision.

11.3 Application to Pasteurized Milk (Class D)

The microscopic method has limited but possibly beneficial application in determining the extent of bacterial contamination in pasteurized fluid milk and cream. Because dead cells somewhat lose their ability to take up stain, and because numbers of bacteria are generally low in pasteurized milk products, this method should not be used to determine compliance with finished-product standards. Rather, the microscopic method in this instance should be reserved as a quality control tool. Large numbers of bacteria in uncultured products give evidence of unsanitary conditions no matter what type of organisms are present and whether or not they are viable (living) cells.

11.4. Application to Dry Milk (Class O)

Most dry milk contains relatively few viable bacteria. Nonetheless, microscopic examination of films prepared from reconstituted samples can provide meaningful information on the previous sanitary history of the product. Such evaluation can serve a quality control program.[10] The Direct Microscopic Clump (DMC) Count has been incorporated into both government,[3] and industry standards.[1,2] Arguments have been made against such application, however,[8] and evidence[21] suggests a potential for wide between-laboratory differences in counts made on the same sample of milk. In addition, research indicates that counts made on skim milk do not necessarily correlate well with counts made on their corresponding dry products.[11] Heat-treated products do not stain well with Levowitz-Weber stain;[17] however, periodic acid-bisulfite-toluidine blue (PST) stain has yielded substantially higher counts than the former.[18]

11.5 Source of Error in the Microscopic Method

As in other methods of enumerating bacteria or somatic cells, results of the direct microscopic method should be considered as estimates only. In general, the most important factors in determining accuracy, precision, and reproducibility are the training and skill of the analysts.[10] Even with exacting techniques, replicate estimates may vary appreciably. Among factors responsible for such variation are:

    A. inaccuracy in measuring 0.01-mL quantities of sample;

B. faulty preparation and staining of slides;
C. failure of some bacteria to stain;
D. the minute amount of milk actually examined in counting;[14]
E. irregularity in distribution of bacteria or somatic cells in the films;
F. failure to count a sufficient number of fields;
G. poor microscopy—inadequate or excessive illumination of the microscope, poor focusing, improper use of colored filters;
H. failure to dry films on a level surface;
I. eye fatigue, and
J. errors in observations and calculation.

11.6 Procedure for Collecting Samples

See applicable section of [Chapter 4].

11.7 Equipment and Supplies

A. Microscope slides:

Use only *clean* microscope slides. Optional sizes include 2.5 × 7.50 cm, 5.0 × 7.50 cm , or 5.0 × 11.25 cm. Each slide should have one or more clear, circular, 1-cm^2 (diameter 11.28 mm) areas delineated by etched surfaces or by painted rings. (Slides of the type described are available from Bellco Glass, Inc., Vineland, NJ 08360, Cat. #5638-01930). Slides may be cleaned for repeated use. Clear-glass slides with a guide are unacceptable.

Several procedures are reported as satisfactory for cleaning slides. Slides are often cleaned by submerging them in hot detergent solution until free of residue, then rinsing thoroughly in flowing tap water, and drying. New slides (or used slides) may be effectively cleaned using appropriate commercial alkaline and/or acid cleaners followed by rinsing in distilled water. To avoid fingerprints on clean surfaces, use steel or glass holders for transferring slides.

Always store slides with minimal exposure to dust, insects, or other contaminants. Slides may be stored dry, or in alcohol. If stored in 95% alcohol, slides can be dried just prior to use by flaming. Otherwise, they must be drained and air-dried. Still another method of slide storage involves an application of a surface coating of BonAmi (Faultless Starch/BonAmi Co., Kansas City, MO. 64104)[16] or other comparable detergent. Just prior to use, the detergent is wiped off with a clean, dry cloth.

B. Instrument for transferring 0.01-mL quantities:

1. Metal syringe: (Applied Research Institute, 141 Lewis St., Perth Amboy, NJ 08861). This instrument allows rapid and convenient transfer of milk or cream. It comes equipped with a semi-automatic, spring-actuated plunger, and a stainless-steel piston fitted closely to a stainless-steel measuring tube (Figure 11.1).

   Before use, and periodically thereafter, determine the accuracy

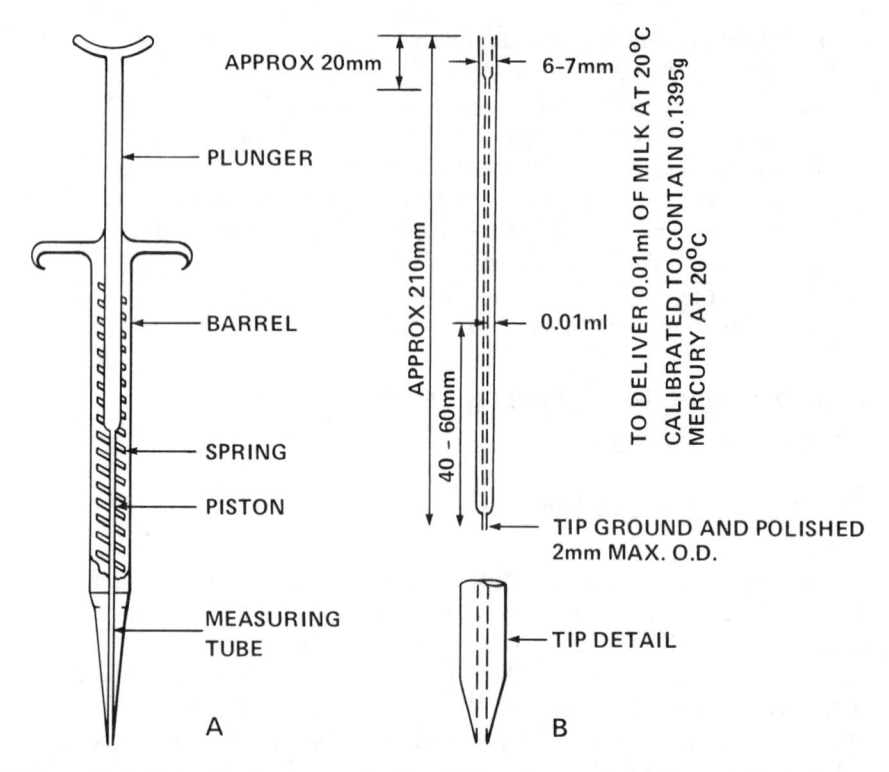

Figure 11.1 Transfer instruments for 0.01-mL amounts. A—metal syringe; B—pipette.

of the syringe adjustment. If certification is required, the appropriate regulatory agency should verify the accuracy.

NOTE: The mechanical principle of the syringe makes possible a *positive* measurement of 0.01-mL portions without manual adjustment in the sample column. It also provides for accurate measurement of cold, viscous cream, an advantage unattainable with 0.01-mL pipettes. Problems may occasionally be encountered in metal syringes due to particles of metal oxide that create artifacts in the milk film.

Do not allow residues to dry on the instrument. Immediately after use, disassemble and clean the measuring tube and piston. *Do not interchange parts from one syringe with parts of another.* Handle parts carefully when they are disassembled. When assembled, the instrument is reasonably sturdy, but should not be dropped or handled carelessly. Except when unavoidable, do not unscrew the barrel at the finger-grip joint. To disassemble, unscrew the tip assembly from the barrel. If necessary to remove the spring, reassemble it with the larger end toward the tip. Clean all residues from piston and lower part of plunger by wiping exposed portions with a dry, paper tissue or cloth. Use soapless detergents and fat solvents *sparingly but as needed* (metal is not as resistant to corrosion as glass). Unless the syringe is properly maintained, it can contribute both bacterial cells and artifacts (metallic oxide particles).

Do not submerge the barrel and finger grip in detergents or solvents. With forced suction or pressure from a close-fitting rubber bulb applied at the *delivery* end of the measuring tube, clean the interior by submerging the opposite end in soapless detergent. Then, in a similar manner, rinse the interior in clean water before air-drying. Do not use acids, alkalis, or ordinary soap solutions for cleaning (as may be the custom with glassware). Use chemical cleaners only in low concentrations at or near pH 7.0. Clean new instruments before first use.

Before assembling the syringe, make certain that both detergent and milk residues have been removed from the instrument. To reassemble, slide the spring (if removed) over the piston and plunger; *carefully* insert the piston into the measuring tube at the countersunk opening, and tighten the threaded joint. Be sure threaded joints are always tightened to identical positions (scribe syringes at each joint).

After repeated use, all syringes tested and found not to conform with milk deliveries as specified above, should be replaced. Syringes should be tested for accuracy biennially.

2. Micro syringes: Many makes of these devices are available and may be used if they meet calibration standards to deliver 0.01 mL of milk.
3. Pipette: The pipette is calibrated to deliver 0.01 mL quantities of milk according to APHA specifications.[5] It consists of straight, thick-walled capillary tubing with a bore of such diameter that a single graduation mark is 40 to 60 mm from the pipette tip. The tip is blunt and formed so as to discharge the milk cleanly [Figure 11.1]. Calibrated to contain 0.1395 g of mercury, it will discharge 0.01 mL of milk at 20°C. Otherwise calibrate with milk as in 11.7B4.

    NOTE: When not in use, keep pipettes submerged in and the bore filled with suitable soapless detergents or strong cleaning solution. When preparing for use, rinse the bore and exterior thoroughly in clean water until free of the detergent or cleaning solution.
4. Calibration: Calibration of the volumetric load of the syringes[19] or pipette[6] is determined indirectly by weight measurements. Withdraw a representative test charge [11.11A, B or C]. Wipe the exterior of the tip. Weigh the charge instrument on an analytical balance. Expel the charge near the center of the 1 cm^2 area of a slide and touch off on a dry spot. Weigh the discharged instrument. Over ten such trial

weighings, the difference in weight of the charged and uncharged instrument should average 0.0103 g ± 0.0005 g.

C. Bent-point needle:

A bent-point biological disecting needle is used for spreading milk or cream delivered from a 0.01-mL pipette or syringe over the square-centimeter area.

D. Drying surface:

A clean, dust-free, level surface should be provided with suitable heat source for drying films at 40 to 45°C. Optionally, use the top of a cabinet or box over an electric bulb or substage microscope lamp. A commercially available drying box (Figure 11.2) may be used for this purpose. (Applied Research Institute, 141 Lewis St., Perth Amboy, NJ 08861). Use of a cross-test level is suggested, to be certain that the surface on which films are made and dried is perfectly level.

E. Forceps and partitioned slide holders:

Forceps and partitioned slide holders are required for dipping and holding slides.

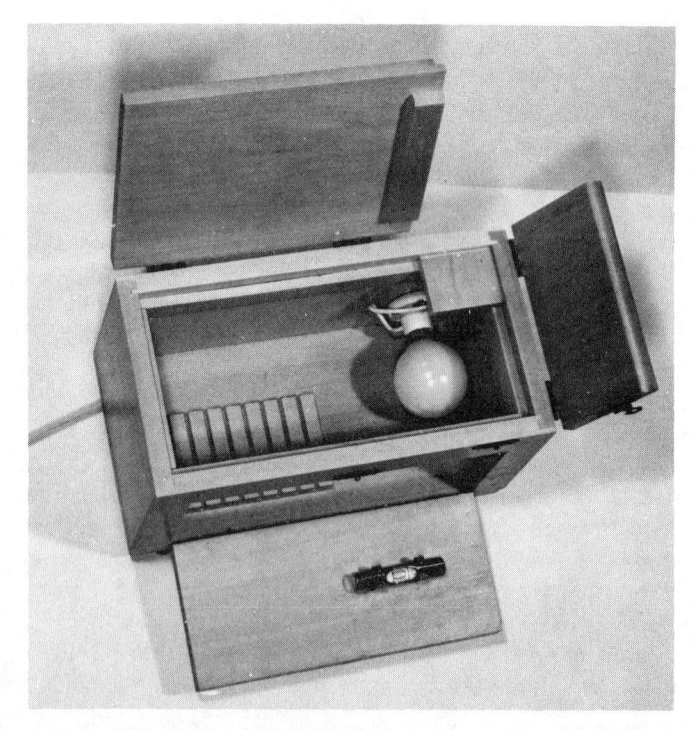

Figure 11.2 Film-drying box, closed type.

11.7 Equipment and Supplies

F. Trays or jars:

Trays or jars should be equipped with tight-fitting covers, and be of convenient size for holding solvents and stains and/or submerging slides held in partitioned racks.

G. Slide storage boxes, cases, or files:

Slide storage boxes, cases, or files must be clean, dust-free, and insect-proof.

H. Compound microscope:

A binocular microscope is preferred, with 1.8-mm oil-immersion objective, a substage actuated by a rack and pinion gear, carrying an Abbe condenser and having a numerical aperture (NA) of 1.25 or higher. It should be equipped with an iris diaphragm, a mechanical stage, and oculars as follows: single for monocular or paired for binocular, 10× magnification recommended (12.5 to 15× permitted). If a mirror is used, it should be flat. For both enumerative and grading purposes, the ocular(s) should be calibrated to provide a microscopic factor of 300,000 to 600,000 [11.18]. Wide field oculars with high eyepoint are advantageous, especially for those who wear eyeglasses.

I. Ocular disc subdivided into quarters:

To use the disc (Figure 11.3), place it in the ocular so that it rests on the edge of the reducing diaphragm. Face the etched surface of the disc according to directions provided by the microscope manufacturer. Only one disc is required for a binocular microscope. Before using such a disc, determine the microscopic factor corresponding to the area to be examined [11.10].

J. Ocular reticle:

For the Strip Reticle Method, special reticles must be used. (Dairy Research Products, Inc., Box 288, Yarmouth, ME 04096. Applied Research Institute, 141 Lewis St., Perth Amboy, NJ 08861)

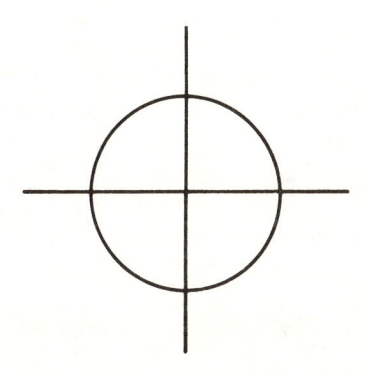

Figure 11.3 Ocular disc, showing general arrangement of guide lines.

K. *Mechanical stage:*
A regular mechanical stage as supplied with the microscope, or a special stage for slides 5.0 × 11.25 cm may be used.

L. *Stage micrometer slide:*
A stage micrometer slide ruled in 0.1- and 0.01-mm divisions.

M. *Hand tally:*
A single or double hand tally may be used.

11.8 Materials

A. *Chemical reagents:*
Chemical reagents must be reagent grade unless otherwise specified. Keep containers of stock reagents tightly closed.

B. *Methylene blue chloride, certified:*
Use either dry stains certified by the Biological Stain Commission for the purpose intended, or liquid stains prepared from dry stains so certified. (Applied Research Institute, 141 Lewis St., Perth Amboy, NJ 08861).

C. *Immersion oil:*
Immersion oil should have a refractive index of 1.51 to 1.52 at 20°C and be of a nondrying type. Type B (medium viscosity) is preferred.

11.9 Adjustment and Calibration of Microscope

When determining numbers of bacterial clumps or somatic cells in milk on a per milliliter basis, the microscopic field area is fundamentally important because it determines the amount of milk that can be examined in any one field. Field diameters providing microscopic factors of from 300,000 to 600,000 are recommended. Microscopic factors are the values by which the average number of bacterial clumps or somatic cells per field are multiplied to calculate the respective counts.

Manufacturers of binocular microscopes normally adjust tube length and paired oculars for use at the most common interpupillary distance (65 mm). Or, they may be specifically adjusted for a given individual. *When estimates are to be made by an individual whose interpupillary distance varies substantially from that for which the microscope was set up, the individual should use paired oculars with diaphragms especially adjusted for that individual; otherwise, counts must be corrected according to the diameter of the field actually observed.*

Presently available standard equipment includes binocular and trinocular microscope bodies of constant tube length. With this feature, the resultant calibrated field and magnification remain constant regardless of changes in interpupillary settings.

11.10 Calculation of the Microscopic Factor (MF)

The area of a single field determines the amount of milk film which can be seen at any one time. Microscopes vary in size of field, and accurate counting can only be accomplished when the field size (area) is known. Field size, number of fields in 1 cm², and amount of milk used (0.01 mL), are the factors considered in calculating the Microscopic Factor (MF).

To determine the MF, first adjust the microscope lamp for maximum optical resolution. Verify the manufacturer's indicated MF. To do so, place a stage micrometer slide, ruled in 0.1- and 0.01-mm divisioins on the stage of the microscope. Proceed as follows:

A. Determine the diameter of one field under the oil immersion lens. Make the reading to the third decimal point, as in 0.146mm.

B. Determine the area of the field by squaring the radius (r) and multiplying by pi or 3.1416. The radius is one-half the diameter.

C. Convert the area of the field as measured in square millimeters to square centimeters; i.e., divide square millimeters by 100.

D. Determine the number of fields in 1 cm², i.e. divide 1 cm² by the area of one field as determined in step C.

E. Determine the number of microscopic fields per milliliter by multiplying the number of fields (as determined in step D) by 100. The factor 100 derives from the fact that 0.01 mL of milk is used and is spread over 1 cm² of surface.

The above calculations may be condensed into the following formulae:

$$\text{MF} = \frac{10,000}{3.1416 \times r^2} \quad \text{or} \quad \text{MF} = \frac{40,000}{3.1416 \times d^2}$$

In the formulae r = radius, and d = diameter.

The MF in each instance depends on a specific tube length and combination of ocular(s) used. Distance from eye to ocular must remain constant. Therefore, each analyst should determine his or her MF. The reciprocal of the MF represents the fraction of 1 mL of milk actually observed in one microscopic field. MF should be expressed to two significant figures (e.g., 480,000 or 4.8×10^5).

11.11 Using Transfer Instruments and Preparing Films

Always keep instruments and glassware scrupulously clean and free of foreign materials. Cleanliness of milk or cream contact surfaces is especially important, but sterility is not necessary.

Shake the test sample vigorously 25 times, as per [6.2F2]. Allow sample to stand until foam disperses (but no more than 3 min) to obtain a virtually foamless test portion. Legibly and indelibly identify each film as it is prepared; i.e., use a number or other symbol and place it on the etched margin of the slide.

Work carefully but as rapidly as possible to avoid bacterial growth while preparing films. Dry the films at 40 to 45°C within 5 min on a *level* surface protected from dust, as per [11.7D]. Heat slowly to avoid cracking and/or peeling of films. Protect films from exposure to dirt, dust, insects, or other damage at all times through actual counting.

After thoroughly mixing sample, as per [6.2F2], measure 0.01 mL portions by the procedures A, B, or C as follows:

A. Use of 0.01-mL syringe:[19]

Before each use and between successive samples, rinse measuring tube in fresh tap water (25 to 35°C) by dipping tip slightly beneath the surface and repeatedly drawing in and expelling water. Before transferring test portions to the slide, dip tip of tube not more than 1 cm below the surface (excluding foam) of well-mixed milk or cream, and repeatedly rinse tube by drawing in expelling portions. Holding tip beneath the surface, fully release plunger and withdraw a test portion. With clean paper tissue or cloth, remove excess milk or cream from exterior of tip. Holding the instrument in nearly vertical position, place tip near center of the area designated for the film, and, while touching the glass, expel the 0.01-mL test portion and touch off on a dry spot.

B. Use of 0.01-mL pipette:

Before each use and between samples, rinse graduated portion of pipette in clean water (25 to 35°C) by dipping the tip slightly beneath the surface and repeatedly drawing in and expelling water. Do not use any pipette in which the tip has been chipped through to the bore. Before measuring test portion, any traces of residual water must be removed from the bore of the pipette. Withdraw portions of well-mixed samples slightly above the graduation mark. Be sure saliva does not get into pipette to contaminate the sample. A cotton plug may be placed in the mouth end for this purpose. Wipe exterior of pipette with a clean, dry, paper tissue or towel and, by absorbing liquid at the tip, adjust length of column to the graduation mark. Place tip near center of area to be covered and expel a 0.01-mL test portion.

Tempering cream samples at 30 to 35°C for not more than 10 min before withdrawal of a test portion promotes more uniform deliveries with the pipette.

C. Micro syringes:

Follow manufacturer's recommendations.

With a clean, bent, dissecting needle, promptly spread the sample over the cm^2 area. Wipe needle between samples on a clean, dry tissue or towel.

11.12 Handling of Slides During Staining

For handling slides containing dried films, use either glass, stainless-steel, or other holders partitioned to permit multiple transfer of slides. Such holders must be sealable and maintained free from residues that might subsequently contaminate the staining solution in which slides are dipped.

11.13 Preparation and Use of Stains

Two different stains may be used for cow's milk. One is the Levowitz-Weber modification of the Newman-Lampert stain.[15] The other is a stain very similar to the modified Newman-Lampert stain, but substitutes xylene, a less toxic compound, for tetrachloroethane. Always avoid use of liquid reagents containing suspended foreign matter. After a stain has been formulated, handle it in a manner that minimizes evaporation of solvents and consequent formation of a precipitate that might introduce artifacts into stained films. Solvent loss must be minimized at all times to allow proper staining.

The Levowitz-Weber modification of the Newman-Lampert stain has produced highly variable results with and is unreliable for staining goat's milk for somatic cell count.[9] Pyronin Y-methyl green stain, however, when used with Carnoy's fixative, was found to be suitable replacement.[9]

A. Preparation of staining solutions:

Use one or the other of the two stain formulations given below. CAUTION—Prepare stain under a fume hood to avoid toxic fumes.

Prepare modified Newman-Lampert stain by adding 0.6 g of certified methylene blue chloride to 52 mL of 95% ethyl alcohol and 44 mL of tetrachloroethane (technical) in a 200 mL flask. Swirl to dissolve. Let stand for 12 to 24 h at 4.4 to 7.2°C, add 4 mL of glacial acetic acid, and filter through fine filter paper (Whatman No. 42 or equivalent). Store in a clean, tightly closed bottle with a closure that is not affected by stain reagents.

Prepare alternative stain using xylene in place of tetrachlorethane in the same manner as the above, but with the following amounts of reagents: 0.50 g methylene blue chloride, 56 mL of 95% ethyl alcohol, 40 mL of xylene, and 4 mL of glacial acetic acid.

Prepare fixative and stain for goat's milk as follows: Carnoy's fixative: Combine 60 mL chloroform, 20 mL glacial acetic acid, and 120 mL 100% ethyl alcohol. Pyronin Y stain: Combine 1.0 g Pyronin Y, 0.56 g methyl green, and 196 mL water.[20] Filter stain through Whatman #1 before use. Note: This stain is sensitive to light and is best stored in brown bottles.

B. Staining procedures:

1. Using Levowitz-Weber stains (Class 0): Place stain in a container that can be covered to prevent evaporation while slides are in stain solution. Repeated use of stain in an uncovered vessel may result in formation of precipitate.

    Submerge or flood slides of the fixed, dried films, singly or in multiples, as per 11.12, into stain for 2 min. Remove and drain off excess stain by resting edge of slide in a near vertical position on absorbent paper. Dry thoroughly (by forced air if available). Rinse the dried, stained slides in three changes of tap water at 35 to 45°C, then drain and air-dry before examining the film under the microscope.

Proceed as above in preparing and staining cream films.

When small numbers of raw milk films are to be stained, flooding of slides may be more practical than submersion. Care must be taken to limit flooding exposure so that evaporation does not progress to the point where precipitation of dye occurs. Use adequate ventilation (fume hood is recommended).

2. Using pyronin Y methyl green stain on goat's milk (Class C): Prepare Carnoy's fixative and fresh Pyronin Y stain [11.13]. Fix smear in Carnoy's fixative for 10 min. Hydrate each for 2 min in 50 % ethyl alcohol, 30% alcohol, and MS water. Stain for 6 min in Pyronin Y-methyl green. Keep stain covered with aluminum foil. Immerse for 20 min in N-butyl alcohol. Immerse for 30 min in a fresh change of N-butyl alcohol. Clear for 10 min in each of two changes of xylol. Mount in 50% piccolyte dissolved in xylol. Protect slides from exposure to light.

C. *Storage of staining solution:*

Discard used stain whenever the solution becomes contaminated or otherwise unsuitable. When not in use, keep containers of staining solution tightly closed to prevent evaporation. This applies to surplus supplies as well as to "working batches" of stain. Avoid use of containers or closures that may dissolve or disintegrate and thus contaminate staining solutions. Store solutions in a relatively dark, cool place but do not refrigerate.

11.14 Storage of Slides

After slides have been examined, remove immersion oil on films by dipping each slide in xylene for 15 to 20 sec and allowing films to air-dry. Store slides in dust-free, slide-storage boxes for future observations. Avoid accumulation of pencil marks on slides, because their presence will interfere with the staining procedure or with future re-examination of milk films.

11.15 Examining Films for Bacteria or Somatic Cells

To obtain estimates of the bacteria or somatic cell count per milliliter, examine each film with an oil-immersion objective after placing one drop of immersion oil on the film. When counting bacteria, count as separate clumps any cell or group of cells (apparently of the same type) separated by a distance equal to or greater than twice the smallest diameter of the two cells nearest each other. Regardless of proximity to each other of cells of different types, count each type as a separate unit. When making Direct Microscopic Somatic Cell Counts (DMSCC), count only those somatic cells with an identifiable stained nucleus. A stained nucleus is identified as follows: For polymorphonucleated cells, count as a cell any that has two or more discernible nuclear lobes; for other somatic cells count any that has a nucleus that appears to be essentially intact. If in doubt about a cell, which may in fact be only a fragment, do not count it.

Cells are counted by one of three methods.

11.16 Field-wide Single Strip Method,[13] (Class O)

A. Determining counting factor:

This counting method uses as boundaries a single strip the width of the microscopic field and running across the diameter of the film of milk. A single-strip factor (SSF) must first be calculated:

Area of a single strip (mm²) = 11.28 mm × D (in mm) Determine the number of single strips in the 0.01-mL milk film by dividing 100 mm² (area of the 0.01-mL milk film) by the strip area:

Number of single strips = 100 mm² − area of single strip. Convert the number of single strips in 0.01 mL of milk to a 1.0-mL basis by multiplying the number of single strips in 0.01 ml by 100.

SSF = Number of strips in 0.01 ml × 100

EXAMPLE:

Diameter of microscopic field = 0.148 mm
Length of strip (diam. of 1-cm² circle) = 11.28 mm
Area of single strip: 11.28 × 0.148 = 1.67 mm²
No. of single strips in area of milk film
(0.01 mL of milk): 100 − 1.67 = 59.9
SSF = 59.9 × 100 = 5,990

B. Counting procedure:

To make a single-strip count, focus on the film edge in the oil-immersion field that appears to be at the maximum horizontal or vertical excursion. Traverse the entire diameter of the milk film, counting those cells within the strip and also those cells touching one edge of the strip. Do not count bacteria or somatic cells that touch the other edge. During scanning of the strip, continually make fine focusing adjustments. The bacteria or somatic cell count is computed as follows:

Total number of somatic and/or bacterial cells in single strip:

Count per mL = Number × single-strip factor (SSF)

EXAMPLE: If an analyst counts 84 cells in a field-wide strip, assuming a single-strip factor of 5,990, the count is computed as follows:

Count per mL = 84 × 5,990 = 503,160

Round off the answer to two significant figures. The reported count becomes 500,000 per milliliter.

11.17 Field Method (Class D)

A. Determining counting (working) factor:

If a constant number of fields is counted in routine work, a working factor (WF) may be calculated. The WF is determined by dividing the microscopic factor (MF) by the number of fields counted. With this factor, estimated microscopic counts can be deterrminined simply by multiplying by the WF the number of clumps of bacteria or numbers of somatic cells observed.

Sample calculation: Assume a MF of 600,000 and a constant of 20 fields examined. The working factor becomes:

$$WF = \frac{600,000}{20} = 30,000$$

Then, if a total of 40 clumps of bacteria were counted in twenty fields, the DMC would be 40 × 30,000 = 1,200,000. Note that the value is again expressed only to two significant figures.

B. *Counting procedure:*

Select successive rows, each perpendicular to the previous row, across the film. Start about midway on any side and two or three fields in from the edge. Proceed to count, accumulating totals of bacteria or somatic cells and of fields examined in accordance with Table 11.1. Then:

$$\text{Count per mL} = \frac{\text{Total cells}}{\text{Number of fields}} \times \text{MF}$$
$$\text{or} = \text{Total cells} \times \text{WF}$$

When it is anticipated that the quality of milk is such that the maximum number of fields will not have to be examined, count cells in 10 to 15 fields to establish an approximate number of cells per field, and thus tentatively establish the count limit within which the sample will fall. From this information the actual number of fields to be counted can be determined for the microscopic factor being used. After examining the number of fields required on the basis of a preliminary count, verify that the cell count falls within limits prescribed for the number of fields to be counted.

Table 11.1 Suggested Microscopic Factors (MF) with Corresponding Field Diameters,* Numbers of Fields to be Examined, and Calculated Working Factors (WF) for DMC or DMSCC per Milliliter†

Dependent values, MF	300,000	400,000	500,000	600,000
Field diameters (mm)	0.206	0.178	0.160	0.146
Count limits per mL:				
30,000–300,000				
No. of fields	30	40	50	60
Working factor	10,000	10,000	10,000	10,000
300,000–3,000,000				
No. of fields	20	20	20	30
Working factor	15,000	20,000	25,000	20,000
Over 3,000,000				
No. of fields	10	10	10	20
Working factor	30,000	40,000	50,000	30,000

* Using 10 × oculars or higher, with 1.8-mm oil-immersion objectives.
† Magnification of at least 900 × is necessary.

11.18 Strip Count Methods[12,28] (Class D)

A. Determining counting factor:

In this method, the unit area examined is a strip of milk film defined in length by the fixed diameter of the circular film of milk and in width by the parallel lines in one of three special eyepiece reticles. Special ocular reticles are needed.[25] They consist of two parallel lines, centrally located, with different spacings between the lines for various somatic cell concentrations. (AO Reichert Instruments, PO Box 123, Buffalo, NY 14240, Catalog No. K2013, for 1.5 million cells per mL; Catalog No. K2022, for 1.0 million cells per mL; Catalog No. K2023, for 0.75 million cells per mL.)

Selection of strip width at the maximum acceptable somatic cell concentration provided locates the region of optimum count precision at this critical concentration.

To calibrate the microscope for the strip-reticle count, insert the appropriate eyepiece of the binocular microscope. Align the oil-immersion objective. Using the stage micrometer, measure the distance in millimeters between the parallel lines seen in the visual field, estimating the third decimal place. This is the strip width. Strip length is fixed at 11.28 mm (diameter of 1 cm^2 circle). Determine area of strip (strip width × 11.28) in mm^2. Divide film area by strip area to get the number of such strips in the milk film. Multiply the result by 100 to get the number of strips in 1 mL of milk. The computations are expressed in the following equation:

$$\text{Strip reticle factor (SRF)} = \frac{10{,}000}{11.28 \times \text{strip width in mm}}$$

The SRF must be determined for each microscope and reticle combination.

B. Counting procedure:

To perform a strip reticle count, locate the horizontal diameter by observing the left or right edge of the milk film through high-dry or oil-immersion magnification while manipulating the vertical travel of the mechanical stage. Follow the film edge to its widest horizontal point. Rotate the eyepiece containing the eyepiece reticle until the parallel lines are horizontally oriented. Place the oil-immersion objective in alignment if not already done. Using the horizontal mechanical-stage control, move the milk film completely across to its far edge, counting all somatic cells. Similarly, locate the vertical diameter of the milk film by observing the top or bottom edge, and count cells found in the vertical strip. When counting films with boundary lines, observe the following rules:

1. Count somatic cells as they pass through the center of the visual field to lessen the need for fine focusing.
2. Count all somatic cells that are entirely within both boundaries.
3. Count all somatic cells that touch one of the boundaries; do not count those somatic cells that touch the other boundary. Note that a cell that just touches one boundary line from either its outside or

its inside is counted; a cell that touches the other boundary from either of its sides is not counted.

Somatic cell counts are made and computed in the reticle strip method in one of two ways:

a. Four-strip reticle procedure: Count four strips, one horizontal and one vertical on each of two films prepared from the same sample. Compute the average somatic count per strip (to two decimal places) and multiply this value by the SRF. The resulting value is the direct microscopic somatic cell count per milliliter:

$$\text{DMSCC per mL} = \frac{H_1 + V_1 + H_2 + V_2 \times \text{SRF}}{4}$$

where H = count obtained on a horizontal strip, V = count obtained on a vertical strip and films are numbered 1 and 2.

EXAMPLE: Using a microscope and reticle yielding an SRF of 15,160, the following counts are made: 72, 80, 78, 69.

$$\text{DMSCC/mL} = \frac{72 + 80 + 78 + 69}{4} \times 15,160 = 1,130,000$$

Round off to two significant figures; thus, the reported DMSCC would be 1,100,000 per milliliter.

b. Single-strip reticle procedure:[12] Since precision increases with the number of strips counted, more strips must be counted to determine compliance as a count approaches the control limit. The following procedure permits a single-strip count for those milks falling into the extreme high or low ranges but provides for counting of additional strips to confirm those values that are close to the control limit. This is a modification of the strip-reticle counting procedure that permits rapid examination of samples by examining only one strip. First, process two milk films as usual. Then begin counting but count only the horizontal strip of the first film. The number of somatic cells found is compared with a low control value (C_L) and a high control value (C_H), specific for the microscope and reticule used. If the count, is less than C_L, the milk sample should be graded acceptable. If the count is greater than C_H, the count may be considered as confirmed (see example below), and the milk sample is graded unacceptable. If the count is within the range C_L to C_H, the count must be confirmed. To accomplish this, count the additional three strips required. Values of C_L and C_H for the microscope and reticle are selected from Table 11.2. They depend on the strip equivalent (SE), which is computed as follows:

$$\text{SE} = \frac{\text{Maximum acceptable somatic cell concentration}}{\text{SRF}}$$

11.18 Strip Count Methods (Class D)

Table 11.2 Values of C_L and C_H Appropriate for a Range of Strip Equivalents (SE)*

G_L	SE	C_H	D_L	SE	C_H	C_L	SE	C_H
44	65	96	64	89	124	84	113	151
45	66	97	65	90	125	85	114	153
46	67	98	66	91	126	86	115	154
47	68	99	66	92	127	87	116	155
47	69	100	67	93	128	88	117	156
48	70	101	68	94	130	88	118	157
49	71	103	69	95	131	89	119	158
50	72	104	70	96	132	90	120	159
51	73	105	71	97	133	91	121	161
51	74	106	71	98	134	92	122	162
52	75	107	72	99	135	93	123	163
53	76	109	73	100	137	94	124	164
54	77	110	74	101	138	94	125	165
55	78	111	75	102	139	95	126	166
56	79	112	76	103	140	96	127	167
56	80	113	77	104	141	97	128	169
57	81	114	77	105	142	98	129	170
58	82	116	78	106	143	99	130	171
59	83	117	79	107	145	100	131	172
60	84	118	80	108	146	101	132	173
61	85	119	81	109	147	102	133	174
61	86	120	82	110	148	102	134	175
62	87	121	82	111	149	103	135	176
63	88	123	83	112	150			

* SE equals the legal maximum cell concentration divided by the strip reticle factor (SRF).

EXAMPLE: If the legal maximum number is 1.5 million somatic cells per milliliter and the SRF is in the neighborhood of 15,000 when the proper eyepiece reticle is used, then,

$$SE = \frac{1,500,000}{15,000} = 100$$

C_L and C_H from the table are, respectively, 73 and 137. The categories defined for a milk sample by the values C_L and C_H may be accepted with at least as much confidence as results attained using the confirmatory procedure. The table of 95% confidence limits designed for interpretation of confirmatory counts is based on the assumption that a coefficient of variation (CV) of 12% will describe the usual laboratory performance; in this screening method an average CV of 19% is assumed.[23]

Therefore, any milk sample that yields a count greater than C_H has more than 95% probability of containing more somatic cells than the legal limit.

11.19 Reporting Counts

Report counts only to the first two left-hand digits of the estimate, followed by the appropriate number of zeroes. If using the method as a tool for broadly classifying milk as to bacterial quality, select only two or three classes representing significant differences in quality, thus assuring appropriately wide ranges between classes. Save slides of reported counts for future reference.

11.20 References

1. AMERICAN DRY MILK INSTITUTE. Recommended sanitary/quality standards for the dry milk industry. Chicago, IL.: ADMI; Bull. 915, 1970.
2. AMERICAN DRY MILK INSTITUTE. Standards for grades of dry milks, including methods of analysis. Chicago, IL: ADMI; Bull. 916; 1971.
3. AGRICULTURAL MARKETING SERVICE, DAIRY DIV. Milk for manufacturing purposes and its production and processing recommended requirements. Fed. Reg. 37 (68):7046–7066. Washington, DC; 1972.
4. BREED, R.S. The determination of the number of bacteria in milk by direct microscopic examination. Centralb. f. Bakt. Part II 30:337–340; 1911.
5. BREED, R.S.; BREW, J.D. Counting bacteria by means of the microscope. Tech. Bull 49. Albany, NY: N.Y. Agr. Exper. Sta.; 1916.
6. BREW, J.D. A comparison of the microscopical method and the plate method of counting bacteria in milk. Tech. Bull. 373. Albany, NY: N.Y. Agr. Exper. Sta.; 1914.
7. CLAIBORNE, F.B.; COX, K.E. The direct microscopic count on preserved milk samples: an effective measure for uniform state-wide control. J. Milk Food Technol. 14:105–108; 1951.
8. COULTER, S.T. Quality considerations. J. Dairy Sci. 40:1012–1015; 1957.
9. DULIN, A.M.; PAAPE, M.J.; WERGIN, W.P. Differentiation and enumeration of somatic cells in goat milk. J. Food Prot. 45:435–439; 1982.
10. FOREST, H.L.; SMALL, E. The direct microscopic clump count as a measure of quality of non-fat dry milk. 15th Proc. Intern. Dairy Cong. 3:1883–1889; 1959.
11. HEINEMANN, B. Factors affecting the direct microscopic clump count of nonfat dry milk. J. Dairy Sci. 43:317–328; 1960.
12. LEVINE, B.S. Effect of formaldehyde on the direct microscopic count of raw milk. Pub. Health Rep. 65:931–938; 1950.
13. LEVOWITZ, D. Reproducible data by microscopic method, p. 198. Bull. Intern. Assoc. Milk Dealers; 1944.
14. LEVOWITZ, D. An inquiry into the usefulness of the standard methods direct microscopic count. J. Milk Food Technol. 20:288–293; 1957.
15. LEVOWITZ, D., WEBER, M. An effective "single solution" stain. J. Milk Food Technol. 19:121 and 127; 1956.
16. MILONE, N.A. Retention of milk and milk-product films on glass slides during staining, defatting and rinsing for microscopic examination. J. Milk Food Technol. 15:195; 1952.
17. MOATS, W.A. Chemical changes in bacteria heated in milk as related to loss of stainability. J. Dairy Sci. 44:1431–1439; 1961.
18. MOATS, W.A.; MARSHALL, R.T. Relationship between direct microscopic counts of nonfat dry milk by two staining methods. J. Dairy Sci. 62:(Suppl 1) 45; 1979. (Abstr.)

11.20 References

19. NEWMAN, R.W. The Smith 0.01-mL syringe in the microscopic grading of milk. J. Milk Food Technol. **15**:101–103; 1952.
20. PAAPE, M.J.; HAFS, H.D.; SNYDER, W.W. Variation of estimated numbers of milk somatic cells stained with Wright's stain or pyronin Y-methyl green stain. J. Dairy Sci. **46**:1211–1216; 1963.
21. PEDRAJA, R.; CHOL, R.; MENGELIS, A.; SMALL, E. Joint collaborative study of the direct microscopic clump count method in dry milks and its statistical considerations. J. Dairy Sci. **46**:810–818; 1963.
22. PRESCOTT, S.C.; BREED, R.S. The determination of the number of body cells in milk by a direct method. J. Infec. Dis. **7**:362; 1910.
23. READ, JR., R.B.; BRADSHAW, J.G.; PEELER, J.T. Collaborative study of confirmatory testing procedures for somatic cells in milk. J. Milk Food Technol. **34**:285–288; 1971.
24. RICHARDS, O. The effective use and care of the microscope. Buffalo, NY: Amer. Optical Co.; 1956.
25. SCHULTZE, W.D. Design of eyepiece reticles for use in the DMSCC method. J. Milk Food Technol. **31**:344–349; 1968.
26. SCHULTZE, W.D.; SMITH, J.W.; JASPER, D.E.; KLASTRUP, O.; NEWBOULD, F.H.S.; POSTLE, D.S.; ULLMAN, W.W. The direct microscopic somatic cell count as a screening test for control of abnormal milk. J. Milk Food Technol. **34**:76–77; 1971.
27. SMITH, J.W. Development and evaluation of the direct microscopic somatic cell count (DMSCC) in milk. J. Milk Food Technol. **32**:434–441; 1969.
28. SUBCOMMITTEE ON SCREENING TESTS. Direct microscopic somatic cell count in milk. J. Milk Food Technol. **31**:350–354; 1968.

CHAPTER 12
METHODS TO DETECT ABNORMAL MILK

R.E. Ginn, V.S. Packard, Jr., R.D. Mochrie, W.N. Kelley, and L.H. Schultz
(R.E. Ginn, Tech. Comm.)

12.1 Introduction

Regulations in most areas of the U.S. require that raw milk for commercial sale be derived from one or more healthy cows or goats; that it be practically free of colostrum; and that milk otherwise abnormal not be offered for sale. This includes milk from cows or goats with mastitis, milk from cows or goats treated with medicinal agents that may be secreted in the milk, and milk containing pesticides, herbicides, or other agents judged hazardous to the consumer.

Concentrations of somatic cells in excess of 500,000 per milliliter are generally considered indicative of mastitis or other abnormality.[17] Cows in very early or very late lactation, or cows with a low-grade or latent udder infection, are likely to produce milk having an excessive number of somatic cells.

Screening and confirmatory tests described herein estimate the number of somatic cells (including leucocytes) in milk. Although polymorphonuclear leucocytes are the cells principally associated with inflammation of mammary glands, enumeration of all nucleated somatic cells has generally been found more feasible when determining normality or abnormality of milk. The tests are not intended for diagnosis of mastitis.

12.2 Screening Tests

A. California mastitis test (Class O):

The California Mastitis Test (CMT) was developed to test milk from individual quarters at the side of the cow.[33,34] The test may also be applied to bulk-tank milk and other blended supplies.

The CMT reagent consists of a detergent plus a pH indicator (bromcresol purple). The extent of the reaction between the detergent and the deoxyribonucleic acid (DNA) of cell nuclei is a measure of the number of somatic cells in the milk. At a concentration of 150,000 to 200,000 cells per milliliter,

a precipitate begins to form. Thicker gels occur in samples with larger numbers of cells. The grading system is based on the amount of gel formation, but assignment of test scores is subjective. This test has the advantages of simplicity, rapidity, low cost and convenience.
1. Equipment and supplies: For programmed screening of tank samples, it is faster and more accurate to dispense milk and reagent with automatic syringes, which can be set to deliver 2-mL quantities. For cowside, bucket, and occasional tank testings, a polyethylene squeeze bottle may conveniently replace the automatic syringe for dispensing the reagent.
   a. Cornwall continuous-pipetting device.
   b. Cornwall 2-mL syringe.
   c. Cornwall metal pipette holder.
   d. Laboratory cannula, 10 cm.
   e. Special CMT plastic paddle.
   f. CMT reagent, (Dairy Research Products, Inc. 5345CR 68, Spencerville, IN, 46788; or NASCO, Inc., Fort Atkinson, WI 58538.)
   g. A source of cold rinse water and a waste bucket or drain
2. Procedure:
   a. Collect and store sample(s) as specified in [4.1C]. Test sample(s) within 36 h after sampling.
   b. Mix sample(s) as specified in [6.2F].
   c. Place 2 mL of milk in a cup of the paddle.
   d. Add 2 mL of CMT reagent.
   e. Gently rotate paddle (or cup) in a circular pattern for 10 sec so that milk and reagent mix thoroughly.
   f. Immediately score reaction.
   g. Rinse the paddle and shake off excess moisture. The paddle need not be dried.
3. Recording and reporting results: Record results in reaction categories as indicated in Table 12.1. Cell count ranges for the various reactions scored are in accordance with published values.[26,27,36,38] Do not report numbers of somatic cells in place of test reaction categories.
4. Precautions:
   a. If milk is kept refrigerated, reliable readings can be obtained for up to 36 h. Thereafter, DNAse activity causes milk to lose its reactivity.[15] Unrefrigerated milk cannot be tested accurately after about 12 h.
   b. Somatic cells tend to migrate with milkfat. Therefore, thorough mixing of milk just before testing is essential to assure a representative sample.
   c. The CMT reaction must be scored between 10 and 15 sec after mixing starts. Weak reactions fade thereafter.

12.2 Screening Tests

Table 12.1 Grading and Interpretation of Milk, California Mastitis Test

Symbol	Suggested Meaning	Visible Reaction	Interpretation (cells per mL)
−	Negative	Mixture remains liquid with no evidence of precipitate formation	0–200,000
T	Trace	A slight precipitate forms, seen to best advantage by tipping paddle back and forth and observing mixture as it flows over bottom of cup. Trace reactions tend to disappear with continued movement of fluid.	150,000–500,000
1	Weak positive	A distinct precipitate forms, but with no tendency toward gel formation. In some milks, reaction is reversible with continued movement of the paddle and the precipitate may disappear.	400,000–1,500,000
2	Distinct positive	Mixture thickens immediately, with some suggestion of gel formation. Upon swirling, mixture tends to move toward the center, leaving bottom of outer edge of cup exposed. When motion is stopped, mixture levels out, once again covering bottom of cup.	800,000–5,000,000
3	Strong positive	A gel is formed that causes surface of mixture to become convex. Usually there is a central peak that remains projected above the main mass after motion of paddle has been stopped. Viscosity is greatly increased, so that the mass tends to adhere to bottom of cup.	Generally >5,000,000
+	Alkaline milk	This symbol should be added to the CMT score whenever the reaction is distinctly alkaline, as indicated by a contrasting, deeper purple color.	An alkaline reaction reflects depression of secretory activity, occurring as a result either of inflammation or of drying-off of the gland.
Y	Acid milk	Bromcresol purple is distinctly yellow at pH 5.2. This symbol should be added to score when mixture is yellow.	Distinctly acid milk in the udder is rare. When encountered, it indicates fermentation of lactose by bacteria.

d. Because of the critical time factor in reading the CMT, use of paddles or other devices containing more than four cups is not recommended. It is not possible to make accurate readings on more than four cups within 5 sec.
e. It is important to use the approved CMT reagent to be sure it is a standardized product.
f. The light must be sufficient to permit clear observation of reactions.

B. *Wisconsin mastitis test (Class O):*

The principle of measuring viscosity after adding detergent to raw milk samples was first used in the CMT. Subsequently, methods for quantitatively measuring the viscosity of a detergent-milk mixture were developed.[13,32,39] One of these, the Wisconsin Mastitis Test (WMT),[39] is widely used in the United States. The WMT is rapid, simple and inexpensive. It provides a more objective means of grading the viscous material obtained. The reaction will diminish significantly in samples that have not been continuously refrigerated at 0 to 4.4°C or that have not been tested within 36 h after collection.[29]

1. Equipment, supplies and reagents:
   a. Test tubes: Clear plastic, flexible, measuring 12.1 × 130 mm ID, with an air vent having an approximate diameter of 3 mm located about 65 mm from the outside bottom of the tube.
   b. Metal caps: With an orifice in the center 1.09 to 1.10 mm in diameter; made of 25-gauge (0.5-mm) brass. (Dr. Z. D. Roundy, 725 South, 600 West, Orem, UT 84057; or Dairy Research Products, Inc. 5345CR68, Spencerville, IN, 46788)
   c. Tube racks: Aluminum, to firmly hold ten plastic WMT tubes.
   d. Automatic syringe: Capacity, 2-mL, for pipetting milk samples [Becton Dickinson (B-D) syringe No. 1250S, or equivalent]. Adjust the syringe to deliver 2 mL of milk and check accuracy by collecting a series of delivered volumes in a graduated cylinder.
   e. Metal syringe holder: B-D No. 1250 MH, or equivalent.
   f. Continuous-pipetting device: 2-mL size, for pipetting reagent, B-D No. 1251, or equivalent. Adjust the syringe to deliver 1.0 mL and check the accuracy by collecting delivered volumes in a graduated cylinder.
   g. Two laboratory cannulas: 14-gauge, 100 mm long, B-D No. 1250 NR, or equivalent. (These may be cut to the desired length.) One cannula is connected to the 2-mL syringe and the other, approximately 82 mm long, is connected to the continuous-pipetting device just described. This 82-mm cannula should extend at least 6 mm below the level of the surface of the milk sample when the pipetting device is resting on the two-way valve.
   h. Timer: With ± 0.1 sec accuracy or better.

12.2 Screening Tests

i. Measuring square: (Roundy or Dairy Research Products op. cit.) Calibrated in 1-mm graduations.
j. Reagent: (Roundy or Dairy Research Products op. cit.) Standardized, commercial WMT reagent concentrate. Prepare a supply of fresh reagent daily diluted with 7 parts of LG water for use in the WMT or as directed by the manufacturer. CAUTION: Old solutions of ready-to-use reagent may contain microbial populations sufficiently high to affect the test.
k. Water bath: To maintain reagent at a temperature which will provide a reagent-milk mixture temperature of 24 ± 2°C.
l. WMT nozzle gauge.

2. Care of samples: Samples must be held continuously at 0 to 4.4°C. Samples should be tested promptly after the first removal from refrigeration. Reliable results will not be obtained on samples that have been warmed above 4.4°C and/or have been cooled slowly in a conventional refrigerator. All samples must be tested within 36 h of collection.

3. Procedure:
   a. Rinse tubes and shake to remove excess water before use (or before calibration). Rinse syringe in water and immerse in well-mixed sample [6.2F]. Dispense 2 mL of well-mixed sample into each tube, running it down the side of the tube to avoid excessive foaming.
   b. Quickly add two 1-mL portions of warmed reagent to each of the tubes in one rack to provide good initial mixing of milk and reagent. Deliver reagent below the surface of the milk.
   c. Promptly cap tubes with caps that had been previously arranged right side up in front of the rack.
   d. Start the mixing action within 30 sec after adding reagent to the first tube in the rack. Mix by holding tubes in nearly horizontal position and tilting them back and forth, permitting the liquid to run forward to the caps 10 times and return. The 10 excursions should be made within 8 to 10 sec. During the forward tilting of the tubes, the most fluid samples should cover about half the cap (180°). During the backward tilt, the butts of the tubes should move down through an arc of about 10 to 15 mm. Avoid vigorous agitation. The temperature of the milk-reagent mixture at the time of inverting must be 24 ± 2°C.

   NOTE 1: The following should be helpful in establishing uniform technique: To guide the tilting movement, place the back of a WMT rack on top of a rod 5 to 6 mm in diameter, or a 1-mL milk pipette that is lying on the laboratory table. The rod acts as a fulcrum and should reduce the tendency to tilt tubes too far backward. Minimize touching the table top with the rack during tilting.

   e. Invert the rack within 30 sec after the mixing action begins. Before inverting, hold the tubes in a horizontal position while waiting for the sweep second hand on the timer to reach a convenient starting point. Then invert the rack rapidly but smoothly and hold it in a vertical position allowing a 15-sec outflow. Return the tubes to an upright position, remove the caps, and let stand for at least 1 min to allow the reagent/milk mixture to drain down before reading.

   f. Use the measuring square to measure the fluid column remaining in each tube. Measure to the top of the meniscus and record readings in millimeters.

     NOTE 2: Clean caps by placing them in a small container of warm water and shaking them in two or three changes of water. After each use, rinse the tubes two or three times in warm tap water (not over 45°C) and shake out excess water before reuse.

4. Precautions:

   a. To calibrate caps, use the special WMT nozzle gauge or equivalent. Correct size of the orifice is indicated when the end of the WMT nozzle gauge wire can be inserted easily into the orifice 1.5 to 2.5 mm.

   b. With continuous use, WMT tubes change shape. Frequent checking of calibration is therefore necessary. WMT tubes have correct inside dimensions when the combined height of the column of 2 mL of milk plus 2 mL of reagent reaches 37 mm. Pipettes must be used to check calibration. Tubes not meeting this standard must be discarded.

C. Rolling ball viscometry (Class B):

The objectivity and speed associated with measurement of viscosity, as achieved in the WMT, has been improved by the Ruakura Rolling Ball Viscometer (RBV).[5,32] This unit was developed in New Zealand and is used to evaluate bulk milk in many nations. The RBV expands the linear scale of the WMT five-fold and allows one sample to follow another in the same measurement tube. The instrument cannot be used on individual cow samples that have extremely high somatic cell counts because the high viscosity may plug the precision-tube. The CMT or conductivity meters would be preferred for individual animal samples. However, blend-milk samples with ranges from 2×10^5 to 3×10^6 somatic cell per milliliter can be estimated at 40 to 80 samples per hour with the RBV screening test. Samples must be held at 0 to 4.4°C[29] and tested within 36 h because the DNA-detergent reaction gradually diminishes with extended storage.[15]

   1. Equipment, supplies and reagents:

12.2 Screening Tests

   a. Viscometer, funnel holder, funnel and tube, spare parts, and instruction manual. (Automation Engineering, 277 Neilson Street, Onehunga, PO Box 12072, Auckland, New Zealand.)
   b. Reagent: Standardized commercial WMT reagent concentrate [12.3B1j]. Prepare the diluted reagent fresh each day the viscometer is used. (Mix one volume of WMT reagent concentrate with 7 volumes of water or as specified by the manufacturer.)
   c. Standard solution: Weigh 28 g of Reagent Grade glycerol into a beaker. Make up to 100 g with MS water. Mix well and store in a stoppered container at ambient temperatures.
2. Installation: Locate and level the instrument on a solid surface to the left of a laboratory sink or similar drain system.
3. Procedure:
   a. Calibration: Pour 15 mL of standard solution into the funnel and press the start button. The solution will push the steel ball toward the end of the tube. Turn the start button in the direction of the arrow to stop solution flow before air enters the measurement tube. In about 20 sec the tube will be rotated counter clockwise and dwell about 3 to 5 sec. The tube will then be returned to the horizontal and the ball will be fixed in position depending upon the viscosity of the solution. The instrument is correctly calibrated if the ball is at 20. Rotate the "calibrate" knob clockwise to increase the reading and counterclockwise to reduce it. Pour another 15 mL of standard solution into the funnel and repeat the reading. Continue to add standard solution and to finely adjust the "calibrate" knob until the instrument reads 20.

   Check calibration every 100 samples, or more frequently if ambient or reagent temperatures change rapidly.

   A standard curve should be prepared comparing the RBV data with those obtained with a confirmation method such as the direct microscopic method [11.16]. Fifty samples with a wide range of somatic cell counts should be run on both methods and a linear regression curve established for possible modification of the RBV data to provide the most accurate estimate of cell numbers.
   b. Measurements: Put 5 mL bulk milk (less than 36 h old and held at 0 to 4.4°C) in a small glass or plastic container. Add 10 mL of reagent using a tilting pipettor or similar device. Do not cause foaming and *DO NOT MIX!* Immediately pour sample and reagent into viscometer funnel. Depress the start switch to allow fluid to enter the viscometer tube and displace the ball to the end of scale. Turn the start switch to stop fluid flow. No bubbles should appear in the tube. The next milk sample can be poured to another container, but *DO NOT ADD REAGENT YET!* Take the instrument reading when the tube returns to the horizontal

(approx. 24 sec after pushing start switch). Multiply the reading by 10^4 to obtain the somatic cell count. If sample viscosity is high (reading over 100), flush viscometer with water before introducing the next sample. Add reagent to the next sample and pour it immediately into the funnel.

   c. Instrument cleaning and storage: Turn off the instrument and depress the start switch. This switch should remain down (and not turned) while the instrument is not in operation because this keeps the valve open. Rinse the viscometer with WMT reagent and water. Leave a 1% iodophore solution in viscometer tubing during extended storage.

D. *Modified whiteside test (Class D):*

The amount and opacity of the precipitate that is formed when milk and sodium hydroxide are mixed together in the prescribed manner, roughly parallels the somatic cell count of the milk. The Modified Whiteside Test (MWT) may be applied both to quarter and to blended herd milk.[13] Assignment of a grade to the test reaction is subjective. A picture guide is available for this grading. Equipment and supplies are easily obtainable and the method is quick, simple and inexpensive.

  1. Equipment and supplies:
    a. Smooth, dark surface such as black bakelite sheets or window glass (not highly polished) enameled on one side; areas about 4 cm² may be marked by lines etched on the surface.
    b. Sodium hydroxide (NaOH), 4%: Weigh out 4 g of NaOH pellets (do not use pellets that have fused together from moisture) and make up to 110 mL with freshly boiled and cooled LG water; store in flint glass or suitable screw-capped polyethylene or polypropylene bottles. Discard the solution if it becomes cloudy.
    c. Medicine droppers: Of good quality, with uniform bore size. Use separate droppers for milk and for NaOH.
    d. Applicator sticks: 15 cm long, for mixing milk and NaOH.
    e. Sample containers: sterile as described in 4.1A3.
    f. Picture guide for scoring test reactions: (C. Hadley Smith, Carey Building, Ithaca, NY 14850) For scoring test reactions.
  2. Procedure: All samples must be stored at 0 to 4.4°C from the time of collection until examination. Testing must be completed within 36 h after collection.
    a. Place 5 drops of cold, properly agitated milk [6.2F] on one spot on the test plate. Rinse the dropper in cold water.
    b. Add 2 drops of 4% NaOH.
    c. In 20 to 24 sec, using an applicator stick held at a low angle, beat the mixture, and then, with a circular motion, spread it over an area of no more than 4 cm² of the test plate.

12.2 Screening Tests

    d. Score the reaction. The light source should be sufficient to permit clear observation of the reaction.

    e. Complete one test before starting the next. For each succeeding sample, rinse the dropper to remove any traces of the previous sample or any moisture. The same dropper may be used throughout.

3. Grading reactions: Negative (−). The mixture is opaque and milky in appearance, and entirely free from precipitate. The somatic cell count is usually under 0.5 million per milliliter.

    Trace (T). The mixture is still opaque and milky, but fine particles of coagulated material are present. These may be few or many and are easily overlooked unless the test is examined closely in good light. The somatic cell count is usually between 0.5 and 1.0 million per milliliter but a few counts may be slightly higher.

    Positive (1+). The background is less opaque but still somewhat milky. Slightly larger particles of coagulated material are present and quite thickly scattered throughout the area. Each of these reactions is easily discernible when compared to a trace reaction, although only very slight clumping will be observed. The somatic cell count is usually between 1 and 2 million per milliliter.

    Positive (2+). The background is slightly watery in appearance, with definite clumps of coagulated material that are easily differentiated from those of the 1+ reaction. Fine strings or threads may form instead of clumps, if stirring has been very rapid. The somatic cell count is usually 1.5 to 2 million per milliliter, although some counts may be higher.

    Positive (3+). The background is definitely watery, with larger clumps of coagulated material than those observed in the 2+ reactions. The somatic cell count is usually in excess of 3 million per milliliter.

    Milk having a high somatic cell count that produces strong reactions, usually thickens shortly after stirring is started and has a tendency to follow the applicator stick in a somewhat gummy mass. As stirring is continued, it breaks up into what is typical for a 2+ or 3+ reaction.

4. Precautions:

    a. Milk must be thoroughly agitated before sampling and testing. No cream plug or sedimentation should occur in individual samples.

    b. The temperature of each sample must be in the range of 0 to 4.4°C, and the storage period between sampling and testing must not exceed 36 h.

    c. Good lighting [6.2B1] is essential when reading the test.

5. Recording and reporting results: Record results in reaction categories (i.e., negative, trace, 1+, 2+, or 3+). For guidance, refer to the grading chart. Do not report numbers of somatic cells in place of test reaction categories.

E. *Conductivity measurement (Class C):*

Inflamed mammary glands produce milk with a reduced amount of lactose.[31] The osmolality of milk must remain comparable to that of blood, thus the mammary membranes allow transfer of salts from the blood to maintain osmotic equilibrium.[3,31] Because lactose is nonconductive and serum salts conduct electrical current, the conductivity of milk increases as it becomes abnormal.[6]

Micro-electronics has allowed the production of small, accurate conductivity meters for cowside measurement of milk conductivity.[19,42] Readings from these instruments correlate with lactose content, somatic cell count, and protein content of milk.[19,37] Because high conductivity readings do not always correlate with the presence of primary pathogens, the technique is for screening only. Recent instrumentation has provided expanded readout scales to allow evaluation of milks from both high- and low-fat breeds.[19] Thus greater precision and objectivity is available at the cowside than in some other screening methods.

1. Equipment and reagents (Wescor Inc. 459 So. Main St., Logan, UT 84321):
    a. Hand-held, portable conductivity meter with approximate readout range of from 5,000 to 12,000 microsiemens (ms) and the scale divided into 10 subranges[19] to allow coverage of high- and low-fat breeds of cows.
    b. Appropriate 9v battery power source.
    c. Standardized salt solution (123 uM NaCl) for checking instrument operation.
2. Procedure: Check instrument at least weekly by pouring the standardizing NaCl solution (at 25°C) into the funnel. The light emitting diode (LED) reading should be 5. If this reading is not obtained, the unit should be recleaned or checked for leaks. Replace the battery when the low-battery indicator shows. If the unit fails to readout the proper values, return it to the manufacturer for repair.

    Manually inject a stream of foremilk from one teat of a cleaned and dried cow's udder into the funnel of the conductivity meter. Push the button and observe the reading (0 to 9) on the LED scale. Immediately follow this with sequential foremilk streams from each of the other three teats. Absolute readings of 5 or above in either low- or high-fat breeds indicate abnormal conductivity. Care in interpretation is needed because some individual animals may produce consistently high readings in all quarters, although appearing normal in other respects. Herd mean conductivity must also be considered during interpretation. Thus this screening test must be followed by confirmatory methods to establish significance of the readings obtained.

After use, the instrument should be rinsed with milk detergent solution and water.

12.3 Confirmatory Tests

Confirmatory tests for somatic cell count may be made using the direct microscopic somatic cell count (DMSCC) [11.18], the Coulter Counter electronic method [12.3A], and the automated Foss fluorescent dye method [12.3B].

Methods to detect somatic cells in goat milk must differentiate between nucleated and cytoplasmic particles. The Pyronin Y stain and method is appropriate in this respect, and is the confirmatory test for somatic cells in goat milk.

A. Direct microscopic somatic cell count (DMSCC) Class O:

All details relevant to the use of the microscope in determining somatic cell count of cow and/or goat milk are given in Chapter 11.

B. Electronic somatic cell counting procedure (Class A1):

Somatic cells in milk have been counted electronically by several procedures that employ either a centrifuge to remove the fat globules[2,13,28] or a chemical dispersion of fat.[2,4,11,12,18,21,22,23,25,41] The chemical method is currently more popular. In the United Kingdom, the procedure is standaridzed by the Milk Marketing Board, and in Europe, by the International Dairy Federation.[12]

A high degree of correlation between results of the chemical procedure and microscope counts has been confirmed.[23,41] Cell size distribution data obtainable from the electronic device offers additional important information.[18] Good correlation has been noted between results of the centrifuge and chemical procedures.[22,24] Results of the DMSCC and electronic counts were compared,[14] and these procedures have also been compared with the WMT.[40] In a collaborative study, the electronic chemical procedure yielded precise results.[7] Results of the automated and semi-automatic Coulter Counters agree well with each other and with those of the DMSCC.[1]

The semi-automated method uses the Electronic Somatic Cell Counter (ESCC). The automated method uses the Milk Cell Counter (MCC). Both pieces of equipment function on the same principle of operation and, with one exception, the same methodology. The major difference is in their degrees of automation. Both instruments require calibration and standardization.

1. Equipment and reagents: The following equipment is required for the semi-automated ESCC method. The same reagents and essentially the same methodology serve the automated milk cell counter (MCC) procedure [12.4B].

a. Coulter Counter: (Coulter Electronics, Inc., 590 W. 20th St., Hialeah, FL 33010) (Model ZB, ZBI, ZF, FN, ZM or equivalent) fitted with 100-μm aperture tube and 100-μL manometer.
b. Automatic diluter: 0.1-mL sampling and 9.9-mL dispensing capacity and tolerant to ethanol.
c. Water bath(s): Either two water baths maintained at 80 and 60°C (± 1°C) or one quick-recovery bath capable of meeting both temperature requirements.
d. Glass medicine vials: (Demuth Glass Works, Div. of Brockway Glass Co., Inc., Parkersburg, WV 26101) 7-dram, 66 mm height, 25 mm outside diameter, thin-glass, snap-cap.
e. Accuvettes: (Coulter Electronics, Inc., op. cit.) 20-mL capped plastic vials.
f. Timer: 10 min.
g. Vortexing mixer.
h. Racks, two: To accept medicine vials and accuvettes, 24-sample capacity.
i. Pipettes: 10-mL, wide-tipped (for dispensing milk sample).
j. Cell-fixing solution: Filter 10% formalin (prepare from 40% w/v formaldehyde) in MS water. Twenty milligrams of eosine may be added to 100 mL of the diluted formalin to provide a visible indication of fixing. An alternative solution is Somafix. (Coulter Electronics, Inc., op. cit.) Note: The pH and concentration of formaldehyde in formalin will change with time.
k. Fat-dissolving electrolyte diluent solution.
   84.5 volumes of 0.9% sodium chloride/water.
   12.5 volumes of ethanol.
   2.0 volumes of Triton X-100. (Rohm & Haas, Philadelphia, PA 19104)
   1.0 volume of formalin (40% w/v formaldehyde).

   Buffer this solution with 1.0 molar tris-(hydroxymethyl) amino methane primary standard to pH 7.0. Filter this solution.

   An alternative solution is Somaton. (Coulter Electronics, Inc., op. cit.)
l. Calibration spheres: Use 3.4-μm diameter monosized latex (Coulter Electronics, Inc., op. cit.) for ESCC equipment and 5.0-μm diameter monosized latex (Coulter Electronics, Inc., op. cit.) for the MCC equipment.
m. The above equipment and reagents are all necessary for making analyses by the semi-automated (ESCC) method. The automated (MCC) procedure utilizes the same reagents, but does not require separate diluter, tempering bath, timer, or mixer. These mechanisms are built into the original equipment.

12.3 Confirmatory Tests

2. Procedure: Paragraphs [a] and [b] include steps needed to analyze for somatic cell count by both the semi-automated (ESCC) and automated (MCC) method, but all of the steps undertaken in [b] are automated in the MCC equipment. Paragraph [c] further explains the latter equipment.
   a. Fixing milk samples:
      1. Hold all milk samples at 0 to 4.4°C before fixing.
      2. Put three drops of Somafix into each medicine vial.
      3. Before sub-sampling, invert milk samples 25 times.
      4. Forcibly expel from a 5-mL syringe, two 5-mL portions (10 mL total) of each milk sample into its medicine vial. Good mixing is needed to prevent coagulation of milk by Somafix.
      5. Repeat steps 2,3,4 until *all samples* have been thus treated.
      6. Rack sub-samples and heat in water bath for 6.5 min at 60 ± 1°C (1.5-min heat-up time and 5-min hold time).
      7. Remove rack and cool samples in ice water for 2 min and then remove to room temperature or refrigerate for later counting.
   b. Diluting and counting fixed samples:
      1. Calibrate the Coulter Counter according to the following instructions and set the lower threshold at 4.4-μm cell diameter (44.6 μm^3). Using an Accuvette and Somaton solution, do five or six blank counts on the Coulter Counter. The average of these counts should be below 20. If it is not, reclean the glassware used, refilter the Somaton solution, or check electrical interference. Fixed samples must be at room temperature for diluting.
      2. Mix each fixed sample (consecutively, 3 to 5 sec on a vortexing mixer).
      3. Invert each sample 2 to 3 times just before dilution.
      4. Dilute the fixed samples as follows:
         a. Draw sample from the medicine vial into automatic diluter.
         b. As medicine vial is taken away from diluter pipette, touch off (with the edge of the vial) the milk drop remaining on the diluter pipette.
         c. Expel milk and diluent from diluter into an Accuvette, being careful to prevent foaming. This can be done by directing flow down the side of the Accuvette at an appropriate angle, while keeping the pipette tip from making contact with the Accuvette and never allowing the tip to project under the surface of the solution. Do not mix in Accuvette.
      5. Cap Accuvette and place in rack. Dilute and rack all samples before going to step 6. Also insert a capped Somaton blank

and carry it through the remaining steps to serve as background count.
6. Place rack of capped Accuvettes in water bath at 80°C for 10 minutes. Be certain that the top of the mixture in the Accuvette is 1.8 cm below the water surface.
7. Cool samples in ice water for 3 min and remove to room temperature. This step is not required in the MCC equipment.
8. Invert each sample four times, allow bubbles to disperse (about 10 sec), and count on the Coulter Counter. At this stage all samples should be analyzed within 1 h. The lower threshold of the Coulter Counter should be set at 4.4 μm cell diameter (44.6 μm^3), and the manometer in the ESCC equipment should be set for 100 μL (0.1 mL) of sample volume. The count readout minus blank counts multiplied by 1000 will be the cell count per mL of milk sample. Round off counts to two significant figures.

c. The basic steps in the automated MCC analysis are essentially the same as for the ESCC procedure. Samples are fixed initially as described in [12.3B4a], and then placed in the MCC rack. Milk samples are automatically stirred, an aliquot is automatically removed, diluted to approximately 1:100 with Somaton, dispensed into a reaction tube, and passed through the MCC equipment. During this time, the diluted samples are heated to 80°C in a polyethylene glycol bath for 10 min. The MCC equipment counts immediately without cooling.[8]

3. Calibration:
   a. Initial calibration will establish the relationship between the threshold setting on the Coulter Counter and spherical particles of known size; 3.40 μm in diameter for the ESCC equipment, 5.0 μm for the MCC equipment. The Coulter threshold is linear, with particle size values in cubic micrometers.
   b. Make a dilute suspension of calibration particles in an Accuvette with Somaton. Place the Accuvette in the counting position of the Coulter Counter and activate the counting cycle. A pulse pattern will appear on the Coulter oscilloscope. Pulses generated from monosized spheres will have uniform amplitudes. Adjust these pulse heights to between ⅓ and ½ the height of the oscilloscope screen. Adjust the lower threshold shadow line to coincide with these pulse peaks. (Inactivate the upper threshold.) Note the lower threshold value. Repeat the lower threshold adjustment three times and average the values. Rotate the lower threshold clockwise until it is set to a value that is ½ of the average value just determined. Take four or five counts and

average them. They should be between 1000 and 4000. If not, dilute or add calibration particles as required.

Adjust the lower threshold counter clockwise to a value of 1.5 times the averaged volume just determined. Take 4 or 5 counts and average them. Add the averages from the ½ and 1½ positions and divide by two. This is the ½ count. Return the lower threshold to the average value and take several counts. These should be within 7% of the ½ count just determined. If they are not, adjust the lower threshold slightly until they are. Record this lower threshold value. Using this value and the cubic micrometer value of the calibration spheres, compute the appropriate threshold, aperture current, and amplification settings for counting at 4.4-μm diameter (44.6 μm^3). Use the computation in the Coulter Instruction Manual. A change of aperture tubes requires a recalibration. Routine maintenance of calibration is performed according to [12.3B3].

4. Precautions:
   a. Before fixing, samples must be stored in a refrigerator or ice bath.
   b. Threshold calibration must be done with stable, near-monosized spheres of known size. They should be dispersed with an ultrasonic bath before use.
   c. Samples should be at room temperature when diluted.
   d. Containers, Accuvettes, test tubes, and other glassware must be clean and nearly particle-free at all stages. Normal cleaning procedures (manual or automated washers) will suffice. Rinse water however, must be essentially free of particles to yield the minimum background count required. Similar cleanliness is required for the aperture tube and the outside electrode. Well-filtered diluent solution [5.3] may be used for flushing this part of the equipment.

   In the shutting down procedure for the MCC, run the equipment on 20 samples each of dairy acid cleaners followed by 30 samples of hot MS water.
   e. Blank samples of diluent solution should show a total background count of less than 20 per 0.1 mL.
   f. The outside electrode, which extends into the sample, should be kept clean (use nitric acid) to prevent gas bubbles from collecting on the electrode surface.
   g. Check the actual volumes of the manometer and diluter used.

C. *Flourescent dye: automated,*[9,16,35] *Foss (Class A1):*
   1. Principle: A preheated milk sample is mixed with preheated buffer and dye. Part of the mixture is transferred to a rotating disc, which serves as the object plane for the microscope. The mixture on the

disc is illuminated by a xenon-arc lamp that excites the somatic cell nuclear-dye complex to fluoresce. Each cell produces an electrical pulse that is counted, and the total cell content is printed out.
2. Apparatus: The Fossamatic 215 (Dickey-john Corp., Box 10, Auburn, IL 62615) consists of 3 functional sections:
   a. A sample preparation section consists of external stirrer, pipette, syringes for buffer and dye, rotary table with 6 glasses, heating coils, internal stirrer, filter, measuring syringe, valves, discharge jet, rotating disc and containers for rinsing liquid, buffer dye and waste.
   b. An optical system consists of a xenon-arc lamp, fluorescent microscope, excitation, and emission filters.
   c. The electronics section consists of photomultiplier, amplifier, oscilloscope, control and alarm functions, and printer with display.
3. Reagents:
   a. Stock solutions:
      1. Dye solution, 0.1% ethidium bromide ($C_{21}H_{20}BrN_3$): 1.00 g of ethidium bromide is diluted in 1 L of distilled water. Heating to 40 to 50°C helps dissolve the powder. Store in a light-proof and air-tight bottle no longer than 60 d.
      2. Triton X-100 (1%): 10 mL of Triton X-100 is dissolved in 1 L of distilled water. Heating to 60°C may be necessary. This solution can be stored in an air-tight container for a maximum of 25 d.
      3. Buffer, 0.025 M potassium acidphthalate ($C_8H_5KO_4$): 51.0 g of potassium acidphthalate plus 13.75 g of potassium hydroxide (KOH) are dissolved in 10 L of LG water. Heating to 50°C helps dissolve the powder. Add 10 mL of 1% Triton X-100 solution. Store no longer than 7 da.
   b. Working Solutions:
      1. Dye solution: 20 mL of 0.1% ethidium bromide solution per liter of buffer.
      2. Diluting solution: Use the buffer solution.
      3. Rinsing liquid: Add 10 mL of 1% Triton X-100 and 25 mL of 25% ammonia solution to 10 L of LG water.
      4. Procedure: Turn on the instrument and allow it to cycle six times without any samples. Check instrument performance and fluid transfer systems.
         Heat samples to 40°C (approx. 5 min for a 10 to 15 mL sample) in a water bath and invert them once. Remove sample bottle caps and place bottles on the automatic feeding rack or transfer the samples to containers of the type required for the

12.3 Confirmatory Tests

instrument rack. Stir samples with the external stirrer while they are still on the rack.

Press the "START" button and the instrument will start its automatic procedure, which will continue until the last sample has been measured and the result printed. Complete systems operational check as suggested in operational manual and read a set of standard solutions before testing the first set of unknowns. Repeat procedure for a duplicate set of samples.

Maintain routine calibration as specified in the operations manual [12.3D].

Contributions to the count by materials other than cells (e.g., bacterial clusters in which partly degenerated cells are found) are of no significance. Individual and intact bacteria will not be dyed and consequently will not contribute to the count. Dead and partly degenerated bacteria will react with the dye if the DNA molecule is intact, but the signals from these are small and are absorbed into the background noise without being counted. Therefore, any inaccuracies in the count are most likely due to:

a. inadequate mixing of the sample, leaving somatic cells unevenly distributed,
b. errors in pipetting,
c. loss of ability to absorb dye. (As cells age or are exposed to formalin, they lose their ability to absorb dye.) and/or
d. degeneration of the somatic cells. This is a primary cause of error. Cell degeneration causes include: bacterial growth, high storage temperatures, excessive agitation, and freezing.

5. Semiautomatic model:[16a] The Fossomatic 90 has been compared to the Fossomatic 215, found to be of comparable precision, approved by the Technical Committee, and classified as A2.
   a. Working solutions:
      Use the stock solutions and working solutions described for the Fossomatic 215 [12.C3] except for the dye solution. Add 26 mL dye stock solution to 2.5 L of buffer.
   b. Procedure: Heat 60 mL sample to 40°C and invert about 10 times. Pipette to 500 mL into intake chamber. Operate instrument as described in the operations manual.
   c. Standardization: Obtain 3 milk samples with a somatic cell count range of between 300,000 and 2,000,000 cells/mL. Subdivide the samples to avoid excessive heating. Determine the cell counts using a reference method (DMSCC). Compare the samples over ⩾ discriminator settings having

increments of 0.25 to 0.5 between settings. Choose initial setting near previous operating point and additional settings to provide ⩾ set of readings above and ⩾ below apparent optimum. Optimum is setting at which deviations of Fos-somatic readings from those of the reference samples are minimal and 1 of opposite sign from rest. Check instrument every 180 samples or after each 2 h of operation against preserved reference samples.

D. Control samples:

Control samples should be used to check calibration of the Coulter and Foss confirmatory procedures. To ensure appropriate adjustment, a control sample should be applied to the given method at least once during each hour of operation. If the method permits use of a preserved sample, a set of controls can be prepared and used over a period of 7 to 10 d or possibly longer, depending upon the quality of the milk and storage conditions. Control samples stored at 4°C have kept for 23 wk.[13a] Otherwise use fresh samples, which will maintain appropriate quality for only a few days because of cell deterioration.

Control samples can be obtained from a commercial source (Pennsylvania State University, 213 Borland Laboratory, University Park, PA 16802) or prepared by the individual laboratory as follows:

1. Use data taken from one day's analyses to select a number of samples to be pooled into the following ranges of somatic cell counts: approximately 200,000 to 300,000; 500,000 to 600,000; and 1.0 million to 1.5 million. Samples that have been previously heated should not be used as controls.
2. Commingle samples of appropriate counts into three lots at the above ranges, each lot ultimately being of such size as to provide controls over a 1-wk period.
3. Sub-divide each lot into 60-mL samples, and to each sample thus obtained add one tablet of commercial potassium dichromate preservative. (NASCO, 901 Janesville Ave., Fort Atkinson, WI 53538) The 60-mL samples can be further sub-divided daily into amounts and vessels appropriate to the given method.
4. Analyze each range of samples in duplicate by the direct microscopic somatic cell count, and average the results.

12.4 References

1. BREMEL, R.D.; SCHULTZ, L.H.; GABLER, F.R.; PETERS, J. Estimating somatic cells in milk samples by the membrane-filter—DNA procedure. J. Food Prot. **40**:32–38; 1977.
2. CULLEN, G.A. The use of electronic counters for determining the number of cells in milk. Vet. Rec. **77**:858; 1965.
3. DAVIS, J.G. The detection of sub-clinical mastitis by electrical conductivity measurements. Dairy Ind. Int. **40**:286; 1975.

12.4 References

4. DIJKMAN, A.J.; SCHIPPER, C.J.: BOOY, C.J.; POSTHUMUS, G. The estimation of the number of cells in farm milk. Neth. Milk Dairy J. **23**:168–181; 1969.
5. DUIRS, G.F.; COX, N.R. A comparison of viscometer, microscopic and Coulter counting of somatic cells in bulk milk. N.Z. J. Dairy Sci. Technol. **12**:107–112; 1977.
6. FERNANDO, R.S.; RINDSIG, R.B.; SPAHR, S.L. Electrical conductivity of milk for detection of mastitis. J. Dairy Sci. **65**:659–664; 1982.
7. GINN, R.E.; THOMPSON, D.R.; PACKARD, V.S. Collaborative study of the Coulter counter-chemical method for counting somatic cells in raw milk. J. Food Prot. **40**:456–458; 1977.
8. GINN, R.E.; PACKARD, V.S.; THOMPSON, D.R. Comparison of the automated with the semi-automated Coulter Counter method and the direct microscopic somatic cell count (DMSCC) on raw milk samples. J. Food Prot. **42**:567–568; 1979.
9. GRAPPIN, R.; JEUNET, R. Premiers essais de l'appreil "Fossomatic" pour la détermination automatique du nombre de cellules du lait. Le Lait (Nov.-Dec.) 627–644; 1974.
10. HORWITZ, W. Official methods of analysis, 13th ed. Washington, DC: Assoc. Offic. Anal. Chem.; 1980.
11. HAUSLER, JR., W.J., ED. Standard methods for the exmaination of dairy products, 13th ed. pp 127–132. Washington, DC: Amer. Public Health Assoc.; 1972.
12. INTERN. DAIRY FEDERATION. Electronic counting of somatic cells in milk: A recommended procedure for milk sample preparation and cell counting with a Coulter Counter. Annu. Bull. Part II:23–31. Brussels, Belgium: Intern. Dairy Federation; 1971.
13. JAARTSVELD, F.H.J. Contribution to the diagnosis of mastitis in cattle in connection with mastitis control. Neth. Milk Dairy J. **16**:260–264; 1962.
14. MACAULAY, D.W.; GINN, R.E.; PACKARD, V.S. Experience in adaptation and comparative evaluation of the Coulter Counter and direct microscopic somatic cell count (DMSCC) method for determining somatic cell counts in milk. J. Milk Food Technol. **39**:250–252; 1976.
15. MARSHALL, R.T.; BRECHBUHLER, J.D. Change in WMT score with time of milk storage. J. Food Prot. **40**:195; 1977.
16. MOCHRIE, R.D.; MONROE, R.J. Collaborative study of somatic cell counting—Fossomatic method. Washington, DC: Assoc. Offic. Anal. Chem. Proc. 91st Ann. Meeting; 1977.
16a. MOCHRIE, R.D.; DICKEY D.A. Comparison of Semiautomated method with official optical somatic cell counting method III for determinig somatic cells in milk. J. Assoc. Offic. Anal. Chem. **67**:615–617, 1984.
17. NATIONAL MASTITIS COUNCIL, WRITING COMMITTEE. Current concepts of bovine mastitis. Washington, DC: Nat. Mastitis Council; 1965.
18. NEWBOULD, F.H.S. A technique for differential somatic cell counts in milk. Brussels, Belgium: Int. Dairy Federation Document 85. Proc. Seminar on mastitis control. p. 136–141; 1975.
19. OKIGBO, L.M.; SHELAIH, M.A.; RICHARDSON, G.H.; ERNSTROM, C.A.; BROWN, R.J.; TIPPETTS, E.L. Portable conductivity meter for detecting abnormal milk. J. Dairy Sci. **67**:1510–1516; 1984.
20. PEAKER, M. The electrical conductivity of milk for the detection of subclinical mastitis in cows: Comparison of various methods of handling conductivity data with the use of cell counts and bacteriological examination. British Vet. J. **134**:308–314; 1978.
21. PEARSON, J.K.L.; WRIGHT, C.L.; GREER, D.O.; PHIPPS, L.W.; BOOTH, J.M. Electronic counting of somatic cells in milk. A recommended procedure. U.K. Government, Belfast, Northern Ireland: Northern Ireland Min. Agr.; 1970.
22. PEARSON, J.K.L.; WRIGHT, C.L.; GREER, D.O. A study of methods for estimating the cell content of bulk milk. J. Dairy Res. **37**:467–480; 1970.
23. PHILPOT, W.N.; PANKEY, JR., J.W. Comparison of four methods for enumerating somatic cells in milk with an electronic counter. J. Milk Food Technol. **36**:94–100; 1973.

24. PHIPPS, L.W.; NEWBOULD, F.H.S. Isolation and electronic counting of leucocytes in cow's milk. Vet. Res. **77**:1377–1379; 1965.
25. PHIPPS, L.W.; NEWBOULD, F.H.S. Determinations of leucocyte concentrations in cow's milk with a Coulter counter. J. Dairy Res. **33**:51; 1966.
26. POSTLE, D.S.; BLOBEL, H. Studies of bulk milk screening procedures for mastitis. Amer. J. Vet. Res. **26**:90–96; 1965.
27. READ, JR., R.B.; BRADSHAW, J.G.; PEELER, J.T. Collaborative study of confirmatory testing procedures for somatic cells in milk. J. Milk Food Technol. **34**:285–288; 1971.
28. READ, JR., R.B.; REYES, A.L.; BRADSHAW, J.G.; PEELER, J.T. Electronic counting of somatic cells in milk. J. Dairy Sci. **50**:669–674; 1967.
29. READ, JR., R.B.; REYES, A.L.; BRADSHAW, J.G.; PEELER, J.T. Evaluation of seven procedures for detection of abnormal milk due to mastitis. J. Dairy Sci. **52**:1359–1367; 1969.
30. REGENTS. U. OF CALIFORNIA. Assignee California mastitis reagent. U.S. Patent 21,998,392. 1956.
31. RENNER, E. Possibility of determination of udder diseases with the infra-red milk analyzer (IRMA). Deutsche Molkeri-Zeitung (Kempten-Allgau) F.**3**:71; 1972.
32. RICHARDSON, G.H.; BROWN, R.J.; CASE, R.; GINN, R.E.; KASANDJIEFF, T.; NORTON, R.C. Collaborative evaluation of Rolling Ball Viscometer for measuring somatic cells in abnormal milk. J. Assoc. Offic. Anal. Chem. **65**:611–615; 1982.
33. SCHALM, O.W.; NOORLANDER, D.O. The California mastitis test. California Vet. (October):Unpaginated; 1956.
34. SCHALM, O.W.; NOORLANDER, D.O. Experiments and observations leading to development of the California mastitis test. J. Amer. Vet. Med. Assoc. **130**:199–207; 1957.
35. SCHMIDT MADSEN, P. Flouro-opto-electronic cell-counting on milk. J. Dairy Res. **42**:227–239; 1975.
36. SCHULTZE W.D.; SMITH, J.W. The cellular content of cows milk. II. Comparison of the CMT and microscopic count for estimating cell concentrations in quarter samples. J. Milk Food Technol. **29**:126–129; 1966.
37. SHELDRAKE, R.F.; MCGREGOR, G.D.; HOARE, R.J.T. Somatic cell count, electrical conductivity, and serum albumin concentration for detecting bovine mastitis. J. Dairy Sci. **66**:548–555; 1983.
38. SMITH, J.W.; SCHULTZE, W.D. The cellular content of cows milk. I. An evaluation of the California mastitis test as a method of estimating the number of cells in milk. J. Milk Food Technol. **29**:84–87; 1966.
39. THOMPSON, D.R.; PACKARD, V.S.; GINN, R.E. Electronic somatic cell count—chemical method, DMSCC, and WMT tests for raw milk. J. Milk Food Technol. **39**:854–858; 1976.
40. THOMPSON, D.I.; POSTLE, D.S. The Wisconsin mastitis test—an indirect estimation of leucocytes in milk. J. Milk Food Technol. **27**:271–275; 1964.
41. TOLLE, A.; ZEIDLER, H.; HEESCHEN, W. Ein Verfahren zur elektronischen Zahlung von Milchzellen. Milchwissenschaft **21**:93–98; 1966.
42. TWOMEY, A.; DUIRS, G.F.; HAGGIE, D. Final report on conductivity type mastitis detector. Hamilton, New Zealand: AHI Plastic Moulding Co.; 1979.
43. WALTER, W.G., ED. Standard methods for the examination of dairy products, 12th ed. pp. 287–291 and Chapter 12. New York, NY: Amer. Pub. Health Assoc.; 1967.

CHAPTER 13

REDUCTION METHODS

J.E. Edmondson, R. Golden, and D.B. Wedle
(R.T. Marshall, Tech. Comm.)

13.1 Introduction

Dye reduction tests indirectly estimate numbers of bacteria in milk in terms of the time required, after starting incubation, for the color of a dye-milk mixture to change. Reduction time is generally inversely proportional to the bacterial content of the sample at the start of incubation.[1,10,11,12,20,21] Different reduction times permit rapid groups of samples into classes or grades.[4]

The method depends on the ability of microorganisms to transfer electrons from constituents of milk to oxygen and to reducible dyes.[7] Methylene blue and resazurin are the dyes used in currently accepted methods.

Methylene blue thiocyanate[19] (1:250,000 concentration) produces a blue color in the dye-milk mixture. When fully reduced the dye is colorless, and the mixture has the white appearance of milk. The reaction is reversible. Resazurin (1:200,000 concentration) has a high tinctorial value, imparting a blue color. As it is reduced, resazurin passes through a spectrum of colors; first, from blue through violet to pink as resazurin is irreversibly reduced to resorfurin; second, from pink to colorless as resorfurin is reversibly reduced to dehydroresorfurin.[16] Reduction of resazurin from blue to pink is less dependent on oxygen content and Eh (oxidation-reduction potential) than is reduction from pink to white.[7,15]

Dye reduction tests are favored over more precise methods because they are easy to perform and economical. They are unsuited, however, for analyses of milk having low numbers of bacteria. In addition, several factors can cause reduction times to fail to compare well with numerical estimates of microbial numbers in samples. These include:

A. Growth of mesophilic bacteria is favored over growth of psychrotrophic bacteria in dye reduction tests by the relatively high incubation temperature (36 ± 1°C).

B. Thermoduric and psychrotrophic bacteria reduce substances more slowly than do mesophiles.[8,9,15]

C. Some bacteria that reduce dyes fail to produce visible colonies on plates at the temperature or in the medium used.

D. Variations occur in the proportions of the microflora carried into the cream layer by rising fat globules.[13] (To lessen the variance caused by this factor, tubes are inverted to redistribute microorganisms and fat globules whenever multiple readings of samples are made.)

E. Reducing activities of bacterial cells are not diminished by clumping, whereas plate and microscopic counts are reduced by clumping.

F. Inhibitors such as antibiotics affect tests.[13,17]

G. Large numbers of somatic cells in milk tend to shorten reduction times.

H. Incorporation of oxygen into samples prolongs reduction time; therefore, caution should be exercised in differentiating milks reduced in less than 1 h.[18]

13.2 Standards

U.S. Department of Agriculture Recommended Requirements for Milk for Manufacturing Purposes[21] contains standards for grading manufacturing milk by the resazurin reduction test. The sanitary codes of two dairy industry associations[2,6] provide standards for the use of either the methylene blue or the resazurin reduction test. Reduction times permitted for acceptable milk are approximately one-half as long for the resazurin as for the methylene blue version of the test.[14]

13.3 Sampling

Provide equipment and supplies and collect samples as described in [4.1].

13.4 Equipment and Supplies

A. Dipper or pipette: To deliver 10 mL.

B. Buret or pipette: With 1-mL graduations.

C. Tubes, vials or suitable nontoxic plastic bags: See [4.1A3].

D. Tube closures: Rubber or other suitable stoppers for individual tubes that will allow inversion of tubes in racks without loss of contents and without contamination of adjacent tubes.

E. Incubator for samples: Water bath of sufficient size designed, 1) with thermostatically controlled heat source to maintain samples at 36 ± 1°C; 2) with volume of water sufficient to bring contents to at least 36°C within 10 min after placing samples therein; 3) to maintain water level slightly above level of contents in tubes; and 4) to protect samples from action of light during incubation.

F. Sheet metal or wire racks [4.1A4].

G. Cooling bath: To hold samples at 0 to 4.4°C until tested.

H. Reagent grade resazurin (11.0 mg) or reagent grade methylene blue thiocyanate (8.8 mg).

I. Light-resistant flask, stoppered: Amber or low-actinic glass bottle (zero transmission at wavelengths less than 380 nm) with screw-cap closure, or other container offering similar protection, capacity 250 mL.

J. Interval timer.

K. Thermometer: [2.3O].

L. Light source and filter [17.4A7]; essential for color comparisons in the resazurin test.

M. Color standards: For triple reading resazurin test, use a single color standard 5 P 7/4 (purple) (Munsel Color Company, Inc., 2441 North Calvert St., Baltimore, MD 21218). For the 1-h resazurin test, use a color grader consisting of four color standards: 5 PB 7/4 (blue); 10 PB 7/5.5 (bluish purple); 5 P 7/4 (purple); and 10 P 7/8 (pinkish purple). To prevent fading, minimize exposure of color standards to strong lighting conditions.

Precautions: Glassware must be chemically clean. Do not use chlorine or other chemical compounds as bactericidal agents on glassware, dippers or closures. Avoid recontamination of that portion of the sterilized stopper that is inserted into the sample tube. Store and cover tubes in racks. Protect sterilized equipment from dust and unnecessary handling.

13.5 Preparing Dye Solution

Autoclave or boil 200 ± 2 mL of MS water in a light-resistant flask [13.4I]. Adjust volume of heated water by adding MS water. Promptly transfer 11.0 mg of resazurin dye or 8.8 mg of methylene blue dye to the flask of hot water. Swirl flask to completely dissolve the dye before cooling the mixture.

Use solution in proportion of 1 mL to 10 mL of sample. Keep dye solution in sterile, light-resistant glass containers in a cool, dark place. Prepare fresh solutions weekly. Mark date of preparation on containers.

13.6 Testing Samples

Label each tube legibly with the producer's name or number, or make a sample identity chart for each rack of tubes.

13.7 Methylene Blue Reduction Method (Class D)

A. With buret or pipette, transfer 1 mL of methylene blue thiocyanate solution to each tube, preferably shortly before (or shortly after) placing 10 mL of sample therein. After adding dye either to empty tubes or to milk in tubes, avoid prolonged exposure of tubes to light, especially direct sunlight.

B. Cover or stopper tubes loosely as soon as the sample has been added and place them in a cooling bath. To monitor the temperature, place one tube containing 11 mL of milk and a thermometer in the rack with other sample tubes.

C. When each rack or batch of samples is complete, either start incubation promptly or transfer rack(s) of tubes or vials from cooling bath to suitable temporary storage at 0 to 4.4°C to await a more convenient time for incubation. When ready to incubate, bring temperature of the test portions to 36°C within 10 min [13.4E]. Check heating rate by means of the thermometer in the monitor tube. Loosen stoppers slightly. To distribute cream when the temperature reaches 36°C, tighten stoppers, slowly invert (do not shake) each rack of tubes three times, and record the time of completion as the beginning of incubation.

D. If the number of samples, or the temperature of tube contents when samples are removed from the cooling bath and placed in the incubating bath, require tempering for longer than 10 min, preheat samples in an auxiliary water bath controlled at a temperature not to exceed 40°C; or place tubes in bath in smaller numbers and at successive periods. When necessary to increase the rate of heat transfer, use more heated water or forced agitation of water, or both.

E. Use an interval timer to remind the analyst when successive examinations should be made.

F. *Reading and recording:*

1. Because of the uneven disappearance of blue color that may be noted as reduction progresses during the test, at each reading, record any sample as decolorized if four-fifths of its column is white.

2. If a sample is decolorized after incubation for 30 min, record methylene blue reduction time for that sample as MBRT 30 min.

3. Following the initial 30-min reading, make all subsequent readings at successive hourly intervals. Record such readings as reduction times in whole hours between completion of the initial triple inversion of tubes and the time when decolorization is noted. For example, if color disappears between 0.5 and 1.5 h, record as MBRT 1 h: similarly, if between 1.5 and 2.5 h, record as MBRT 2 h.

4. Immediately after each reading, remove and record all decolorized samples and then slowly make one complete inversion of the remaining tubes.

13.8 Resazurin Reduction Method (Class D)

Proceed as in [13.7A through D], except substitute resazurin solution for methylene blue thiocyanate solution. When tube contents reach 36°C, slowly invert (do not shake) each rack of tubes three times, start timing, and incubate. Avoid making tests in bright sunlight, which can cause colors to change rapidly.[9]

A. *Reading the test:*

13.8 Resazurin Reduction Method (Class D)

Because color comparisons—whether by daylight or by artificial light—may be unsatisfactory, examine tubes under a fluorescent tubular daylight lamp against a neutral gray background[5] [17.4A7]. Assign examination of tubes only to persons who can make the necessary color comparisons accurately.

B. *Resazurin triple-reading test:*[14]

Adhere to the following procedure during the incubation period:

1. Incubate at 36 ± 1°C for 1 h. Record color at an appropriate time before 1 h has elapsed or at the end of 1 h, in accordance with the grading system used. Compare each tube with single color standard 5 P 7/4 [13.4M].

2. After the initial incubation period of 1 h, invert all non-reduced tubes once. Then continue incubation at 36 ± 1°C for 1 h.

3. During or at the end of the second hour of incubation, make color comparisons at appropriate time stipulated by grading system followed.

4. After the second hour of incubation, invert tubes once and continue incubation for 1 h.

5. Make final color comparison at appropriate time during or at end of third hour, as dictated by grading system used.

C. *Resazurin 1-h test:*

Proceed as in [13.7 A through D]. After incubation at 36 ± 1°C for 1 h, compare each tube with color standards under standard lighting conditions [13.4M]. Assign class designation to samples according to grading system used.

D. *Interpreting and reporting reduction times:*

Application of the grading principle permits 1) classification of samples into two or more major groups, according to appreciable differences in bacterial quality, and 2) use of descriptive grading terms to indicate whether or not the product is "acceptable." When reduction times are shorter than those permitted by regulations or industry standards, appropriate grading terms that clearly indicate lack of compliance should be used. Under no circumstances report results as numbers of bacteria.

The 1-h test is recognized in the sanitary codes of two dairy product associations.[6,2] Three classes of milk are established on the basis of purple, lavender and pink end points after 1 h of incubation. Dual standards—one for milk handled in cans and one for milk handled in farm bulk tanks—should be applied in the 1-h test to provide uniform classification of milk supplies.[3,4]

The triple reading test, because of its longer incubation period, has advantages over the one hour test. Some properly cooled milks containing many dormant bacteria,[11,16] and other milks containing large numbers of somatic cells may show little color change during the first hour of incubation. Such milks may be missed by the 1-h test but often are detected during further incubation.[14,16]

13.9 References

1. ABELE, C.A. The methylene-blue reduction test as a means of estimating the bacterial content of milk, to determine its suitability for pasteurization or as a basis for grading. J. Milk Technol. 8:67–79; 1945.
2. AMERICAN DRY MILK INSTITUTE. Recommended sanitary/quality standards code for the dry milk industry. Bull. 915 Chicago, IL: Amer. Dry Milk Inst.; 1970.
3. DABBAH, R.; MOATS, W.A.; TATINI, S.R.; OLSON, JR., J.C. Evaluation of the resazurin reduction one-hour test for grading milk intended for manufacturing purposes. J. Milk Food Technol. 32:44–48; 1969.
4. DABBAH, R.: TATINI, S.R.; OLSON, JR., J.C. Comparison of methods for grading milk intended for manufacturing purposes. J. Milk Food Technol. 30:71–76; 1967.
5. DAVIS, J.G.; NEWLAND, L.G. The use of the resazurin comparator in artificial light. Dairy Ind. 8:555–563; 1943.
6. EVAPORATED MILK ASSOCIATION. Evaporated milk industry sanitary standards code and interpretations. Washington, DC: Evaporated Milk Assoc.; 1970.
7. FRAYER, J.M. Dye reduction in milk related to Eh, pH and dissolved gases. Bull. 498 Burlington, VT: Vermont Agr. Exper. Sta.; 1942.
8. GARVIE, E.I.; ROWLANDS, A. The role of micro-organisms in dye-reduction and keeping-quality tests. II. The effect of micro-organisms when added to milk in pure and mixed culture. J. Dairy Res. 19:263–274; 1952.
9. GREENE, V.W.; JAMISON, R.M. Influence of bacterial interaction on resazurin reduction times. J. Dairy Sci. 42:1099–1100; 1959.
10. JOHNS, C.K. Place of the methylene blue and resazurin reduction tests in a milk control program. Amer. J. Pub. Health and the Nation's Health. 29:239–247; 1939.
11. JOHNS, C.K. Some aspects of the resazurin test. p 173. New York: NY Assoc. Dairy and Milk Inspectors; 1941.
12. JOHNS, C.K. Relation between reduction times and plate counts of milk before and after pasteurization. J. Milk Food Technol. 17:369–371; 1954.
13. JOHNS, C.K.; DESMARAIS, J.G. The effect of residual penicillin in milk on the dye reduction tests for quality. Canad. J. Agr. Sci. 33:91–97; 1953.
14. JOHNS, C.K.; HOWSON, R.K. A modified resazurin test for the more accurate estimation of milk quality. J. Milk Technol. 3:320–325; 1940.
15. JOHNS, C.K.; HOWSON, R.K. Potentiometric studies with resazurin and methylene blue in milk. J. Dairy Sci. 23:295–302; 1940.
16. RAMSDELL, G.A.; JOHNSON, JR., W.M.T.; EVANS, F.R. Investigation of resazurin as an indicator of the sanitary condition of milk. J. Dairy Sci. 18:705–717; 1935.
17. ROWLANDS, A. The effect of penicillin on the methylene blue reduction time and keeping quality of milk. Proc. Soc. Appl. Bacteriol. 14:7–14; 1951.
18. THORNTON, H.R. The use of the methylene blue reduction test. Canad. J. Pub. Health 24:192–196; 1933.
19. THORNTON, H.R.; SANDIN, R.B. Standardization of the methylene blue reduction test by the use of methylene blue thiocyanate. Amer. J. Pub. Health and the Nation's Health. 25:1114–1117; 1935.
20. THORNTON, H.R.; STRYNADKA, N.J.; WOOD, F.W.; ELLINGER, C. Milk contamination and the methylene blue reduction test. Canad. J. Pub. Health 25:284–294; 1934.
21. U.S. DEPARTMENT OF AGRICULTURE, CONSUMER AND MARKETING SERVICE. Milk for manufacturing purposes and its production and processing. Requirements recommended for adoption by state regulatory agencies. Fed. Reg. 37:7046–7066. Washington, DC: U.S. Dept. of Agr.; 1972.

CHAPTER 14

DETECTION OF ANTIBIOTIC RESIDUES IN MILK AND DAIRY PRODUCTS

J.C. Bruhn, R.E. Ginn, J.W. Messer, and E.M. Mikolajcik
(J.C. Bruhn, Tech. Comm.)

14.1 Introduction

Widespread use of antibiotics has significantly contributed to the control of diseases in dairy cattle. Antibiotics and drugs are administered by infusion, by injection, and orally. Unfortunately, all three routes can permit the chemicals to reach the milk supply, thus creating problems for the producer, processor, and consumer.

The need for qualitative and quantitative tests to detect antibiotic residues and drugs in milk and dairy products has long been recognized. The cylinder plate assay method was described in 1941.[1] The filter paper disc method was developed in 1944–1945.[19,30] Concurrent but independent work by many investigators led to the application of both methods to milk from 1950 to 1955. The principle of the reverse phase disc assay technique developed in 1960[16] served as the basis of the agar diffusion technique described in 1975,[29] which was successfully used in some laboratories.[23]

Chemical tests have been developed to assay commercial antibiotic preparations. Recently developed radio-immunological techniques detect the presence of penicillin (beta-lactam) residues. Other chemical tests that are being developed are claimed to be rapid, easy to run, sensitive, and economical. One such method, the penzyme test has been reviewed by the technical committee and given a Class C status (being introduced for antibiotic testing). (Smith, Klein Animal Health Products P O Box 2650, Westchester, PA 19380) Other methods as they are developed may be submitted to the technical committee for review.

The *Bacillus subtilis* disc assay method[28] with modifications[2,4,8,13,15,20,21,24,27] has been used more extensively than the cylinder assay method[3,9,17,31] to determine the presence of residual antibiotics in fluid milk. Recently, disc and tube assay methods that use *Bacillus stearothermophilus* as the test organism have gained in popularity.[11,14,21]

Use of standardized materials will reduce variability in results of all assay procedures. With adequate care and suitable controls, reliable assays can be made.

14.2 Bacillus subtillis Disc Assay (Class D)

A. *Equipment and supplies:*
1. Seed agar: Antibiotic Medium No. 1[30] [Chapter 5].
2. Petri dishes: Flat-bottomed glass or plastic dishes.
3. Spore suspension: A standardized suspension of spores of *Bacillus subtilis* (ATCC 6633). Store the spore suspension at 0 to 4.4°C.

    Each lot of spores (from commercial sources or produced in the laboratory) must be tested to determine the amount of inoculum to be used in Methods A, B, or C [14.2B1,2,3]. To determine the number of spores per mL of stock suspension, prepare dilutions of 10^{-4}, 10^{-5}, 10^{-6}, and 10^{-7} of the stock spore suspension with sterile, phosphate-buffered water [5.3A]. Plate 1.0 mL of each of the four dilutions in triplicate by using Antibiotic Medium No. 1. Incubate the inverted plates for 24 ± 2 h at 32°C. Obtain the mean count of the three plates for the dilution that has 25 to 250 colonies per plate. Multiply by the appropriate dilution factor to obtain the number of spores per mL of the stock suspension. Dilute the stock suspension as necessary to obtain the required number of spores per mL of the "seeded" agar for Methods A, B, and C [14.2B1,2,3]. Preparation of a proper dilution of spore suspension for seeding agar will permit development and easy recognition of zones when the sample contains the equivalent of, or less than, 0.05 IU (International Unit) of penicillin.
4. Filter paper discs, penicillin-impregnated: Diameter, 1.27 cm, 0.05 IU, with high absorbance, for use as reference standards. Refrigerate discs as directed by manufacturer. Additional reference discs that contain 0.01, 0.025, 0.1, 0.25, 0.5, and 1.0 IU may be used if more exact information on the level of antibiotic present is needed. When variability is suspected among commercial filter discs impregnated with penicillin, new lots of discs must be checked against discs dipped in solutions with known penicillin concentrations.
5. Standardized crystalline penicillin G, sodium or potassium: Store at −10°C. Do not use beyond the expiration date.
6. Filter paper discs, nonsterile blanks: Diameter, 1.27 cm, nontoxic to *B. subtilis*, with high absorbance, for sample application. Sterilized blank discs may be used if they give results equivalent to those obtained with non-sterile discs.
7. Filter paper discs, penicillinase (beta-lactamase)-impregnated: Diameter 1.27 cm, with high absorbance, to identify penicillin. NOTE: All discs used in testing must be of the same size and absorbance to insure comparable results.
8. Penicillinase: Available from many biological supply houses.
9. Forceps: Dissecting, with fine points.
10. Water baths: Thermostatically controlled.

14.2 Bacillus subtillus Disc Assay (Class D)

11. Incubators: 32 ± 1°C; 35 ± 1°C, 37 ± 1°C.

B. *Procedure:*
1. Method A (32°C overnight incubation):
   a. Prepare and sterilize a suitable amount of Antibiotic Medium No. 1 [Chapter 5]. Cool the sterile medium to 50 to 55°C and prepare a "seed" agar suspension [14.2A3]. Inoculated agar should contain *approximately* 10^5 spores per mL.[21] Mix well but do not incorporate air bubbles into the agar. Add the amount of agar specified in Table 14.1 to each petri dish and allow the agar to solidify.

   The amounts of medium specified in the table assume uniform agar thickness. Zones are usually larger and more clear, and lower concentrations of inhibitory substances can be detected, with a thin layer of medium.[13,21] Special care must be exercised to maintain uniformity of inoculum and depth of medium.

   b. Divide the outside bottom of the petri dish with a marking pencil into four or more approximately equal sections so the distance between discs will be at least 20 mm (on centers) and label the sections with identifying marks. With clean, dry forceps remove a paper disc from a vial or other suitable container and touch the edge into a well-mixed [6.2F2] sample of raw milk. Allow the milk to wet the disc by capillary action; avoid an excess of milk on the disc. Immediately place the disc, flat side down, on the agar surface near the center of a marked section. Place a .05-IU penicillin reference standard disc [14.1B4] on each plate. Vary the location of the reference disc so that in each series of samples the reference disc is placed in representative portions of the plate. Do not wet reference discs, because freshly poured plates have enough moisture on the agar surface to saturate the discs. Invert

Table 14.1 Quantity of Medium Required per Petri Dish to Insure Uniform Thickness of Agar Layer*

Petri Dish Size (mm)	Material and Type	Diameter of Agar Layer (mm)	Area of Agar Layer (cm^2)	Amount of Agar Required (mL)
100 × 22	Glass, flat-bottom	85	57	6.0
100 × 15	Plastic, round standard†	85	57	6.0
100 × 15	Plastic, square, phage typing	89	79	8.5

* Based on the ratio of volume of agar medium to surface area, using the glass, flat-bottom dish as the standard.

† As designated in catalog. Round, plastic dishes are actually 85 mm OD.

the plates and incubate at 32°C for 14 to 24 h until growth becomes apparent. Examine plates and interpret results as in [14.2C]. Report as in [14.2D].

d. Samples of raw milk which give a positive test must be heated to 82°C for 2 to 3 min,[17] cooled, and retested as in [14.2C1]. This heat treatment inactivates naturally occurring inhibitory substances in milk and eliminates the possibility of a false-positive test result. Alternatively, heat all samples before testing; however, fewer than 1% of the samples are then likely to give a positive result.

2. Method B (35°C short incubation):

Prepare seed agar and plates as in [14.2B1] by inoculating agar with approximately 10^6 spores per mL. Invert plates with discs in place and incubate at 35°C for 5 to 7 h. Examine plates and interpret results as in [14.2C]. Report as in [14.2D].

3. Method C (37°C short incubation):

Melt Antibiotic Medium No. 1, cool to 70°C, and inoculate with a quantity of spore suspension [14.2A3] predetermined to give approximately 5.0×10^6 spores of *B. subtilis* per milliliter.[21] Rotate the agar flask to distribute spores, hold 15 min in a thermostatically controlled water bath at 70°C, mix the agar again, and prepare plates as in [14.2B1].

Invert plates with discs in place and incubate at 37°C until growth becomes apparent: normally, observations can be made after 3 to 4 h. Optionally, hold poured plates at 37°C for 1 h before "spotting" with discs wet with milk.[12] Examine plates and interpret results as in [14.2C]. Report as in [14.2D].

C. *Interpretation:*

Examine plates for clear zones of inhibition. Any clear zone surrounding a disc prepared with a non-heat-treated sample represents a positive test. Such tests must be replicated using a heated sample to eliminate false-positive results. Any clear zone surrounding a disc prepared from a heat-treated sample is a positive test. To determine if the clear zone resulted from penicillin, or another inhibitor or both, proceed as in [14.2C1].

1. Tests that identify the inhibitor:

a. To determine whether inhibition resulted from penicillin, place a penicillinase-impregnated disc [14.2A7], a blank disc [14.2A6] moistened with the heated positive test sample, and a 0.05-IU penicillin control disc on the same plate. Alternatively, add 0.05 mL of penicillinase [14.2A8] to 5 mL of the heated milk sample and shake well. Wet one blank disc [14.2A7] with the penicillinase-treated sample and one blank disc [14.2A7] with a heated untreated portion of the same sample. Place both discs and an 0.05-IU penicillin control disc on the same plate. Incubate as in [14.2B methods a, b, or c]. If no clear zone appears around the penicillinase disc, but one does surround the disc containing the test sample, penicillin is present. If clear zones equal in size surround both discs, an inhibitor other than penicillin is present. If the clear zone surrounding the penicillinase disc is 5 mm or more smaller than the zone surrounding the blank disc, both penicillin and another inhibitor are present. If the clear zone surrounding the penicillinase disc is less than 5 mm smaller than the zone surrounding the blank disc, an inhibitor other than penicillin is present.

b. Alternatively, place one disc moistened with the heated milk and one moistened with penicillinase (or commercially prepared with penicillinase) near each other (edges 2 to 3 mm apart) on a plate. Place an 0.05-IU penicillin control disc on the plate. Incubate as in [14.2B methods A, B, or C]. If penicillin is in the milk, a characteristic neutralization zone appears at the juncture. If penicillin is not the inhibitor, a neutralization zone will not appear. This test demonstrates only the presence of penicillin or another inhibitor. It does not demonstrate both inhibitors simultaneously.

D. *Reporting:*

Report the presence of inhibitor only in heated milk samples. If tests in [14.2C1] were done, report the results demonstrated. If tests in [14.2C1] were not done, report as positive for inhibitor. To estimate the concentration of penicillin, use reference discs [14.2A4] as a guide. Report absence of inhibitors as "Not Found" or "Negative."

14.3 Qualitative *Bacillus stearothermophilus* var. *calidolactis* Disc Assay[22] (Class A1)

The sensitivity of this method is equivalent to the AOAC method (16.131–16.136)[9] and is applicable to levels ≥ 0.008 IU penicillin/mL.[22]

A. *Equipment and supplies*:
1. a. Antibiotic medium 4 [Chapter 5].
   b. Agar medium P: Dissolve 3.0 g beef extract, 5 g peptone, 1.7 g tryptone, 0.3 g soytone, 5.25 g glucose, 0.5 g NaCl, 0.25 g anhydrous $K_2H\ PO_4$, 1.0 g polysorbate 80, 0.06 g bromcresol purple, and 15 g agar in MS water and dilute to 1 L. Adjust pH after sterilization to 7.8 ± 2. (Difco PM Indicator Agar is satisfactory.)
   c. Agar medium M: Dissolve 15.0 g pancreatic digest of casein, 5.0 g papaic digest of soybean, 5.0 g NaCl, and 15.0 g agar in MS water and dilute to 1 L. Adjust pH after sterilization to 7.3 ± 2. (BBL Trypticase Soy Agar is satisfactory.)
   d. Broth medium D: Dissolve 17.0 g pancreatic digest of casein, 3.0 g papaic digest of soybean, 5 g NaCl, and 2.5 g anhydrous K_2H

PO$_4$ in MS water and dilute to 1 L. Adjust to pH 7.3 ± 2 after sterilization. (BBL Trypticase Soy Agar without dextrose is satisfactory.)
2. Culture of *Bacillus stearothermophilus* var. *calidolactis* (ATCC 10149)
3. Seeded agar plates: Inoculate a portion of liquified agar medium 4[14.3A1] or P[14.3Alb], cooled to 55°C or 64°C, respectively, with the previously prepared spore-suspension [14.4B] or with a commercial spore suspension. Inoculate the medium with the spore suspension to achieve a spore density of 400,000 or 1,000,000 per mL medium, depending on whether an incubation temperature of 55°C or 64°C, respectively, is being used. Pour 6 mL of the inoculated agar medium 4 or P, into each plate and let harden on flat, level surface. Use within 5 days.
4. Penicillin stock solution: [14.3A5]. (Difco PM Positive Controls, Penicillin G and Penicillin Assays, Inc. penicillin standards are satisfactory.)
5. Penicillinase (beta-lactamase): [14.2A7].
6. Filter paper discs, blank: Use S and S 74OE, 1.27-cm discs or discs of equivalent absorption performance, qualities, and purity.
7. Control discs: Prepare fresh daily from positive control milks containing 0.008 IU penicillin/mL.
8. Petri plates: [14.2A2].
9. Forceps: [14.2A9].
10. Incubators: 55 ± 2°C or 64 ± 2°C.
11. Water bath: To heat sample to 82 ± 2°C for at least 2 min.
12. Sterile isotonic saline solution: Dissolve 9.0 g NaCl in distilled or deionized water and dilute to 1 L. Sterilize by autoclaving for 30 min at 121°C.
13. Vernier calipers: [14.3A12].

B. Preparation of stock culture of test organism:
Commercially available spore suspensions are satisfactory.
1. Maintain the *B. stearothermophilus* on agar medium M [14.3Alc], transferring weekly.
2. Inoculate a slant of agar medium M with the test organism and incubate overnight at 55 ± 2°C or 64 ± 2°C. Following incubation, inoculate 3, 250-mL Erlenmeyer flasks, each containing 150 mL broth of medium D [14.3A1d], with 1 loopful of the test organism per flask. Incubate at 55 ± 2°C or 64 ± 2°C and periodically make a spore stain to determine the extent of sporulation.
3. When about 80% sporulation has occurred (usually after 72 h), sediment the cell-spore suspension at 5,000 rpm for 15 min. Decant the supernatant liquid, resuspend cells and spores in a saline solution [14.3A12], and recentrifuge. Repeat saline solution washing. Resus-

14.3 Qualitative *B. stearothermophilus* var. *calidolactis* Disc Assay (Class A1)

pend washed spores in 30 mL of saline solution and store at 0 to 4°C. Spore suspension will remain viable 6 to 8 mo. Check viability of spore preparation by use of trial test plates.

C. *Procedure*:
1. Screening:
   a. With clean, dry forceps, touch a paper disc to the surface of a well-mixed sample of milk and let the milk be absorbed by capillary action. Drain the excess milk by touching the disc to the inside surface of the sample container. Immediately place the flat side of the disc on the seeded agar surface, pressing with the tip of the forceps to insure good contact. Identify each disc or the section on which it is placed [14.2B1b].

   Place a control disc containing 0.008 IU penicillin/mL on the plate. Invert the plate and incubate at 55 ± 2°C or 64 ± 2°C until well defined zones of inhibition (17 to 20 mm) are obtained around the 0.008 IU/mL control disc.

   b. Examine plates for clear zones of inhibition surrounding the discs. Clear zones > 14 mm indicate the presence of inhibiting substances. Zones of ≥ 14 mm are read as negative. To confirm the presence of inhibitors, proceed as in [14.4C].

2. Confirmation:
   a. Heat samples to 82°C for ≤ 2 min. Cool promptly to room temperature.
   b. With clean, dry forceps, touch a paper disc to the surface of a well-mixed sample of milk and let the milk absorb by capillary action. Also fill a penicillinase-impregnated disc with the milk, or add 0.05 mL of penicillinase to 5 mL of the sample and then fill a paper disc with that mixture. Drain excess milk by touching each disc to the inside surface of the sample container. Immediately place each disc on the seeded agar surface and press gently with the forceps to insure good contact.

   Place a control disc containing 0.08 IU/mL on the plate. Invert the plate and incubate at 55 ± 2°C or 64 ± 2°C until well defined zones inhibition (17 to 20 mm) are obtained around the 0.008 IU/mL control disc. Examine the plate for clear zones of inhibition (≥ 14 mm) surrounding the disc, which indicate the presence of inhibitory substance(s)

C. *Interpretation and reporting:*
Assay of test milk in screening and confirmatory tests may produce the following results:
1. No zone < 14 mm around the disc containing untreated milk in a screening test is reported as no zone for the inhibitory substances.
2. A zone > 14 mm around the disc containing untreated milk but no zone around the disc containing penicillinase-treated milk in a

confirmatory test is a positive test. Record the zone size and report the sample as positive for beta lactam residue.
3. Clear zones equal in size around both discs in a confirmatory test indicate presence of inhibitors other than beta-lactam residues.
4. A clear zone around penicillinase-treated milk disc that is at least 4 mm smaller than the one around the untreated milk disc in a confirmatory test indicates the presence of beta-lactam residues as well as another inhibitor(s). Record zone size and report sample positive for beta lactam residues and other inhibitors.

A penicillin-positive control disc at 0.008 IU/mL should produce a clear, well-defined zone of inhibition (17 to 20 mm).

14.4 Quantitative *Bacillus stearothermophilus* var. *calidolactis* Disc Assay[6,7,10] (Class A2)

This test is based on the Difco Disc Assay Procedure (Difco Laboratories, P.O. Box 1059A, Detroit, MI 48232) and uses the paired-t analysis of data, rather than a regression curve to allow the quantitation of beta-lactam antibiotic residues that are 0.003 IU on either side of a reference standard of about 0.016 IU.

A. *Equipment and samples:*
 1. Seed agar: Antibiotic Medium 4 (Difco or BBL Yeast Beef Agar, or equivalent).
 2. Petri dishes: Flat-bottomed glass or plastic dishes.
 3. Standardized suspension of spores of *Bacillus stearothermophilus* (ATCC 10149) available from commercial sources. Store the spore suspension at 0 to 4.4°C.

 One ampule of spores per 100 mL of medium provides approximately 10^6 spores mL. Occasionally confirm spore concentrations in ampules.
 4. Crystalline USP Na penicillin G for standardization: Store at $-10°C$. Do not use beyond expiration date.
 5. Standard penicillin solution: Prepare a standard reference solution of penicillin containing 1000 IU USP Na penicillin G/mL in phosphate buffer [14.3A4]. This stock solution may be frozen for later use.
 6. One percent phosphate buffer: Dissolve 8 g anhydrous KH_2PO_4 and 2 g anhydrous K_2HPO_4 in MS water and dilute to 1 L with distilled or deionized water. Adjust the pH to 6.0, if necessary.
 7. Filter paper discs, penicillinase (beta-lactamase)-impregnated: [14.2A7].
 8. Filter paper discs, non-sterile blanks: [14.2A6].
 9. Standard reference milk containing 0.016 IU/mL penicillin G.

 For each assay, first dilute the standard penicillin solution [14.3A5] to 1.0 IU/mL in phosphate buffer. Then prepare a 0.016 IU/mL *standard reference milk by diluting* an aliquot with previously tested penicillin-free milk.

14.4 Quantitative *B. stearothermophilus* var. *calidolactis* Disc Assay (Class A2)

10. Microliter pipettor: 90 μL with disposable tips. (Eppendorf Model 4700, Curtis Matheson Scientific, Inc., P.O. Box 1546, Houston, TX 77001, or equivalent)
11. Vernier calipers: Readable to 0.1 mm.
12. Forceps: [14.2A9].
13. Incubator: 64 ± 2°C.
14. Water baths: 65 ± 1°C; 82 ± 1°C; 0 ± 1°C.

B. *Procedure:*

1. Prepare and sterilize [Chapter 5] no more than a 30-day supply of Antibiotic Medium 4 in flasks of 100 mL each. Use the medium fresh or store in refrigerator until used. Add 1-mL ampule of *B. stearothermophilus* spore suspension to each 100-mL lot, which has been previously tempered to 65°C. Final spore concentration in medium should be 10^6 spores per mL. Preferably seed the medium fresh each day of testing. Work on an absolutely flat surface because even a slight variation will cause an uneven distribution of the medium and irregularly shaped zones.

   Using a warm 10-mL pipette (pipettes may be stored in the 65°C incubator), transfer of 6 mL of seeded medium to the center of the petri dish. Cover the dish and swirl gently to cover entire bottom surface. Let medium solidify at room temperature.

2. Prepare a 0.016 IU/mL standard reference milk [14.4A9]. Heat the standard reference milk and sample at 82°C for 2 min in a temperature control milk sample. The thermometer should be inserted during heating. Cool in an ice water bath. It is essential that the standard reference milk samples and the unknown be treated identically during the heating step. At the very low concentration of penicillin being tested, some evaporation loss may occur during heating.

3. Using clean, dry forceps place six filter discs evenly around the periphery of the culture plate, approximately 1 cm from the edge. Bond touch each disc to the surface of the medium, gently, by using the tips of the forceps. *Using separate* forceps, place the penicillinase disc in the center of the plate. Within 45 to 60 sec after the final disc is placed on the culture plate, add 90 μL of the standard reference milk or unknown by using the 90-μL pipettor. On each plate, alternate the reference standard milk with the unknown sample, and add the unknown sample to the penicillinase disc. Thus, on each plate, triplicate reference standards and triplicate unknowns are alternated on the six peripheral discs. Any given standard must always fall adjacent to an unknown. To use the pipettor correctly when adding sample, hold it over the center and 3 to 5 mm above each disc. Eject the sample in an even flow, avoiding splashing. Use one disposable tip for the triplicate standard reference samples and another for the triplicate unknown samples.

In the above manner, a total of *three* such plates are prepared for each unknown sample. Invert plates and place immediately in the incubator at 64 ± 2°C. Incubate for 2 h and 45 min. Remove plates in sets as prepared and place on laboratory bench at room temperature while measuring.

4. Using Vernier calipers, measure the zone diameter of the standard reference milk and the sample to the nearest 0.1 mm. Include both the disc and zone of inhibition in the measurement. If a zone of inhibition is present around the penicillinase disc, record as "other inhibitor" and proceed no further with the test.

To measure zone diameter: (a) Turn petri dish upside down. (b) Using a fluorescent light source, maneuver the dish to show a clearly defined zone. (c) With the pointed tip of the caliper (the shortest of the two measuring ends of the caliper), anchor lightly one point of the tip on the plastic dish at the outermost edge of the zone. Spread the caliper directly opposite, crossing the center of the disc, to the outermost edge of the zone on the other side. (d) Measure diameter to nearest 0.1 mm.

If no zone of inhibition is present around the penicillinase disc, and if the zones of inhibition of the unknown samples are greater than those of the reference standard, proceed with the calculation.

5. Determine the relationship between the reference standard milk and the unknown sample zone by using a paired-t analysis, noting paired differences that occur between the standard reference and the clockwise adjacent unknown. Each plate thereby yields three differences for use in paired-t analysis.

Calculate t-value, $t = \sqrt{n} \times \bar{d}/S_d$, when n = number of observations; d = mean differences, and S_d = standard deviation of the difference. The simplified formula for calculation of t for 9 replicates (3 plates) is as follows:

$$t = \frac{2.828 \times D_1}{\sqrt{9 \times D_2 - (D_1)^2}} \quad \text{where}$$

D_1 = sum of differences of 9 replicate determinations, and

D_2 = sum of squares of differences of 9 replicate determinations.

C. Interpretation:

A t-value of > 1.860 indicates with a 95% confidence that the sample contains ≥ 0.016 IU penicillin/mL. Therefore if the t-value calculated is larger than the given t-value, then the difference between the zone size of the reference standard and unknown is confirmed. If the zone size is larger, *and is confirmed to be so*, the concentration of beta-lactam residue may be considered higher than the reference, or vice versa, at the specific level of confidence taken.

The method can discriminate, at a 95% confidence level, the concentration of beta-lactam that is 0.003 units different from a reference of about 0.016 IU.

D. Reporting:

Report the concentrations of beta-lactam as: present in a concentration > 0.016 IU/mL; ≤ 0.016 IU/mL; or not found, "negative." If inhibitor other than a beta-lactam is detected, report as "other inhibitor" without reference to concentration.

14.5 Delvotest® -P- Ampule and Multi Test Procedures.[14] (Class A2)

These tests are based on the rapid growth and acid production of *Bacillus stearothermophilus* var. *calidolactis*, with the associated subsequent change in the bromcresol purple dye from purple to yellow in the absence of beta-lactam inhibitors. The test detects and confirms levels of beta-lactam residues ≥ 0.005 IU/mL in processed fluid milk products (except for the Multi Test on chocolate-flavored milk products) and raw milk.[14]

A. Equipment and supplies:

1. Delvotest® -P- Ampule test kit: (GB Fermentation Industries, Inc., P.O. Box 241068, Charlotte, NC 28224). Each kit contains 100 test ampules, seeded with *B. stearothermophilus* var. *calidolactis* (ATCC 10149) in plain, solid agar medium. Each kit contains 100 nutrient tablets composed of tryptone (0.5 mg), glucose (5.0 mg), nonfat dry milk (2.0 mg), and bromcresol purple (0.025 mg). Store at 4 to 15°C until opened; store opened packs at room temperature. Plastic forceps to transfer the nutrient tablets and a plastic syringe with 100 disposable tips for sampling and dispensing 0.1-mL portions of milk are also included in each kit.

2. Delvotest® -P- Multi test kit: (GB Fermentation Industries, Inc.) consists of three plates in a hermetically sealed aluminum bag. Each plate contains 96 cups and each cup contains *B. stearothermophilus* var. *calidolactis* sealed in solid medium supplemented with bromcresol purple indicator. Each plate can be divided into six blocks of 16 cups each. Similar aluminum bags contain three similar plates with one nutrient tablet per cup, and sealing tapes to cover all blocks individually. Store all materials at 4 to 15°C until opened; opened kits are to be stored at room temperature.

3. Heaters: Block heater and water bath, both thermostatically controlled at 65 ± 2°C. Check daily.

4. Dispensing pipettes: Disposable, as in ampule test, or micropipettor (Micro/Pettor, Scientific Manufacturing Industries, 1399 64th St., Emeryville, CA 94608) or equivalent.

5. Phosphate buffer: 1%, pH 6.0. Dissolve 8.0 g of anhydrous KH_2PO_4 and 2.0 g of anhydrous $K_2H\,PO_4$ in MS water and dilute to 1 L with water.
6. Penicillinase (beta-lactamase): Concentrate, store at 0 to 4.4°C (Difco, BBL, or Penicillin Assays, Inc., materials are satisfactory).
7. Penicillin stock solution: Accurately weigh 30 mg USP K penicillin G reference standard and dissolve in pH 6.0 buffer to give known concentrations of 100 to 1000 IU/mL. Store diluent at 0 to 4.4°C for not more than 2 d.
8. Inhibitor-free milk: Any fluid milk product (milkfat content 0.00 to 3.50%, total solids < 13%) may be used after being tested with this method to verify that it is inhibitor-free. Use milk for diluting standards and as the negative control.

B. Controls:

As a check on proper functioning of reagents, both penicillin-positive and -negative control samples must be included with each series of samples tested by the screening and confirmatory procedures. A single positive and a single negative control will suffice for each series of samples tested.

1. Negative control: Any fluid milk product (milkfat content 0.00 to 3.5%, total solids < 13%) may be used after being tested with this method to verify it is inhibitor-free.
2. Positive control: From the stock penicillin solution [14.3A4] prepare a reference concentration of 0.006 IU per mL in inhibitor-free milk. Dispense 5-mL quantities into 15 × 100 mm tubes. Cap the tubes and freeze them at −10°C. Do not store at −10°C for longer than 6 mo. For use, thaw at room temperature. When tested, this milk must not allow any change in the purple color of the entire solid test medium.

C. Check trial procedures:

Perform the procedure on each new lot number of ampules or test kits and on new batches of prepared reagents.

Prepare 10-mL volumes of penicillin standards in inhibitor-free milk, that contains concentrations of 0.002, 0.004, 0.006, 0.008, and 0.010 IU/mL, 10 mL water, and 10 mL inhibitor-free milk in test tubes. Transfer a 5-mL portion from each tube into a corresponding tube. Heat both sets of tubes to 82°C for 3 min in a water bath. Remove; cool rapidly to room temperature. Add 0.2 mL of penicillinase to one series of tubes. Shake well; let stand for 15 min at room temperature.

For the Ampule Test, remove one ampule for each test tube from the test kit and label the ampules. Break off the neck of each ampule and place ampule in lid of ampule box, or other suitable rack. With clean, dry forceps, place one nutrient tablet in each ampule. For the Multi Test, remove one plate from each of the two foil bags. Score the foil covering at the cutline on the end block of each plate and break off one block of agar cups and

14.5 Delvotest-P-Ampule and Multi Test Procedures (Class A2)

one block of cups containing nutrient tablets. Open both blocks by carefully tearing back the foil covers. Place the block containing agar upside down, exactly on top of the block containing nutrient tablets. Holding both blocks together, invert them; the tablets should fall into each corresponding cup. Light tapping of tablet cups may be needed to dislodge all of the tablets. Arrange test tubes of controls according to cup locations in the block. Attach a dry sampling pipette to a plastic syringe. Completely depress the plunger, place the end of the pipette into a tube, ca 1 cm below the top of the sample level. Allow the plunger to return slowly under pressure of the spring. The level of sample should reach the wide part of the pipette. If an air bubble appears, slowly expell the sample back into the tube and let the plunger slowly return again. Do not contaminate the syringe. If a drop of milk clings to the outside of the pipette tip, gently touch it off on the edge of the tube. This volume is ca 0.1 mL. Empty the pipette into an appropriate ampule or cup. Remove the pipette tip and replace it with a new tip for each sample, control, or standard. If using a micropipettor, wipe the outside with tissue and rinse three times in sample before removing a test aliquot. Continue this pipetting procedure until all tubes have been sampled.

For the Ampule Test, place the ampules into appropriate holes in the heater block. Incubate at 65 ± 2°C for exactly 2.5 h, then remove the ampules. Read the penicillinase-treated ampule and its corresponding untreated ampule side by side, looking through the agar against a white, reflective background; compare and record colors as yellow, purple, or yellow-purple. Disregard any intense color surrounding the nutrient tablet.

For the Multi Test, carefully seal a block of cups with the strip of adhesive tape enclosed in each kit. Very carefully float the sealed block in a 65°C waterbath. Incubate at 65 + 2°C for exactly 2 h and 45 min, remove the block from the waterbath and read and record the colors developed. Read from the bottom side of block. Compare and record colors as yellow, purple, or yellow-purple.

The following colors are satisfactory: yellow for water, inhibitor-free milk, and 0.002 IU/mL standard; yellow-purple for 0.004 IU/mL standard; purple for 0.006, 0.008, and 0.010 IU/mL standards.

NOTE: Occasionally kits of a particular Lot Number may require a longer incubation time for color to fully develop. If water and inhibitor-free milk samples are not truly yellow and/or 0.006 and 0.008 IU/mL standards are not completely purple after 2.5 h of incubation (2 h and 45 min for the Multi Test) continue incubating until proper colors are developed. Check color development at 10-min intervals and record the optimum incubation time required for each Lot Number.

    D. Screening test procedure:
1. Use one negative and one positive control (0.008 or 0.010 IU/mL). Samples may be preheated as in the check trial.

2. Place the required number of ampules or blocks for the samples to be tested in a suitable rack.
3. Label each ampule or block of cups legibly and indelibly.
4. Shake the sample 25 times through an arc of 30 cm in 7 sec.
5. Remove the tip of each ampule with clean forceps, add one nutrient tablet to each ampule.
6. With the syringe, add 0.1 mL of a well-mixed sample, or of the positive or negative control sample, to each identified ampule. Use a clean disposable tip for each sample and control transferred. Extreme care must be taken not to contaminate the multi-use syringe when sampling and dispensing.
7. Place the rack containing the samples in the block heater or the blocks of cups in a water bath and incubate as in the check trial [14.5C].
8. After incubation, remove the ampules or blocks of cups and observe the color of the solid medium.
9. An entire or partial purple coloration of the solid test medium in any of the milk sample ampules indicates the presence of substances inhibitory to the test organism.
10. Samples of raw milk that give a purple color to part or all of the solid medium must be confirmed [14.5E] before reporting.

E. *Confirmatory test procedure:*
1. Heat-treat (82°C for 3 min, cool) two 5-mL portions of each sample to be confirmed and the 0.002, 0.004, and 0.006 IU/mL standards. Do not treat the negative control or 0.008 IU/mL standard.
2. Add 0.2 mL of penicillinase, to one portion of each heat-treated sample and to the three low-concentration standards. Mix well.
3. Prepare the ampules or block of cups, add samples and incubate as in the check trial [14.5C].

F. *Interpretation and report:*
All results are reported as positive or negative for beta-lactam residue.
1. Negative: solid yellow in screening test.
2. Negative: heated sample yellow, penicillinase-treated sample yellow.
3. Negative: heated sample purple, penicillinase-treated sample purple.
4. Negative: heated sample yellow-purple, penicillinase-treated sample yellow-purple.
5. Positive: heated sample yellow-purple, penicillinase-treated sample yellow.
6. Positive: heated sample purple, penicillinase-treated sample yellow.

Samples containing heat-stable natural inhibitors give a true negative test for beta-lactam residues. Samples containing heat-stable natural inhibitor plus penicillin may result in false-negative tests for beta-lactam residues. Samples containing other inhibitory substances (e.g., tetracycline) will give a true negative test for beta-lactam residues. Samples of chocolate-flavored products are difficult to read using the Multi Test kit because of light-distorting colors from adjacent tubes and must not be reported as positive until they are confirmed.

14.6 The Charm Screening Assay for Beta-Lactam Residues[5] (Class A1)

This assay is based on the specific, irreversible affinity of beta-lactam antibiotics for certain enzyme sites on the cell walls of microorganisms. ^{14}C-labeled penicillin and *Bacillus stearothermophilus* are added to milk samples. Antibiotic in the sample competes with ^{14}C-penicillin for binding sites. The amount of bound carbon-14 is counted and compared with the control to determine the presence or absence of beta-lactam antibiotics.[5]

This qualitative assay is positive or negative for the detection of beta-lactam antibiotics at concentrations of ≥ 0.01 IU/mL. Levels of radioactive carbon-14 are sufficiently low so that they are exempt from Federal licensing, but the user should check to see that local or state regulations recognize this Federal exemption. The Charm test has also been recently adapted to detect other antibiotics, including tetracyclines, streptomycin, erthromycins, sulfa drugs, and chloramphenicol using H^3 labeled reagents. Check with the manufacturer for further information.

A. *Equipment and supplies:*
1. Charm test apparatus (Penicillin Assays, Inc., 36 Franklin Street, Malden, MA 02148): Includes dry-well tube heater for 13 × 100 mm glass tubes, cold plate that accommodates 13 × 100 mm tubes, radiation counter (penicillin analyzer), aluminum counting planchets, and semi-automatic pipettors or equivalents.
2. Diluent A (Penicillin Assay, Inc.).
3. Diluent B (Penicillin Assay, Inc.).
4. Penicillin-free skim milk powder: Reconstitute the milk powder with distilled water and store at 4°C. Five-mL portions may be pipetted into tubes, corked and frozen at $\leq -20°C$ for later use.
5. Penicillin-free milk: [14.5A8].
6. Carbon-14 penicillin (Reagent A): Containing 103 micro Caries/μ mole. Dilute with Diluent A [14.6A2] according to the manufacturer's directions. Use 0.0027 μCi in each assay. The freeze-dried ^{14}C-penicillin is stable at $-20°C$ for at least 6 mo, but will deteriorate 10% in 12 h at 37°C. Reconstituted, it is stable at $-20°C$ for at least 6 mo, but will also deteriorate 10% in 12 h at 37°C. Reconstituted, it is stable for 12 h at less than 4°C; 24 h at $-2°C$ and for at least 1 mo at $-20°C$. Do not refreeze this reagent.
7. *Bacillus stearothermophilus:* About 47 mg of vegetative cells (Reagent B) are used for each test. When reconstituting Reagent B according to the manufacturer's directions, be sure to disperse all clumps by shaking the vial and stirring the contents with a pipette

tip. Allow Reagent B to stand at ambient temperature for 15 min before using. This insures thorough reconstitution. Dilute Reagent B with Diluent B [14.6A3], according to the manufacturer's directions.
8. Vortex mixer.
9. Centrifuge: Suitable for centrifugation of 13 × 100 mm glass tubes at 1200 × g.
10. Cotton swabs (2 per test).
11. Hot plate capable of holding the aluminum planchets at 400°C.

B. *Procedure:*
1. Pipette 5 mL of sample into a 13 × 100 mm glass tube. Add, by pipette, 200 μL of reconstituted ^{14}C-reagent (Reagent A) and mix in a vortex mixer.
2. Add, by pipette, 200 μL of the *B. stearothermophilus* suspension of reconstituted Reagent B and mix in vortex mixer.
3. Incubate tube for 3 min at 90°C in a dry-well heater. A tube, filled with water and inserted in one of the walls of the heater, may be used as a thermometer well to insure that the heater is set at 90°C (5 mL of cold, 15°C or less, milk in a tube should reach 60 to 65°C in 3 min).
4. Remove the tubes and sediment the cells by centrifugation for 4 min at 1200 × g. Decant the milk and carefully swab out the fat ring by using two cotton swabs. This insures removal of the fat ring. Fat prevents penetration of the ^{14}C-radiation and its subsequent measurement by the analyzer of the readout system.

    Rinse the tube twice with deionized or distilled water expressed from a wash bottle without disturbing the precipitate at the bottom of the tube.
5. Add about 300 μL of distilled or dionized water and resuspend the precipitate by using the vortex mixer. Place an aluminum planchet on a hot plate maintained at 400°C. A surface thermometer placed on the surface of the hot plate will aid in monitoring the temperature. Pour the suspension into the planchet and touch the mouth of the tube to the planchet to remove the last drop. Rinse the tube twice using 300 μL of water each time and add the washings to the planchet. Ensure that the entire bottom of the planchet is covered uniformly. Let the material in the planchet dry.
6. Place the dry planchet into the penicillin analyzer [14.6A1] and measure radiation from the ^{14}C for 8 min. Compare the count with the predetermined control point [14.6C] to determine if the sample is positive or negative.

C. Determination of the control point:

The control point establishes the count below or above which one can determine that beta-lactam residues are present or absent. The control point is set at least three standard deviations from the 0.01 IU/mL mean.
1. Analyze 10 samples of milk that are devoid of beta-lactam residues by the procedure outlined [14.6B1 through 6]. Average the counts. If any of the zero-standard samples exceed ± 20% of the average, replace the standard with a new one and determine a new average. Calculate the control point: the control point = 0.80 × the average count. Test samples having counts below the control point contain beta-lactam antibiotic residues or indicate a test failure [see 14.6C2]. Test samples with counts exceeding the control point contain no detectable beta-lactam antibiotic residues.
2. To identify a test failure, make a second determination of the positive sample, the zero standard and an 0.01 IU/mL standard at the same time. If beta-lactam is present, the positive sample and standard should again have counts below the control point, and the zero standard should have counts above the control point.

D. Reporting:

Samples with counts lower than the control point [14.6C] tht are not the result of a test failure [14.6C2] are reported positive for a beta-lactam residue ≥ 0.01 IU/mL. Samples with counts above the control point should be reported as not containing beta-lactam residues in concentrations > 0.01 IU/mL.

14.7 Modified *Sarcina lutea* Cylinder Plate Method for Detection of Penicillin in Nonfat Dry Milk[3,9] (Class A1)

A. Equipment and supplies:
1. Agar: Medium No. 1 (Penassay Seed Agar) and Medium No. 4 (Yeast Beef Agar).
2. Petri dishes: Glass, 20 × 100 mm, equipped with porcelain covers, glazed on the outside, or glass cover lids with filter pads capable of absorbing water of syneresis. Comparable plastic dishes and lids may be used if the lids are raised slightly to allow the escape of water.
3. Cell suspension: Use *Sarcina lutea* (ATCC 9341), which is maintained as a stock culture on agar slants of Medium No. 1 and transferred to a fresh slant every 2 wk. Prepare the suspension as follows: Streak an agar slant heavily with the test organism and incubate for 18 to 24 h at 32°C. Wash growth from the slant with 1 to 2 mL of sterile physiological saline solution and transfer to the dry surface of a Roux bottle containing 300 mL of Medium No. 1. Spread the suspension evenly over the entire surface of the medium with the

aid of sterile glass beads. Incubate 18 to 24 hours at 32°C. Wash growth from the agar surface with 50 mL of saline solution. NOTE: Before the actual assay, determine by trial plates the optimum amount of the bulk suspension to be added to seed agar to obtain the best zones of inhibition. Store this stock suspension at 0 to 4.4°C for ≤ 2 wk.
4. Phosphate buffer: 1%, pH 6.0 ± 0.1, [14.5A5].
5. Cylinders: Stainless steel with an outside diameter of 8 ± 0.1 mm and an inside diameter of 6 ± 0.1 mm and 10 ± 2 mm long.
6. Dispenser: For placing cylinders on plate.
7. Water bath: Thermostatically controlled.
8. Incubator: 30 to 37 ± 1°C.
9. Roux bottle.
10. Glass beads: Sterile, for spreading suspension over the agar surface in the Roux bottle.
11. USP Sodium Penicillin G reference standard: Follow the label directions for storage of the standard. Prepare a stock solution by dissolving an accurately weighed portion of the penicillin standard in buffer [14.5A5] to obtain a solution containing 1000 IU/mL. Store at 0 to 4.4°C. This stock solution may be used for 2 da.
12. Nonfat dry milk: Antibiotic-free [14.5A13].
13. Penicillinase: Available from biological supply houses. Store at 0 to 4.4°C.

B. Procedure:
1. Assay plate: Prepare suitable amounts of Antibiotic Media Nos. 1 and 4, sterilize and cool to 48°C. Add 10.0 mL of Medium No. 1 to each sterile petri dish. Distribute the medium evenly and allow to harden on a flat, level surface. Add the appropriate amount of culture [14.7A3] to each 100 mL of Medium No. 4 cooled to 48°C. Mix well, taking care to avoid incorporating bubbles into the agar. Add 4.0 mL of this inoculated agar to each of the plates receiving 10.0 mL of the base layer. Distribute the agar evenly by tilting the plate from side to side in a circular motion. Allow agar to harden and use the plates on the same day they are prepared.
2. Standard response line:
   a. Prepare this line each day samples are confirmed for penicillin. Prepare a control diluent from antibiotic-free nonfat dry milk [14.5A8] and phosphate buffer [14.7A4] by combining the two materials at the rate of 9 mL of buffer for every 1.0 g of nonfat dry milk. Use this diluent to dilute the penicillin stock [14.5A11] to obtain concentrations 0.00625, 0.0125, 0.025, 0.05, 0.1, and 0.2 IU/mL. The reference concentration is 0.05 IU/mL.
   b. In 15 assay plates [14.7B1], place six cylinders [14.7A5] equally spaced (60° intervals on a 2.8-cm radius) on the inoculated agar

14.7 Detection of Penicillin in Nonfat Dry Milk (Class A1)

surface. Fill three alternate cylinders on each of the 15 assay plates with the 0.05 IU/mL concentration. Fill three alternate cylinders on three assay plates with each of the other concentrations. The three plates containing the lowest concentration of the standard are intended to produce negative results. The other 12 plates are used to construct the standard response line. This will give forty-five 0.05 IU/mL determinations and nine determinations for each of the other points on the line.

c. Incubate the 15 plates at 30 ± 1°C for 16 to 18 h. After incubation, invert the plates to remove the cylinders; measure the diameter of each zone of inhibition as accurately as possible.

d. Average the readings of the 0.05 IU/mL concentration and the nine readings for each of the other reference concentrations tested on each set of three plates. Also average the 45 readings of the 0.05 IU/mL concentration. This average is the correction point for the other concentration points on the curve. Correct the average value obtained for each point (0.00625, 0.0125, 0.025, 0.1 and 0.2 IU/mL concentration) to the figure it would be if the average for the nine 0.05 IU/mL readings for the specific concentration was the same as the correction point. Thus, if in correcting the 0.025 IU/mL concentration, the average of the 45 readings of the 0.05 IU/mL concentration is 20 mm, and the average of the 0.05 IU/mL concentration of this set of three plates is 19.8 mm, the correction is +0.2 mm. If the average reading of the 0.025 IU/mL concentration of these same plates is 17.0 mm, the corrected value is 17.2 mm. Plot the five corrected values and the average of the 0.05 IU/mL concentration on two-cycle, semi-log graph paper, placing the concentration in IU/mL on the logarithmic scale and the diameter of the zone of inhibition on the arthmetic scale. Construct the best straight line through these points, either by inspection or by means of the following equations:

$$L = \frac{3a + 2b + c - e}{5} \qquad H = \frac{3e + 2d + c - a}{5}$$

Where L and H = calculated zone diameters for the lowest and highest (0.0125 and 0.2 IU/mL) concentrations for the standard response line; c = average zone diameter of the 45 zone diameters for the reference concentration of 0.05 unit/mL; a,b,d,e = corrected average zone diameters for each of the other concentrations used for the standard response line.

3. Controls: In using this assay, the lowest standard concentration (0.00625 IU/mL) is intended to be a control, which may produce negative results. This lowest concentration represents control diluent

to which penicillin has been added at a level that is normally below the limit of detectability. On occasion, the lowest concentration (0.00625 IU/mL) will produce a measurable zone of inhibition. The next highest concentration (0.0125 IU/mL) should always produce results. The sensitivity of this assay is normally 0.01 IU/mL. The nonfat dry milk plus phosphate buffer should always produce negative results.

4. Sample examination:
   a. Screening procedure: Accurately weigh 10.0 g of nonfat dry milk and add 90 mL of buffer [14.7A4]. Mix thoroughly. If the concentration of penicillin is expected to be greater than 0.2 IU/mL in this mixture (2.0 units/g of original nonfat dry milk), dilute a portion of the reconstituted milk with control diluent to an estimated 0.05 IU/mL.

   Place six cylinders [14.7A5] on each assay plate [14.7B1] at 60° intervals on a 2.8-cm radius. Use three plates per test. Two samples can be examined per test. On each of the three plates, fill two cylinders [14.7A5] with the 0.05 IU/mL penicillin reference standard and two cylinders with each of the two untreated samples. Incubate plates for 16 to 18 h at 30 ± 1°C. After incubation, invert the plates to remove the cylinders and then measure the diameter(s) of the clear zones of inhibition produced. Average six control readings for the reference concentration and six readings for each sample. Any test sample producing an average zone of inhibition of 9.1 mm or greater is a presumptive positive test and must be confirmed as in [14.7B4b].

   b. Confirmatory procedure: To identify the activity as penicillin, add penicillinase conc. [14.7A13] at the rate of 0.5 mL per 10 mL of sample. Mix, and incubate for 30 min at 37°C. Use three assay plates [14.7B1] per test sample. On each of the three plates, fill two cylinders [14.7A5] with the 0.05 IU/mL penicillin reference standard, two cylinders with the untreated sample, and two cylinders with the pencillinase-treated sample. Incubate the plates for 16 to 18 h at 30 ± 1°C. After incubation, invert the plates to remove the cylinders.

   Measure the diameter of each zone of inhibition as accurately as possible. Determine the average for each of the three types of zone readings (the standard, untreated sample, and penicillinase-treated portion) on the three plates.

   c. Interpretation: A zone of inhibition with the untreated sample and no zone with the penicillinase-treated portion of the sample is a positive test for penicillin. If the average zone of inhibition of the penicillinase-treated portion of the sample is the same size as the average zone produced by the untreated portion of the

same sample, penicillin is not present. If the average zone of inhibition produced by the penicillinase-treated portion of the sample is ⩾ 0.1 mm less than the average zone of inhibition produced by the untreated portion of the same sample, penicillin plus another nonspecific inhibitor is present.
   d. Calculation of penicillin concentration: To calculate the antibiotic content of a sample, average the zone readings of the six 0.05 IU/mL control zone readings and the six zone readings of the sample on the three plates. If the sample gives a larger average zone size than the average of the control, add the difference between them to the zone size of the 0.05 IU/mL reference standard of the response line to obtain the concentration of the sample from the response line [14.7B2]. If the average sample value is lower than the average of the 0.05 IU/mL control, subtract the difference between them from the 0.05 IU/mL reference standard of the response line to obtain the concentration of the sample from the response line [14.7B2]. Multiply the concentration in IU/mL obtained from the response line by the dilution factor of 10× to obtain the final concentration in terms of IU/gram. If additional dilution of the sample has been made, the appropriate dilution factor must be taken into account when calculating the final potency.

C. *Reporting:*
Report samples found positive for penicillin. Report the absence of penicillin as "no zone." Samples found positive for nonspecific inhibitors must be assayed by other methods before reporting.

14.8 Specific Antimicrobial Cylinder Methods

Methods to assay milk and milk products (cheese, buttermilk, ice cream, dry milk, and evaporated milk) for specific antibiotics (bacitracin, chlortetracycline, oxytetracycline, tetracycline, streptomycin, dihydrostreptomycin, neomycin, penicillin, polymyxin, erythromycin, and novobiocin) have been described.[16] If the antibiotic present must be identified, that publication should be consulted.

14.9 Detection of Sulfa Drugs and Antibiotics in Milk (Class C)

This disc assay procedure[26] uses the spore-forming *Bacillus megaterium*. This organism is sensitive to eight sulfa drugs, to bacitracin, and to eight other antibiotics commonly used in mastitis therapy.
   A. *Equipment and supplies:*
      1. Test organism: *B. megaterium* (ATCC 9855).
      2. Prescription bottles: Capacity of 180 mL.
      3. AK Sporulation Medium #2.

4. Incubator: Providing 37°C temperature.
5. Mueller-Hinton agar: [Chapter 5].
6. Filter paper discs: Diameter 1.27 cm (Carl Schleicher and Schuell Company, or various supply houses).
7. Spectrophotometer.
8. Plastic petri dishes: ID 90 mm.
9. Centrifuge: Refrigerated.
10. Refrigerator.
11. Water bath: Capable of 50°C temperature.

B. *Preparation of spore suspension.*
1. Transfer cells of *B. megaterium* ATCC 9855 by streaking the entire surface of sterile AK sporulation medium #2 in a 180-mL prescription bottle previously prepared.
2. Incubate inoculated prescription bottle of agar for 48 h at 35°C.
3. After incubation, wash the spores and vegetative cells from the agar surface with buffered MS water. Sediment the spores and cells by centrifugation at 5,000 × g for 15 min at 3°C.
4. Store the spore suspension in buffered MS water at 0 to 4.4°C until used.

C. *Testing procedure:*
1. Prepare Mueller-Hinton agar and keep it liquid in a water bath at 50°C. Fresh agar should be prepared each day of testing.
2. Adjust the inoculum to an optical density calculated to give a final concentration of about 5×10^4 spores per milliliter of agar.
3. Add the spore suspension to give the approximate concentration desired to the liquid agar at 50°C.
4. Mix the agar and spore suspension by swirling gently to avoid air bubble formation.
5. Pipette 4 mL of inoculated medium into petri dishes and distribute over dishes by rotation. Allow the agar to harden.
6. Proceed to test milk samples by using 1.27-cm discs dipped in milk samples and placed on the agar surface. Space the discs so that zones will not overlap. Incubate the plates at 37°C for 4 to 5 h. This incubation temperature affords maximum sensitivity to sulfa drugs.

D. *Results:*

Any clear zone around the periphery of the 1.27-cm disc is considered a positive test, provided that proper controls are used. NOTE: A1.27-cm disc impregnated with 50 μg of para-aminobenzoic acid (PABA) may be used to identify sulfa drug inhibitors. Discs with 50 μg of PABA overcome the inhibitory properties of sulfa drugs up to concentrations of 5 μg per disc. Identification discs are prepared by adding PABA in aqueous solution to them, followed by drying of the discs at 40 to 44°C.

14.10 References

1. ABRAHAM, E.P.; CHAIN, E.; FLETCHER, C.M.; FLOREY, H.W.; GARDNER, A.D.; HEATLEY, N.G.; JENNINGS, M.A. Further observations on penicillin. Lancet **241**:177–189; 1941.
2. ARRET, B.; KIRSHBAUM, A. A rapid disc assay method for detecting penicillin in milk. J. Milk Food Technol. **22**:329–331; 1959.
3. CARTER, C.G. Detection of penicillin in dry powdered milk by the *Sarcina lutea* cylinder plate method. Nat. Center for Antibiotic Analysis, FDA, USDHEW, Washington, D.C.; 1974.
4. CERNY, J.; MORRIS, R.L. A modified disc assay method for detecting antibiotics in milk. J. Milk Food Technol. **18**:281–283; 1955.
5. CHARM, S.E.; CHI, R.K. Rapid screening assay for beta-lactam antibiotics in milk: collaborative study. J. Assoc. Offic. Anal. Chem. **65**:1186–1192; 1982.
6. GINN, R.E.; CASE, R.A.; PACKARD, V.S.; TATINI, S.R. Quantitative estimates of beta-lactam residues in raw milk around a reference standard: Collaborative study. J. Assoc. Offic. Anal. Chem. **65**:(6)1407–1412; 1982.
7. GINN, R.E.; CASE, R.A.; PACKARD, V.S.; TATINI, S.R. Quantitative assay of beta-lactam residues in raw milk using a disc assay method. J. Food Prot. **45**:571–573; 1982.
8. GROVE, D.C.; RANDALL, W.A. Assay methods of antibiotics. Medical Encyclopedia, Inc., N.Y.; 1955.
9. HORWITZ, W., ED. Official methods of analysis, 13th ed. Assoc. Offic. Anal Chem.; 1980.
10. HORWITZ, E., ED. Official methods of analysis, 13th ed. (Suppl. 3). Assoc. Offic. Anal. Chem.; 1982.
11. IGARASHI, R.T.; BAUGHMAN, R.W.; NELSON, F.E.; HARTMAN, P.A. Rapid antibiotic assay methods using *Bacillus stearothermophilus*. J. Milk Food Technol. **24**:143–146; 1961.
12. JOHNS, C.K. Further observations on testing milk for penicillin. J. Milk Food Technol. **23**:266–268; 1960.
13. JOHNS, C.K.; BERZINS, I. Observations on the determination of antibiotics in milk. J. Milk Food Technol. **19**:14–17; 1956.
14. KELLEY, W.N. Qualitative ampule and multitest for beta-lactam residues in fluid milk products. Collaborative study. J. Assoc. Offic. Anal. Chem. **65**:1193–1207; 1982.
15. KOSIKOWSKI, F.V.; HENNINGSON, R.W.; SILVERMAN, G.F. The incidence of antibiotics, sulfa drugs and quaternary ammonium compounds in the fluid milk supply of New York State. J. Dairy Sci. **35**:533–539; 1952.
16. KOSIKOWSKI, F.V.; LEDFORD, R.A. A reverse-phase disc assay test for antibiotics in milk. J. Amer. Vet. Med. Assoc. **136**:297–301; 1960.
17. KOSIKOWSKI, F.V. Induced and natural inhibitory behavior of milk and significance to antibiotic disc assay testing. J. Dairy Sci. **46**:95–101; 1963.
18. KRAMER, J.; CARTER, G.G.; ARRET, B.; WILNER, J.; WRIGHT, W.W.; KIRSHBAUM, A. Antibiotic residues in milk, dairy products, and animal tissues. Nat. Center for Antibiotic and Insulin Analysis. FDA, USDHEW, Washington, D.C.; 1968.
19. LOO, Y.H.; SKELL, P.S.; THORNBERRY, H.H.; EHRLICH, J.; MCGUIRE, J.M.; SAVAGE, G.M.; SYLVESTER, J.C. Assay of streptomycin by the paper-disc plate method. J. Bacteriol. **50**:701–709; 1945.
20. MACAULAY, D.M.; PACKARD, V.S. Evaluation of methods used to detect antibiotic residues in milk. J. Food Prot. **44**:696–698; 1981.
21. MARTH, E.H.; ALEXANDER, F.J.; HUSSONG, R.V. Studies on disc assay methods for detection of antibiotics in milk. J. Milk Food Technol. **26**:150–155; 1963.
22. MESSER, J.W.; LESLIE, J.E.; HOUGHTBY, G.A.; PEELER, J.T.,; BARNETT, J.E. *Bacillus stearothermophilus* disc assay for detection of inhibitors in milk: collaborative study. J. Assoc. Offic. Anal. Chem. **65**:1208–1214; 1982.
23. PACKARD, V.S.; TATINI, S.; GINN, R.E. An evaluation of methods for detecting and

comparative incidence of penicillin residues in different types of raw milk supplies. J. Milk Food Technol. **38**:601–603; 1975.
24. Parks, O.W.; Doan, F.J. Sensitivities of the disc assay and triphenyltetrazolium methods for antibiotics in milk. J. Milk Food Technol. **22**:74–76; 1959.
25. Pater, B. A collaborative study of the Delvotest-P method to detect low concentrations of penicillin in milk. J. Food Prot. **40**:23–24; 1976.
26. Read Jr., R.B.; Bradshaw, J.G.; Swartzentruber, A.A.; Brazis, A.R. Detection of sulfa drugs and antibiotics in milk. Appl. Microbiol. **21**:806–808; 1971.
27. Siino, F.A.; Czarnecki, R.B.; Harris, W.K. The incidence of penicillin in the market milk supply of a local New England area. J. Milk Food Technol. **21**:211–214; 1958.
28. Silverman, G.F.; Kosikowski, F.V. Systematic testing of inhibitory substances in milk. J. Milk Food Technol. **15**:120–124, 137; 1952.
29. Van Os, J.L.; Lameris, S.A.; Doodewaard, J.; Oostendorp, J.G. Diffusion test for the determination of antibiotic residues in milk. Neth. Milk Dairy J. **29**:16–34; 1975.
30. Vincent, J.G.; Vincent, H.W. Filter paper disc modification of the Oxford cup penicillin determination. Proc. Soc. Exper. Biol. Med. **55**:162–164; 1944.
31. Wright, W.W. Overnight microbial plate assay for penicillin in milk. J. Assoc. Offic. Agric. Chem. **45**:301–306; 1962.

CHAPTER 15

MICROBIOLOGICAL TESTS FOR EQUIPMENT, CONTAINERS, WATER AND AIR

R. Y. Cannon, C. E. Beckelheimer, and R. Burt Maxcy
(R. T. Marshall, Tech. Comm.)

15.1 Introduction

Good sanitation practices require that contamination of dairy products from equipment, containers, water and air be minimized. Following pasteurization of the product, contamination with pathogenic microorganisms has direct public health significance. In addition, contamination with psychrotrophic organisms has a direct bearing on the shelf life of refrigerated dairy products. Even one psychrotrophic organism in a package of fluid milk could be significant in reducing shelf life.[22]

15.2 Sample Collection [Chapter 4]

15.3 Tests for Equipment Sanitation

A. Rinse solution methods:

Effectiveness of cleaning and sanitizing containers and equipment can be determined by repeated flushing of a measured volume of sterile, buffered sanitizer-neutralizing solution or nutrient broth over the product contact surface and then determining the bacterial population of the rinse solution by plating or membrane filter techniques.[19]

Large equipment assemblies can be checked by circulating water that has been chlorinated and neutralized, or sterilized by membrane filtration.[31] The increase in bacterial population is determined by membrane filter techniques.[19] Sanitation of assemblies may also be checked by sampling milk at successive points of access along processing lines.[38]

1. Equipment and supplies:
    a. Transfer pipettes, sterile [2.4].
    b. Petri dishes, sterile [2.4].
    c. Standard Methods Agar, Violet Red Bile Agar, MF Endo Broth.
    d. Stock phosphate solution [5.3].

e. Buffered rinse solution, sterile: Add 1.25 mL of stock phosphate buffer solution, 5 mL of 10% aqueous sodium thiosulfate, 4 g of Azolectin (Associated Concentrates, 32-60 Sixty-First St., Woodside, NY 11377) and 10 g of Tween 20 (ICI Americas, Inc., Wilmington, DE 19897) to MS water and make to 1 L. Adjust to pH 7.2 and dispense sufficient quantities into screw-capped vials made up to contain 20 mL, 100 mL or other suitable volumes, following sterilization. Azolectin is hygroscopic and should be stored in a desiccator. Weigh the powder and rapidly dissolve by heating over boiling water. (Neutralizing Buffer is available in dehydrated form from Difco Laboratories, Detroit, MI 48201. A solution similar to the above rinse solution is produced on rehydration.)

f. Rinse water for large volume rinsing of "cleaned in place" (CIP) equipment: Large volume of water may be treated or sanitized by chlorinating to a residual concentration of 25 mg/L, holding for 10 min., and then neutralizing by adding an excess of 10% sodium thiosulfate solution.

Tap water may also be sterilized by membrane filtration followed by addition of sodium thiosulfate (1 mL of 10% solution per L of water) to inactivate residual disinfectant.[31]

g. Sodium thiosulfate solution, 10%: Dissolve 100 g of $Na_2S_2O_3 \cdot 5H_2O$ in MS water and make up to 1 L; filter; store in the refrigerator in a dark place, or in an amber bottle.

h. Hypodermic syringe and needles, sterile.

i. Ethyl alcohol, 70%.

2. Rinsing containers (Class 0): Sample containers, obtained as directed in [4.1C4] will be handled as follows:

a. Firm-walled capped items: Remove containers from the conveyor line without touching lips or interiors before containers reach the filler valves. Aseptically introduce 20 mL of sterile, buffered rinse solution [15.3A1e] into each container, then return the containers to the conveyor lines beyond the filler valves, to be capped mechanically. For 3.78-L or larger containers, introduce 100 mL of rinse solution. Should inspection show that mechanical cappers are in an unsanitary condition, apply laboratory-sterilized closures to three other containers taken from the conveyor before they reach the filler valves. Proceed as in [15.3A2c].

b. Firm-walled uncapped items: Paper, laminated or plastic containers that are formed, filled, and sealed in a single device must be obtained after sanitization but before the product to be packaged is introduced. Swab a portion of the side surface of containers with 70% ethanol, aseptically fill a sterile hypodermic syringe with sterile, buffered, rinse solution, and inject into the

15.3 Tests for Equipment Sanitation

container. For 0.5- to 1-L containers use 20 mL; for 1- to 2-L containers use 50 mL; and use 100 mL for up to 4-L containers. Aseptically apply a piece of cellophane tape to cover the area punctured by the needle. Proceed as in [15.3A2c].

   c. Shaking firm-walled containers: Grasp the container firmly with its long axis in horizontal position. Shake vigorously 10 times, each shake being a "to-and-return" thrust of about 20 cm. Turn the container 90° and repeat the horizontal shaking treatment; turn the container 90° twice more, repeating the horizontal shaking treatment each time. Swirl the container vigorously 20 times in a small circle with its long axis in vertical position, then invert and repeat. Stand container upright before removing the sample.

   d. Flexible-walled items: Swab all caps on fill tubes of collapsed liners with a soft paper towel moistened with 70% ethanol, then remove and introduce, aseptically, 20 mL of buffered rinse solution [15.3A1e] with a sterile pipette. For 3.78-L or large containers, introduce 100 mL of rinse solution. Where tubes are sealed, swab an area of the tube adjacent to the liner with 70% ethanol and introduce rinse solution with a sterile hypodermic syringe; apply cellophane tape to the puncture area.

     Arrange the collapsed liner on a smooth, clean, firm horizontal surface as flat as its construction permits. With hands or a roller, "work" the rinse solution back and forth 10 times, so it contacts all interior surfaces completely. With sterile scissors, cut into an end of uncapped containers to obtain contents. Lift fill tube of flexible liner before removing closure (or cutting tube with sterile scissors) to avoid losing rinse solution. Then lower tube to permit aseptic transfer of rinse to sterile container.

   e. Equipment assemblies: The rinse solution method may be employed on individual items of equipment or on hookups designed for cleaning in place. A control sample (approximately 1 L) of treated rinse water is obtained before rinsing. A sufficient volume of treated rinse water [15.3A1f] is added to a storage vat or tank at the upstream end of the assembly, then pumped or allowed to flow by gravity through the assembly. Samples are collected at the equipment's point of discharge from the first, middle, and final portions of rinse water. Samples may also be collected at various points throughout the assembly.

3. Plating rinse solutions and counting plates:
   a. Container sample measurements: Distribute a 10 mL sample of the 20-mL, 50-mL, or 100-mL rinse solution equally between two sterile petri dishes; seed three other petri dishes with 1-mL portions of the rinse. Pour dishes that received 10 mL of rinse solution with 15 to 18 mL of Standard Methods Agar; pour others

in the usual manner. Incubate plates at 32°C for 48 ± 3 h and count all colonies. To calculate the Residual Bacterial Count (RBC) per specified container capacity, multiply the total number of colonies by the volume of rinse solution divided by the volume of sample plated. For example, if the volume of rinse is 20 mL, the volume of sample plated is 10 mL and the number of colonies is 10, the RBC is 10 x (20 ÷ 10) = 20.

Numbers of coliforms in rinse solutions are determined by dividing 5 mL of rinse solution between two plates and culturing as directed in [8.8]. To calculate Coliform Count per specified size of container, multiply number of coliform colonies by volume or rinse solution divided by volume of sample plated.

Membrane filter procedures[19] may also be used with the rinse solutions to enumerate coliforms, fecal coliforms and fecal streptococci.

Yeasts, molds, proteolytic bacteria, or other specific microorganisms may be determined similarly by using appropriate differential media and incubation temperatures and times.

b. Equipment samples: Rinse solutions obtained from cleaned in place (CIP) systems or processing assemblies and control samples require use of the membrane filter procedure[19] for analysis. Average the yields of rinse samples taken at the beginning, middle and end of drainage and subtract the yield (if any) of control samples. The result is the "corrected yield." Calculate ratio of sample volume to rinse volume and multiply by corrected yield to obtain an indication of numbers of organisms present in the entire system. Presence of specific types of organisms can be determined by employing appropriate differential media and incubation conditions, e.g., MF-Endo Broth would be used for coliform counts [Chapter 8].

Milk samples obtained from points along processing lines [Chapter 4] may be plated directly to determine where significant changes in counts occur. Presence of psychrotrophic bacteria may be determined by incubating samples at 7°C for 5 da and then replating. A substantial increase in count indicates presence of psychrotrophs. Points of entry of other types of microorganisms can be determined by plating with appropriate differential media.

4. Screening method (nutrient broth modification) for retail milk containers (Class 0): Plates poured from three containers, employing either buffered rinse solution or nutrient broth, sometimes yield no colonies. To determine the frequency with which such containers are encountered, a large number of samples must be examined. Inoculate 20-mL portions of nutrient broth into 50 containers and shake them to rinse their surfaces thoroughly. Preferably incubate

15.3 Tests for Equipment Sanitation

containers at 32°C; otherwise room temperature is acceptable. After 48 h, note the percentage of containers with turbid broth, an indication of bacterial growth. Calculate the percentage of containers exhibiting growth.

B. *Disintegration method to determine the microbial content of paper container materials:*

The disintegration method is applicable to paper, paperboard or molded-pulp items in intermediate or complete stages of conversion. The method is not applicable to items fabricated from plastic, dry-waxed or wet-waxed paper, metal and/or combinations of these materials, where the plastic or metal is the food-contacting surface. For these latter items the rinse or swab method should be used.

Procedures for obtaining and testing finished single-service containers and closures are contained in the *Manual of Recommended Methods for Sampling and Microbiological Testing of Single Service Food Packages and Their Components.*[35]

1. Equipment and supplies:
   a. Pipettes, sterile: Capacity 10 mL, with tip opening cut to about 3 mm for passage of pulp. A 20-mL large-bore pipette is preferable for plating paper pulp.
   b. Petri dishes, sterile.
   c. Plating medium: Standard Methods Agar.
   d. Phosphate-buffered dilution water, sterile.
   e. Scalpels or scissors, sterile.
   f. Forceps, sterile.
   g. Balance.
   h. Kraft paper or envelopes, sterile: For protection of samples in transit to laboratory.
   i. Disintegrator blender: Sterile interior, high-speed, electrically operated, with corrosion-resistant metal (stainless steel preferred) cup, capacity about 500 mL. Optionally, use 500 mL in a 1,000-mL cup.

2. Sampling procedure for paper stock: Sample paper stock before it enters into any converting operation. Partially converted items may also be sampled in intermediate stages of conversion. Sampling methods for these items shall, in general, comply with sampling methods for paper stock or finished products. Items fabricated from plastic or plastic-laminated paper or paperboard, dry-waxed paper, wet-waxed paper, metal and/or combinations of these materials are usually not sampled before conversion; therefore, sampling methods for finished products only would apply.
   a. Roll-stock sampling: Either of the following two methods is acceptable.

1. Roll stock shall be sampled as a butt roll, with a minimum radial thickness of 2.54 cm of paper on the core and a minimum roll width of 30.48 cm, except where mill roll width is less than 30.48 cm, except where mill roll width is less than 30.48 cm and where the full width of the mill roll constitutes the sample. The cut web shall be firmly taped before removal of the butt roll from the roll stand. If the width of the roll is over 30.48 cm, for convenience the butt roll may be cut to a minimum of 30.48 cm, by using either a handsaw or a power band saw, using care to avoid contamination. The roll-stock sample shall be wrapper carefully in wrapping paper and sealed with tape.
2. Samples that are to be cut from the roll of paper shall be cut with a clean sharp knife as follows: Cut into roll two laps deep and discard these first two layers. Then cut into roll six laps deep to a size of approximately 20.22 × 25.40 cm, grasp sample by top and bottom sheets near corners, remove and tape in three or four places to keep sheets in alignment, thus protecting inner sheets, which will be tested. When cutting samples from the roll of paper, three sides of the sample may be cut first, thus forming a flap. This flap may then be stapled or taped to prevent undue exposure and riffling and the hinge side of the sample finally cut as assembled sample is removed. Place in a clean paper envelope, seal and insert into a stout manila or brown kraft envelope. For sample protection, envelopes should be a type that can be sealed, preferably with a pressure-sensitive adhesive rather than by moistening.
   b. Sheeted-stock sampling: Final samples shall consist of a minimum of six sheets, each approximately 20.22 × 25.40 cm. Sheeted paper selected for sampling shall be taken from any portion of the stock except the bottom or top 5.80 cm. Large sheets shall be reduced to a final sample size by using a guillotine-type cutter or equivalent means not involving direct handling of any sample sheets. Stacks of cut sample sheets shall be taped and packaged as in paragraph 2 above. Top and bottom sheets shall be considered as protective covers during sampling and shipping and are to be discarded before testing.
3. Sample preparation: With sterile scalpel or other sterile cutting device, cut representative portions that equal 100 g from rolls of paper or paperboard, sheeted stock, nested containers, blanks or closure discs. With sterile forceps, transfer cut portions to sterile kraft wrappers or envelopes. Using sterile forceps, place bottle caps, hoods and other similar closures directly into sterile envelopes. Protect samples from moisture or other contamination during transit

15.3 Tests for Equipment Sanitation

to the laboratory. Because some papers contain a few bacteria, take great precautions to avoid dust, sputum or other contamination of samples. With sterile scalpel or scissors, cut the middle portion of the sample sheet into pieces [maximal size 8.6 cm], and collect the pieces in a sterile petri dish. Using sterile forceps to handle stock, weigh 3 g of sample directly into a sterile petri dish. Put 300 mL of sterile, buffered dilution water into a large size sterile mixer cup, add 3 g of paper, and re-cover the mixer. In larger cups, use 500-mL portions of sterile, buffered dilution water and 5 g of paper. Place the cup on the blender driving mechanism, operate at low speed for 15 sec and then at high speed so that total time of operation usually does not exceed 2 min. Time of operation will depend on type of paper, however, and may require 5 min. After 30 sec of high-speed operation, stop the machine. By revolving the cup at a slight angle, rinse all pieces of paper or paperboard from cup walls into water. Resume disintegration at high speed, but at intervals re-examine the interior to be sure that no pieces of undisintegrated material remain on sides of the cup.

4. Plating: Because a few bacteria per gram are found in high-quality paperboard used in making bottle caps, closures and milk containers, and because these bacteria are largely spore-forming types that usually produce spreading colonies on agar plates, accurate counts are often difficult to obtain. Protect plating operations from all dust or other forms of contamination. For high-grade virgin-stock paperboard, no further dilution is necessary.

    With a sterile, 10-mL pipette having a large bore at the tip, divide 10 mL of distintegrated paper or paperboard approximately equally among three sterile petri dishes. Pour with Standard Methods Agar, incubate plates at 32 ± 1°C for 48 ± 3 h, and count colonies. When introducing medium, rotate the plate so medium is thoroughly mixed with the distintegrated paper or paperboard. When analyzing board that contains secondary stock, higher dilutions are usually required. Keep the pulp sufficiently concentrated to yield five or more colonies per plate. Optionally, use five instead of three plates per sample.

5. Results, reporting, and standards: Multiply the sum of colonies developing on three or five plates from 0.1 g of paper or paperboard stock by 10 and report the result as the number of colonies per gram of stock. Some industry specifications allow not more than 250 colonies per gram of disintegrated paper or paperboard stock in three of the last four analyses of these products taken at the mill from different batches. A specification of not more than 250 colonies per gram has also been used. This is calculated as the logarithmic average of plate counts of the last four analyses and is therefore somewhat more lenient than the "three of the last four" specification.

6. Test for coliforms: The procedure to be used is outlined in the manual previously cited.[35]

C. *Surface contact methods:*

Microbiological examination of surfaces requires selection of the proper method. RODAC Replicate Organism Direct Agar Contact (RODAC) (Becton Dickinson and Company, Paramus, NJ 07652) plates and the swab procedure are the methods of choice. The swab technique should be used for areas such as cracks, corners, or crevices, i.e., areas of such dimensions that the swab will be the most effective way to recover organisms from them. The RODAC procedure is better used on flat surfaces such as walls, floors, ceilings and equipment surfaces. Selection of the proper technique is essential for meaningful results.[1,17]

1. Swab contact method (Class A1): The swab technique consists of brushing a sterile swab moistened in an appropriate solution over a measured surface area. The swab is then rinsed thoroughly in a measured amount of sterile solution and the liquid examined microbiologically. Select areas for swabbing that appear to be unclean or difficult to clean. Equipment surfaces that are regular, smooth, ground, polished, large or accessible are easier to clean than those that are irregular, rough-seamed, ridged, small, rounded on small radii, or less conveniently reached. If these latter areas are not properly cleaned, residues will accumulate, and surfaces will not be sanitized when chemical solutions are applied.

   a. Equipment and supplies:
      1. Screw-capped vials: Dispense rinse solution [15.3A1e] in small screw-capped vials (7 to 10 cm long) prepared so they will contain 5 mL of solution after autoclaving. (If swabs with wood applicator sticks are autoclaved in the rinse solution, ascertain that there is no change in pH.)
      2. Cotton swabs: Sterile nonabsorbent cotton, firmly twisted to approximately 5 mm in diameter by 2 cm long over one end of a wood applicator stick 12 to 15 cm long. Package as individuals or multiples in convenient, protective containers (glassine envelopes, test tubes, or vials), with swab heads away from the closure, and autoclave. If toxicity of cotton swabs is suspected, autoclave a representative sample of swabs in MS water (5 mL per swab) at 121°C for 15 min and do a water quality test for MS water [Chapter 5] to show the absence or presence of toxicity. Davron or rayon swabs may be used.[26]
      3. Calcium alginate swabs: Swabs made of calcium alginate fibers[3,7,10,25,40] are soluble in aqueous solutions (rinse, culture media, etc.) containing 1% of sodium hexametaphosphate or sodium glycerophosphate, or sodium citrate. All organisms

dislodged from swabbed surfaces are thus liberated. Then using calcium alginate swabs (Spectrum Diagnostics, Inc., Glenwood, IL 60425), prepare rinse solution vials so they will contain 4.5 mL (rather than 5.0 mL) after sterilization.
4. Solubilizing solution for calcium alginate swab: Prepare a 10% solution of sodium hexametaphosphate or other solubilizing compound, package in appropriate vials and autoclave.
5. Mechanical shaker (optional).

b. Procedure:[9,39] Open the sterile swab containers; grasp the end of a stick, being careful not to touch any portion that might be inserted into a vial, and remove aseptically. Open a vial of buffered rinse solution; moisten the swab head and press out excess solution against the interior wall of the vial with a rotating motion. Hold the swab handle to make a 30° angle with the surface. Rub the swab slowly and thoroughly over approximately 50 cm^2 of surface three times, reversing direction between successive strokes. Move swab in path 2 cm wide by 25 cm long or other dimensions to cover equivalent area. Return the swab head to the solution vial; rinse briefly in solution, then press out the excess. Swab four more 50 cm^2 areas of sanitized surface, as above, rinsing the swab in solution after each swabbing; remove excess liquid. After the fifth area has been swabbed, position the swab head in the vial and break or cut it with sterile scissors, leaving the swab head in the vial. Replace screw cap; put vial in a waterproof container packed in cracked ice and deliver to the laboratory at 0 to 4.4°C. With calcium alginate swabs follow the standard procedure, but after swabbing the final (fifth) 50 cm^2 area and depositing the swab head in the rinse vial add 0.5 mL of sterile solubilizing solution.

c. Preparing and plating rinse solutions from swabs: At the laboratory, remove vial from refrigerated storage. Shake vigorously, making 50 complete cycles of 15 cm in 10 sec., striking the palm of the other hand at the end of each cycle. Groups of vials may be put into appropriate racks and shaken manually; or mechanical shaking may be employed for a time interval that treats cotton swab heads similarly to the manual procedure. Plate 2-, 1- and/ or 0.1-mL portions of rinse solutions [Chapter 6]. Plate additional decimal dilutions, if deemed necessary. Pour plates with Standard Methods Agar. Incubate plates, count colonies, and then calculate the number of colonies recovered from 50 cm^2 (equivalent to 1 mL of rinse). When other than aerobic plate counts are sought, plate with appropriate selective or differential media and incubate as required.

2. Cellulose sponge method (Class C):[37] Food processing equipment and environmental surfaces may be sampled using the cellulose sponge swab technique. Cellulose sponges are cut into 5 cm³ pieces, placed in individual kraft paper bags, and steam sterilized at 121°C for 20 min. The sterile sponge is removed from the bag using sterilized or sanitized crucible tongs and moistened with 10 mL of buffered rinse solution or a 10% solution of Tergitol Anionic-7 (Union Carbide, Chicago, IL 60638). It may be desirable, depending on the organism being sought, to replace the Tergitol Anionic-7 with another sampling fluid (0.1% peptone water or buffered rinse solution) because of the possible deleterious effects of Tergitol on gram positive bacteria.

Instead of using crucible tongs, the hands may be covered with sterile or sanitized rubber or plastic gloves to handle the sponge. When sampling for salmonellae, it is sufficient to thoroughly wash the hands with soap, followed by sanitization in a solution of 200 ppm available chlorine.

The surface is sampled by vigorously rubbing the sponge over the designated area until the soil is removed. An area of several square meters may be efficiently swabbed. If the surface is flat, the rinse solution may be applied directly to the surface and then taken up into the sponge by the rubbing action. After sampling, the sponge is introduced into a sterile plastic bag, and transported under refrigeration to the laboratory.

Because of the large areas that can be sampled with cellulose sponge, this technique is particularly useful in detecting salmonellae on food plant equipment and in the environment. When testing for salmonellae, the the sponge is introduced directly into lactose broth, which is then tested by conventional techniques.

The sponge sample may be subjected to a variety of analytical procedures such as the aerobic plate count or counts of coliforms, yeasts and molds, staphylococci. For quantitative analyses, 50 to 100 mL of diluent is added to the bag containing the sponge. The sponge is then vigorously massaged with the diluent for 1 min or more. Portions of the diluent are then removed from the bag and plated in the desired medium, followed by appropriate incubation time and temperature. The numbers of microorganisms per unit surface area are calculated on the basis of the area swabbed, the amount of diluent used, and the size of the sample plated. For example, if 50 colonies are obtained from a 1-mL sample derived from a sponge in 100 mL of diluent that swabbed 1 m², the aerobic plate count/m² will be 5,000.

3. RODAC plate (agar contact) method (Class 0):[4,8,21,41] The RODAC plate method provides a simple agar contact tool for sampling

15.3 Tests for Equipment Sanitation

surfaces. It is of value in estimating sanitary quality of surfaces encountered in dairy operations, food-processing plants, and hospitals. Collaborative studies[4] have indicated that the mean percent recovery of aerosolized *Bacillus subtilis* spores from stainless steel surfaces was slightly less than 50%, both for the swab (47%) and RODAC plate (41%) methods. Results from the RODAC plate method were more reproducible, however, than those from the swab technique. The RODAC plate method is particularly recommended when quantitative data are sought from flat, impervious surfaces. It should not be used for pervious, creviced, or irregular surfaces.

Product contact surfaces amenable to RODAC sampling necessarily will be flat and impervious, and thus relatively easy to clean and disinfect. Other environmental surfaces, such as floors, are more likely to be heavily contaminated. In either event, a sufficient number of spots should be sampled to obtain representative data.

a. Apparatus and materials: Disposable plastic RODAC plates (Becton Dickinson Labware, Oxnard, CA 93030) may be filled in the laboratory or may be purchased prefilled with test medium (Difco, Inc., Detroit, MI 49233 or BBL Microbiology Systems, Cockeysville, MD 21030). They should be prepared with 15.5 to 16.5 mL of Standard Methods Agar, and the meniscus of the agar should rise above the rim of the plate. Where surfaces to be sampled have been subjected to germicide treatment, incorporate appropriate neutralizers into the medium. Use 0.5% polysorbate (Tween 80) to neutralize substituted phenolic disinfectants and 0.07% soy lecithin to neutralize quaternary ammonium compounds.

   Dey-Engley medium may be used.[14] It contains a variety of ingredients able to neutralize germicides likely to be used on surfaces.

   If qualitative data are sought, differential media may be used, or, preferably, colonies should be transferred to differential media from the original broad-spectrum medium. Following preparation, plates should be incubated at 32°C for 18 to 24 h as a sterility check. They should then be used within the succeeding 12 h unless appropriately wrapped and refrigerated to slow dehydration.

b. Procedure: Remove plastic cover and carefully press agar surface to surface being sampled. Make certain that entire agar meniscus contacts the surface, using a rolling uniform pressure on the back of the plate to effect contact. Replace the cover and incubate plate in an inverted position for 24 to 48 h at 32°C. When heavy contamination is present, incubate 24 h to prevent overgrowth. When contamination is light, 48 h should be allowed for growth of colonies. Count colonies and record as either number of colonies per RODAC plate or number of colonies per cm^2.

15.4 Standard Tests for Water Supplies

To be suitable for use in dairy operations, water must be of a safe sanitary quality and practically free of microorganisms that could initiate product spoilage. Spoilage of refrigerated milk and milk products has been caused by contamination with water-borne organisms (primarily psychrotrophs).[42] This contamination has occurred either directly, through product contact with the water itself (e.g., by milk traversing a surface that was freshly rinsed with water, or by cheese curd or butter having been washed in water); or indirectly, by microbes metabolizing nutrient residues on incompletely cleaned equipment surfaces. Since most of the microorganisms of water are destroyed by chlorination or heat, in many plants all water is treated to keep spoilage at a minimum.

Procedures to sample and test water supplies to determine conformance with requirements for coliform concentration by both most probable number (MPN) and membrane filter methods are completely detailed in the most recent edition of Standard Methods for the Examination of Water and Wastewater.[19] Although limitations for aerobic bacterial counts are not included in National Interim Primary Drinking Water Regulations, information on numbers of organisms developing at various incubation temperatures is helpful both in classifying untreated waters and in assaying the efficiency of treatment procedures. The membrane filter technique is particularly useful to determine numbers of organisms in supplies containing so few that the usual plating procedures do not yield meaningful results.

15.5 Tests for Microbiological Quality of Air

The demand for improved shelf life of dairy products has put increased emphasis on the microbiological quality of air in dairy environments. Entrance of a very few viable contaminants from air into pasteurized and cultured dairy products can cause spoilage.[11] Microorganisms must be prohibited from entering sterilized dairy products. The requirement that dry milk products be free of salmonellae emphasizes the need to reduce the microbial content of air used for cooling and/or instantizing of dry milk.

Sources of airborne contamination in dairy plants include people, various dust, drains, insects, rodents and products and materials received into the plants.[20,24] The objectives of an air sampling program should be defined before initiation of the sampling. For example, is information required on 1) the number of viable organisms suspended in a sample of air, 2) the number of particles bearing viable organisms, 3) the size distribution of the particles, or 4) all of these factors? A detailed review of air sampling

15.5 Tests for Microbiological Quality of Air

techniques is presented in *Air Sampling Instruments* published by the American Conference of Governmental Industrial Hygienists.[5]

A. Selection of method:

Basic methods to determine viable particulate matter in air include such techniques as sedimentation, impingement in liquids, impaction on solid surfaces, centrifugation and filtration.[15]

In sedimentation, results are influenced by particle size and by the velocity and direction of air currents in relation to the agar plates. Microorganisms that do not settle onto agar from air are not included in the count obtained by the sedimentation method.

Air samplers to determine the number of microorganisms are chiefly impactors,[28,30] impingers, centrifuges,[36] filters,[16] and electrostatic precipitators.[29]

With impaction methods, organisms from a specific volume of air are deposited on agar or on coated or dry surfaces. Under satisfactory conditions, the agar method is relatively efficient, but it has limitations: too high a count causes excessive colony growth and hence counting difficulties, and temperature must be above freezing. Also, too long an exposure to airflow during sampling causes dehydration of the agar surface, which reduces adhesion of particles and hence fewer colonies will develop. In addition, microorganisms may not adhere to dry surfaces. To overcome the problem evaporation causes during sampling, cover poured media with a monolayer of oxyethylene docosanol.[33] Dry the surfaces of agar in poured plates before pouring onto them a sterile 0.2% emulsion of oxyethylene docosanol. The emulsion may be poured from plate to plate. Plates should be drained a few seconds.

Although treatment with oxyethylene docosanol markedly reduces drying during impaction sampling, it does not significantly reduce the rate of evaporation during storage in static air.[23]

Liquid impingement consists of depositing organisms from air into a liquid such as buffered distilled water or nutrient broth. The method is not suited to small numbers of airborne organisms. Not all organisms are retained, and some may be destroyed with high impingement velocity. With extended sampling, air movement tends to reduce the volume of liquid in the sampler. This problem is aggravated when the liquid contains materials that tend to produce foam.

The viability and number of microorganisms in air may be influenced by temperature, humidity and light rays. Air currents, the number of microorganisms on various surfaces, ambient air conditions surrounding the area, the number and activity of nearby people, and other activities and sanitary conditions also affect viable counts.[18] Sunlight passing through transparent air samplers may reduce counts. These factors should be recorded when air is sampled.

B. *Agar plate preparation:*

Using equipment, supplies and procedures described for the SPC, for psychrotrophic counts, or for yeast and mold counts, aseptically pour 10-mL portions of sterilized tempered agar into sterile petri plates. Replace covers and allow agar to solidify on a level surface. Hold plates at 2 to 7°C in a manner that minimizes dehydration of the agar. These plates with agar are used for sedimentation or impaction methods of estimating numbers of microorganisms in air.

C. *Sedimentation (Class D):*[34]

Place a new paper towel or parchment paper at the location where air is to be sampled, and set the petri plate horizontally on a flat surface. Expose Standard Methods Agar or selective medium by removing the plate cover and, without inversion, place it on the paper beside the bottom of the petri plate. After exposing the agar medium for 15 min, replace the cover and incubate the plate according to the appropriate procedure. Colonies are then counted and results expressed as the number per square foot per minute. Since a standard petri plate bottom area is approximately 60 cm², the number of colonies divided by 60 gives the estimated count per cm². Standard Methods Agar or Antibiotic Supplemented Agar [9.8] are generally used. Other selective media are occasionally substituted,[27] but not all species for which a selective agar was designed will grow on it after their recovery from air.

D. *Impaction onto agars (Class B):*[1,2,6,13,32]

Several types of commercial samplers are available (Andersen 2000, Inc., P.O. Box 20769, Atlanta, GA 30320; Spiral System Instruments, Inc., 4853 Cordell Avenue, Suite A-10, Bethesda, MD 20014; Flow General, Inc., 7655 Old Springhouse Road, McLean, VA 22102). Two common designs are the single-stage slit and the multi-stage sieve sampler. A newly developed air sampling device incorporates into its operation the principles of air centrifugation and agar impaction.[6] Air is drawn toward an impeller that is housed within an open, shallow drum, and it is then accelerated by centrifugal force onto the surface of an agar medium contained within a plastic strip. To obtain optimum results, attention must be given to correct airflow rate, distance of agar from slit, volume of air sampled, and concentration of viable microorganisms.

E. *Liquid impingement (Class B):*[12]

Connect impinger (Ace Glass, Inc., P.O. Box 688, Vineland, NJ 08360) to vacuum pump by means of adequate tubing. Adjust airflow rate through impinger as specified for the unit. Place sterile, buffered, MS water that contains silicone antifoam into the impinger and sterilize. Place the sampling probe in the proper location for procurement of the air sample. Connect impinger to the suction tube and conduct sampling for 0.5 h or for a specific time in accordance with the number of microorganisms expected in the air.

Prepare dilutions, plate and incubate in accordance with appropriate procedures. Count colonies and record as the number per volume of air.

15.6 References

1. ANDERSEN, A.A. New sampler for the collection, sizing and enumeration of viable airborne particles. J. Bacteriol. 76:471–484; 1958.
2. ANDERSEN, A.A.; ANDERSEN, M.R. A monitor for airborne bacteria. Appl. Microbiol. 10:181–184; 1962.
3. ANGELOTTI, R.; FOTER, M.J.; BUSCH, K.A.; LEWIS, K.H. A comparative evaluation of methods for determining the bacterial contamination of surfaces. Food Res. 23:175–185; 1958.
4. ANGELOTTI, R.; WILSON, J.L.; LITSKY, W.; WALTER, W.G. Comparative evaluation of the cotton swab and RODAC methods for the recovery of *Bacillus subtilis* spore contamination from stainless steel surfaces. Health Lab. Sci. 1:289–296; 1964.
5. ANONYMOUS. Air sampling instruments for evaluation of atmospheric contaminants, 5th ed. Cincinnati, OH: Amer. Conf. Governmental Ind. Hygienists; 1978.
6. ANONYMOUS. Biotest RCS Centrifugal Air Sampler. Publication M. B-10D1 4-79 B10M. Miami, FL: Hycon, A Div. of Serlac Corp.; 1979.
7. BARNES, J.M. The removal of bacteria from glass surfaces with calcium alginate, gauze and absorbent cotton-wool swabs. Proc. Soc. Appl. Bacteriol. 15:34; 1952.
8. BOND, R.G.; HALBERT, M.M.; KEENAN, K.M.; PUTNAM, H.D.; RUSCHMEYER, O.R.; VESLEY, D. Development of a method for microbial sampling of surfaces, with special reference to reliability. Final report under contract PH-86-62-182. Washington, DC; U.S. Dept. of Health, Educ. and Welfare; 1963.
9. BUCHBINDER, L.; BUCK, JR., T.C.; PHELPS, P.M.; STONE, R.V.; TIEDEMAN, W.D. Investigations of the swab rinse technic for examining eating and drinking utensils. Amer. J. Public Health and the Nation's Health. 37:373–378; 1947.
10. CAIN, R.M.; STEELE, H. The use of calcium alginate soluble wool for the examination of cleansed eating utensils. Canad. J. Pub. Health. 44:464–467; 1953.
11. CANNON, R.Y.; REDDY, K.K. Contamination of milk with airborne microorganisms through the vacuum defoamer. J. Milk Food Technol. 33:197–201; 1970.
12. COWN, W.B.; KETHLEY, T.W.; FINCHER, E.L. The critical-orifice liquid impinger as sampler for bacterial aerosols. Appl. Microbiol. 5:119–124; 1957.
13. DECKER, H.M.; KUEHNE, R.W.; BUCHANAN, L.M.; PORTER, R. Design and evaluation of a slit-incubator sampler. Appl. Microbiol. 6:398–400; 1958.
14. DEY, B.P.; ENGLEY F.B. JR., Methodology for recovery of chemically treated *Staphylococcus aureus* with neutralizing medium. Appl. Environ. Microbiol. 45:1533–1537; 1983.
15. DIMMICK, R.L.; AKERS, A.B. An introduction to experimental aerobiology. New York, NY: Wiley Interscience, A div. of John Wiley & Sons; 1969.
16. EHRLICH, R. Technique for microscopic count of microorganisms directly on membrane filters. J. Bacteriol. 70:265–268; 1955.
17. FAVERO, M.S.; McDADE, J.J.; ROBERTSEN, J.A.; HOFFMAN, R.K.; EDWARDS, R.W. Microbiological sampling of surfaces. J. Appl. Bacteriol. 31:336–343; 1968.
18. FEDORAK, P.M.; WESTLAKE, D.W.S. Effect of sunlight on bacterial survival in transparent air samplers. Canad. J. Microbiol. 24:618–619; 1978.
19. GREENBERG, A.E.; CONNORS, J.J.; JENKINS, D., EDS. Standard methods for the examination of water and wastewater, 15th ed. Washington, DC: Amer. Pub. Health Assoc.; 1980.
20. GREENE, V.W.; VESLEY, D.; BOND, R.G.; MICHAELSEN, G.S. Microbiological contamination of hospital air. I. Quantitative studies. Appl. Microbiol. 10:561–566; 1962.
21. HALL, L.B.; HARTNETT, M.J. Measurement of the bacterial contamination on surfaces in hospitals. Public. Health Rep. 79:1021–1024; 1964.
22. HARPER, W.J.; HALL, C.W. Dairy Technology and Engineering. p. 545. Westport, CT: The AVI Publishing Co., Inc.; 1976.

23. HARTMAN, P.A.; WEBER, J.A. Negligible evaporation retardation by oxyethylene docosanol under static conditions. Appl. Microbiol. 24:508–509; 1972.
24. HELDMAN, D.R. Significance and control of air-borne contamination in milk and food plants. J. Milk Food Technol. 30:13–17; 1967.
25. HIGGINS, M. A comparison of the recovery of organisms from cotton-wool and calcium alginate wool swabs. Great Britain: Monthly Bull. of Ministry of Health and Pub. Health Lab. Serv. 9:50–51; 1950.
26. HOLLINGER, N.F.; LINDBERG, L.H. Delayed recovery of streptococci from throat swabs. Amer. J. Public Health and Nation's Health. 48:1162–1169; 1958.
27. KINGSTON, D. Selective media in air sampling: A review. J. Appl. Bacteriol. 34:221–232; 1971.
28. KOTULA, A.W.; GUILFOYLE, J.R.; EMSWILER, B.S.; PIERSON, M.D. Comparison of single and multiple stage sieve samplers for airborne microorganisms. J. Food Prot. 41:447–449; 1978.
29. KRAEMER, H.F.; JOHNSTONE, H.F. Collection of aerosol particles in presence of electrostatic fields. Ind. Eng. Chem. 47:2426–2434; 1955.
30. LUNDHOLM, I.M. Comparison of methods for quantitative determinations of airborne bacteria and evaluation of total viable counts. Appl. Environ. Microbiol. 44:179–183; 1982.
31. MARSHALL, R.T.; APPEL, R. Sanitary conditions in twelve fluid milk processing plants as determined by use of the rinse filter method. J. Milk Food Technol. 36:237–241; 1973.
32. MAY, K.R. The cascade impactor: an instrument for sampling coarse aerosols. J. Sci. Instr. 22:187; 1945.
33. MAY, K.R. Prolongation of microbiological air sampling by a monolayer on agar gel. Appl Microbiol. 18:513–514; 1969.
34. OLSON, H.C.; HAMMER, B.W. Numbers of microorganisms falling from the air in dairy plants. J. Dairy Sci. 17:613–623; 1934.
35. PARROW, T. Manual of recommended laboratory methods for sampling and microbiological testing of single-service food packages and their components. pp. 22; Syracuse, NY: Syracuse U. Res. Corp.; 1978.
36. SAWYER, K.F.; WALTON, W.H. The conifuge—a size separating sampling device for airborne particles. J. Sci. Inst. 27:272; 1950.
37. SILLIKER, J.H.; GABIS, D.A. A cellulose sponge sampling technique for surfaces. J. Milk Food Technol. 38:504; 1975.
38. SING, E.L.; ELLIKER, P.R.; CHRISTENSEN, L.J.; SANDINE, W.E. Effective testing procedures for evaluating plant sanitation. J. Milk Food Technol. 30:103–111; 1967.
39. TIEDEMAN, W.D. Chairman. Technic for the bacteriological examination of food utensils. Part 2 pp. 68–70. New York, NY: Amer. J. Pub. Health Yearbook; 1947–48.
40. TREDINNICK, J.E.; TUCKER, J. The use of calcium alginate wool for swabbing dairy equipment. Proc. Soc. Appl. Bacteriol. 14:85; 1951.
41. WALTER, W.G.; POTTER, J. Bacteriological field studies on eating utensils and flat surfaces. J. Environ. Health 26:187–197; 1963.
42. WITTER, L.D. Psychrophilic bacteria—A review. J. Dairy Sci. 44:983–1015; 1961.

CHAPTER 16

SEDIMENT IN MILK

J.C. Bruhn, E.O. Wright, and W.L. Arledge
(J.C. Bruhn, Tech. Comm.)

16.1 Introduction

A sediment test is designed to determine the amount of insoluble matter in a sample of milk. The findings help determine the cleanliness of milking techniques. The sediment test consists of filtering a specific volume of milk through a white filter paper disc and observing the type and amount of residue. The appearance and quantity of residue serves to roughly measure the insoluble matter in the sample. The discs are compared in appearance with discs prepared by filtering milk with known amounts of standard sediment mixtures. The milks can then be classified according to sediment content. Measured portions of either coarse or fine sediment mixtures filtered directly onto a set of standard reference discs, or official photographs of standard discs are used for comparisons with tests on raw milk samples.[1,3,4] Discs prepared from a fine sediment mixture are preferred for comparison with tests of bulk-tank milk and of processed and consumer-package samples. Special procedures, often involving chemical treatment and heating of samples may be required to make cream and other fluid dairy products filterable.[2]

Sediments found in milk from different regions and at different seasons of the year may vary in composition, specific gravity, color, and other physical characteristics. Therefore, a quantity of sediment collected in one locality may differ in appearance from equivalent amounts obtained elsewhere when all samples are on sediment discs.

Efficient use of single-service strainers on dairy farms has reduced the value of sediment tests on milk as delivered to receiving plants. Although the presence of sediment in milk suggests unsanitary methods during production or handling, its absence does not prove that sanitary conditions have prevailed. When appreciable sediment is present in milk when received by the processing plant, efficient use of filters and clarifiers at the plant reduces the value of subsequent sediment tests on its retail products. Despite their limitations, these simple tests do permit a rapid measurement of sediment introduced by careless practices on farms or in processing plants. The visual nature of the test can effectively help motivate a farmer to improve farm sanitation and cleaning practices.

When sediment is a problem, testing of milk at milking time before filtering will assist in determining the source of the problem.

16.2 General Methods

Two general methods ("mixed-sample" and "off-bottom") are used to determine sediment in milk.

The *mixed-sample method* generally uses standard discs prepared from fine sediment mixtures for comparisons. This method commonly is used to determine sediment in milk from farm bulk tanks, transportation tanks, and storage tanks, during processing, and in the consumer package.

The *off-bottom method* uses discs prepared from coarse or fine sediment mixtures for comparisons (choice depends on type of sediment for which samples are being tested). This method was developed to measure the amount of sediment that collects on the bottoms of cans of unstirred milk as offered by dairy farmers. The off-bottom method effectively demonstrates any carelessness in production to dairy farmers and offer evidence of unsanitary methods to control officials. Off-bottom sediment tests may not always reveal true conditions, however because: a) some unstirred milk, as delivered, contains considerable suspended fine sediment; b) the milk may have been agitated shortly before testing; c) surface contours of some can bottoms vary; or d) the technique of operating the sediment gun may vary (for example, air-operated guns may not allow enough time for them to be drawn across the bottom of the can).

The mixed-sample method is recommended for milk in farm bulk tanks. Field experience has shown no uniformity among tanks as to dimensions, amount of milk, or location of sediment.[6] It has been demonstrated that for practical purposes sediment recovered from 3.8 L of mixed sample is equivalent to the sediment from 0.47 L of off-bottom sample.[6,8]

Methods for determining extraneous matter in dairy products, exclusive of sediment in milk, are laboratory procedures, whereas methods for detecting sediment in raw milks are platform tests. Procedures to detect and determine extraneous matter in cheeses are described in [Chapter 18].

16.3 Equipment and Supplies

A. *Tester* (Sediment Testing Supply Co., 1512 West Jarvis Ave., Chicago, IL 60626; Clark Dairy Supply Co., Inc., 430 East Main St., Box 157, Greenwood, IN 46142; NASCO, Ft. Atkinson, WI 53538; and NASCO, 1524 Princeton Ave., Modesto, CA 95352):

The tester shall be simply constructed, easily cleaned, and quickly adjustable between testing of samples to permit sanitary removal of the

16.3 Equipment and Supplies

used disc and replacement with a clean disc. Before using, check tester for repeatability of results [16.3C]. Milk or sediment must not bypass filter disc nor be impeded in its flow. Select type of tester suited to method of sampling and sample size.

For proper sampling procedure, [4.3D].

B. Mixed-sample method, pressure or vacuum-type tester:

Equip a single-unit, off-bottom type of tester, or other suitable tester, with a special head having a filtering diameter appropriate to sample size.

C. Off-bottom tester:

The off-bottom tester may be a single-unit type, for intake 0.47 L of milk on upstroke of plunger and discharge through disc on downstroke, or a two-unit type containing one unit for removal of 0.47 L of milk from bottom of can and another for filtering the sample. Filtering diameter should be 1⅛ in (2.86 cm).

NOTE: Directions for use of standard sediment material in checking sediment testing devices for repeatability have been published.[2] There are two basic requirements: 1) measure quantity of milk delivered to assure that appropriate sample size is withdrawn; and 2) be sure the entire sample passes through disc, with no portion of sample or sediment bypassing the disc.

D. Sediment discs: (Kendall Company, Fiber Products Division, Wapole, MA 02081; Food Filters Corp., Eation, OH 45320; and Sediment Testing Supply Company, 1512 West Jarvis Ave., Chicago, IL 60626.)

Standard sediment discs, 1¼ in. (3.18 cm) in diameter, are used over the flat screen in the tester, and expose a filtration diameter of 1⅛ in. (2.86 cm), 0.40 in. (1.02 cm), 0.20 in. (0.51 cm), 0.14 in. (0.35 cm), or 0.10 in. (0.25 cm) as applicable.[10,11] The disc must not contain phenolic resins or other chemicals that may contaminate milk.

NOTE: To test sediment discs, filter 12 mg of fine standard sediment mixture (60-mL portion) through the disc, using a clean flask to catch the filtrate.[2] Transfer the filtrate to a beaker. Rinse flask 3 times with water and add rinsings to the beaker. Refilter the filtrate through a 7- or 9-cm S & S No. 589 White Ribbon paper, or equivalent, that has been washed with about 200 mL of water, dried to constant weight at 100°C, and cooled in a covered dish in a desiccator before weighing. Rinse beaker and paper thoroughly with water. Dry paper to constant weight as above. Test no fewer than three discs. Average weight of sediment per disc (based on three or more discs) should not exceed 2.8 mg. A standard disc prepared from a fine sediment mixture should not appear to have sediment buried beneath the surface.

E. White cardboard or special cards:

White cardboard or special cards for mounting discs with a semi-gloss to dull finish are preferred.

16.4 Preparation and Use of Standard Sediment Discs

Directions to prepare coarse and fine sediment standard discs have been published,[2] and laboratories wishing to prepare standard discs may refer to these directions. It may be more practical, however, to use photographs of standard discs.

The grading method is as follows: When sediment in a sample is equal to or less than that on the first disc, read it as the value of the first disc. If the sediment exceeds that shown on a standard disc or photograph, read it to the next higher classification and report sediment as "not exceeding _____ mg," or in the event of a rejection, report it as "exceeds _____ mg." Stipulate whether a mixed or an off-bottom sample was used. Rapid screening for sediment content may be accomplished by comparing sample sediment disc to nearest standard disc. The screening method should not be used for milk exceeding the highest standard sediment disc on recognized official charts.[7]

A. Photographs of standards:

Photographs of standard discs may be used as guides when grading sediment tests. Do not use photographs that have become faded, stained, soiled, or otherwise damaged. Photographs of standard discs have been prepared by the U.S. Department of Agriculture (Standardization Branch, Dairy Division, Agricultural Marketing Service, U.S. Department of Agriculture, Washington, DC 20250) in cooperation with the U.S. Food and Drug Administration and the American Public Health Association.

16.5 Procedure (Class 0)

A. Mixed Samples:[5,9]

Pass the sample through a disc [16.3D] that is held in correct position in the tester. Temper 0.47 L of milk to 35 to 38°C and filter through a disc 1.02 cm in diameter [16.3A]. If a single-unit, off-bottom tester with a special head is used, temper a larger sample to 35 to 38°C, withdrawing 0.47 L with tester while stirring; or, draw 0.47 L into tester and warm milk by holding tester under running hot water before discharging milk through discs. (Milk varies in its rate of flow through discs. Pasteurized milk may be more difficult to filter than raw milk. Other factors influencing rate of flow are temperature, fat content, freezing, high acidity, degree of clumping of fat globules, stage of lactation, presence of abnormal milk, and amount of sediment in sample.)

Remove disc from tester and store in a protected manner. If disc is placed on paper while still moist, the milk will act as an adhesive. Grade by comparing with standard discs [16.4A] and indicate on report whether disc was graded wet or dry. Character of sediment may be determined by microscopic examination.

To prevent decomposition on storage, spray used discs with an alcoholic solution containing 2.5 g each of menthol and thymol in 100 mL, or 40% formalin. Do not use glue to affix disc to paper. If disc becomes detached, moisten it with a few drops of water and remount. Protect from contamination.

B. Off-bottom samples:

Take sample from bottom of can while moving sampling device diametrically across bottom of can, or around circumference if can has a high center bottom. Synchronize withdrawal of 0.47-L sample with movement of tester head as follows: Pull plunger up in tube as head of tester is moved once across bottom diameter of can, or around its bottom circumference.

If the tester is of the single-unit type, discharge milk with a complete downstroke of the plunger. After tester is removed from can, draw plunger up 15 to 20 cm and quickly push plunger to its full length. Remove disc from tester and proceed as in [16.5A].

C. Equivalent-diameter disc methods:

A 0.51-, 0.35-, or 0.25-cm diameter sediment disc can be substituted for the 1.02-cm diameter disc in the mixed sample method. A suitable tester, with special head having appropriate filtering diameter, is substituted for the tester having the 1.02-cm diameter filtering disc.

The following mixed-milk sample sizes are to be used with the appropriate filtering diameter:

120-mL sample—0.51-cm filtering diameter
60-mL sample—0.35-cm filtering diameter
30-mL sample—0.25-cm filtering diameter

Follow procedures described in [16.5A].

16.6 References

1. CLARKE, J.O. Determination of filth and extraneous matter in dairy products. Amer. J. Pub. Health 37:728–732; 1947.
2. HORWITZ, W. (ED). Official methods of analysis. 13th ed. pp. 785–788. Assoc. Offic. Anal. Chem., Washington, DC; 1980
3. JOINER, C.R. Report on sediment tests in milk and cream. J. Assoc. Offic. Agric. Chem. 35:99–100, 340–344; 1952.
4. JOINER, C.R. Report on sediment tests in milk and cream. J. Assoc. Offic. Agric. Chem. 36:87–88, 310–315; 1953.
5. KIHLSTRUM, E.E.; DELHEY, R.W. Farm bulk tank sediment testing. J. Milk Food Technol. 23:108–112; 1960.
6. LISKA, B.J.; CALBERT, H.E. A study of mixed milk sample methods of sediment testing using off-the-bottom standards and equipment, for possible use with farm bulk milk holding tanks. J. Milk Food Technol. 17:82–85; 1954.
7. USDA. General specifications for dairy plants approved for USDA inspection and grading service. Federal Register, Vol. 40, N. 198, pp. 47910–47940, Washington, D.C.; October 10, 1975.
8. WATSON, N.E. A sediment pump for farm tanks. J. Milk Food Technol. 15:43; 1952.
9. WATSON, N.E. Determination of sediment in milk in farm tanks and/or plant storage tanks and tank trucks. J. Milk Food Technol. 19:228–229; 1956.
10. WRIGHT, E.O.; HOTCHKISS, D.K.; CLARK, W.S. JR. Preliminary study to determine the

feasibility of using a 0.20-inch diameter disk to measure sediment in fluid milk. J. Milk Food Technol. 37:409-410; 1974.
11. WRIGHT, E.O.; CLARK, W.S. JR.; WEBBER, R.W.; ARLEDGE W.L.; ROMAN, M.H.; HOTCHKISS, D.K. A collaborative study to determine the feasibility of using 0.40-, 0.20-, 0.14-, and 0.10-inch-diameter discs to measure sediment in fluid milk. J. Food Prot. 40:41-42; 1977.

CHAPTER 17

PHOSPHATASE METHODS

I. Druckrey, D.H. Kleyn, and G.K. Murthy
(M.W. Wehr Tech. Comm.)

17.1 Introduction

The phosphatase test is applied to dairy products to determine whether pasteurization was done properly and also to detect the possible addition of raw milk to pasteurized milk. Studies[3] have indicated that all raw milk contains the enzyme phosphatase and that its thermal resistance is greater than that of nonsporeforming pathogens that could possibly be present in milk. The phosphatase test detects improper pasteurization on the basis of residual phosphatase activity.

A negative phosphatase reaction cannot be interpreted as an absoute index of proper pasteurization. The test is based on inactivation of phosphatase to a certain level of activity. It is possible, therefore, that a blend of overheated and underheated milk, or a blend of overheated milk and a very minute amount of raw milk might yield a negative phosphatase test.

Though process controls such as temperature recorders, flow diversion valves, sealed pumps of known delivery capacity, etc., have been designed to assure proper pasteurization, plant operators and regulatory agencies should use the phosphatase test as a routine control procedure. The test is simple and is sensitive to exposures less than those established for pasteurization.

The phosphatase tests described in this chapter are based on the principle that the alkaline phosphatase enzyme in raw milk liberates phenol from a disodium phenyl phosphate substrate or phenolphthalein from a phenolphthalein monophosphate substrate when the tests are conducted at suitable temperature and pH. The amount of phenol or phenolphthalein liberated from the substrate is proportional to the activity of the enzyme. Phenol is measured colorimetrically after its reaction with 2,6-dichloroquinonechloroimide (CQC) to form indophenol blue. Phenolphthalein is detected by addition of sodium hydroxide.

Since publication of the original Kay-Graham[10] procedure for estimation of phosphatase activity, many changes have been made in the method. The test has been simplified and made more rapid, more sensitive, and more quantitative.[7,21,23] A rapid colorimetric test[17] has been developed for use as a confirmatory procedure.

Several substances other than disodium phenyl phosphate, have been suggested for estimating phosphatase activity.[1,11,12,15,17,18] The Rutgers method[11] employs phenolphthalein monophosphate. Only the substrate specified in a particular method should be used in that phosphatase method.

17.2 General Precautions for Methods Employing Disodium Phenyl Phosphate

Because of the extreme sensitivity of phosphatase tests employing this substrate to minute amounts of phenol and phenolic compounds, it is essential that all pipettes, glassware, stoppers, etc., and all reagents be phenol-free to avoid false positive tests.

A. Glassware:

All glassware should be rinsed with warm water, washed with a detergent that contains no phenolic compounds, rinsed in tap water and then distilled water or other suitable reagent-grade treated water, air- or oven-dried, and protected from contamination during storage. It is desirable to follow this procedure immediately after use.

Test representative pieces to determine whether they are free of phenol.

B. Closures:

Use of plastic closures should be avoided since phenolic contamination from them has been encountered. Gum rubber stoppers generally are free of phenolic compounds, but they should not be used in containers of butyl alcohol.

C. Reagents:

Keep all reagents tightly stoppered to prevent deterioration and contamination.

Presence of free phenol in buffered substrate containing disodium phenyl phosphate frequently is a cause of blue color in control tests. This color is caused by deterioration of the buffer substrate or by disodium phenyl phosphate that is not phenol-free. Deterioration of buffer substrate can be delayed by heating to 85°C for 2 min.[10] Phenol-free disodium phenyl phosphate should be used, if possible. To test for phenol in buffered substrate, add 2 drops of CQC to 10 mL of buffered substrate and incubate for 5 min at room temperature. If any color develops, the buffered substrate should be discarded and a fresh solution prepared.

If the disodium phenyl phosphate is not phenol-free, a satisfactory buffered substrate can be made as follows: In a separatory funnel, dissolve 0.5 g of disodium phenyl phosphate in 25 mL of buffer [17.4B2]. Add 2 drops of CQC, mix well, and allow 10 min for color development. Extract the indophenol blue color by shaking vigorously with 10 mL of n-butyl alcohol

[17.4B7]. Allow to stand until the alcohol layer separates completely. Remove the alcohol layer and dilute to 500 mL with water.

If Phos-Phax tablets (Applied Research Institute, 90 Brighton Avenue, Perth Amboy, NJ 08861) are used in the Scharer rapid test [17.4B4], dissolve 1 Phos-Phax tablet in 5 mL of water. Add 2 drops of CQC, shake well, and allow 5 min for color development. Extract color with 2 mL of n-butyl alcohol. Allow to stand until the alcohol separates, remove the alcohol with a pipette, and discard. Dilute the aqueous solution to 50 mL with water. These solutions should be prepared shortly before use, as they decompose.

CQC powder should be stored refrigerated in tightly closed bottles. When preparing CQC solutions, bring the powder to room temperature before opening the bottle to avoid moisture condensation on the reagent. Refrigerate the CQC solution and discard it when it deteriorates (turns brown).

17.3 Controls Applicable to All Phosphatase Procedures

Controls are used to: detect deterioration or contamination of reagents; detect presence of interfering substances such as flavoring materials having a phenolic structure; account for variations in color of the indophenol blue that results from interfering coloring materials and fats, particularly where direct extraction with butyl alcohol has been done; and check for possible presence of microbial phosphatase.

A. Negative control:

A negative control should be used with each type of dairy product being tested. Place approximately 5 mL or 5 g of product in a test tube containing a thermometer and heat at 90°C for 1 min followed by rapid cooling. Stir during heating to insure complete inactivation of phosphatase. Test the heated sample for phosphatase. Any color developed in the negative control is not due to residual phosphatase, but indicates contamination of reagents or presence of interfering coloring materials or both. Ideally, the negative controls should show no color.

B. Positive control:

As a check on the proper functioning of reagents, a single positive control will suffice for each series of samples tested. Add 0.2 mL of fresh, raw, mixed-herd milk to 100 mL of raw milk which has been heated at 90°C for 1 min, followed by rapid cooling to room temperature. When the phosphatase test is done on this sample, a positive result should be obtained.

Alternatively[21,22] prepare sample as follows: Dilute two parts of raw skim milk with one part of the corresponding skim milk heated to 95°C for 1 min. Place about 100 mL of standardized skim milk in a 14-cm plastic or glass petri dish. Immerse a 12.5-cm diameter Whatman No. 40 filter paper for 10 sec. With a tweezer, invert the filter paper in the milk and soak for 10 sec. Remove the filter paper and drain excess milk for 10 sec and then blot lightly between folds of blotting paper. Place the filter paper in 1.5- to 2.5-

cm stacks over blank filter paper inside a vacuum desiccator containing silica gel. After 24-h aspiration, replace blank filter papers and continue aspiration for an additional 4 to 5 da. Remove the filter paper from the desiccator and punch out 0.64 cm disks using a standard punch. Store the disks in screw-cap test tubes at room temperature in the dark. These disks are suitable for use up to 8 to 12 mo. When tested, each disk will show 2 to 3 µg phenol or its equivalent.

C. *Interfering substance control:*

Interfering substances may result in false-positive or false-negative tests. Vanillic acid resulting from the oxidation of vanillin, a flavoring agent used in many dairy products, will yield a false-positive test when the substrate is disodium phenyl phosphate. Sterile milk products processed by the HT-ST procedure may contain substances that react in the phosphatase test to produce false-positive results.

When positive results are obtained on samples, the test should be repeated using the same amount of sample but substituting for buffered substrate the same amount of buffer solution. In the absence of substrate, any color in this control is caused by interfering substances and not by residual phosphatase. However, if the color of the test using buffered substrate is greater than the color obtained in the interfering-substances control, under-pasteurization and/or contamination with raw product is indicated.

Chocolate milk spiked with raw milk has been found to yield lower alkaline phosphatase activity than expected (58 to 60%), thus indicating the presence of enzyme inhibiting substances.

D. *Microbial phosphatase:*

The possibility of phosphatase production in dairy products by microorganisms, particularly in cream, has been demonstrated.[2,7] Microbial phosphatases are considerably more heat resistant than is alkaline milk phosphatase,[2] so differentiation is possible with the following technique:

If the original phosphatase test is positive, repasteurize a 5-mL portion of the sample at 62.8°C (66°C for samples with a fat content greater than 10%) for 30 min in a test tube. Completely immerse the tube or immerse so that the water line is approximately 4.0 cm above the level of the milk in the tube. The temperature of the milk should reach at least 62.3°C within 5 min. Keep a thermometer immersed in the sample or in a suitable control. Cool, and do the phosphatase test on the repasteurized sample, the original sample, and the negative control. If the repasteurized sample does not show a significant reduction in color, the original result was due to microbial phosphatase. It should therefore be classified as false-positive and reported as negative.

E. *Collection of samples [Chapter 4]:*

All samples should be placed in clean containers and refrigerated and tested within 36 h. It is desirable to take samples for phosphatase testing before samples are admixed with nuts, fruits, candy and other such additives.

If flavored products are to be tested, however, it is recommended that such additives be removed by straining before testing.

17.4 Scharer Rapid Phosphatase Test[25 to 29] (Class O):

This method should be limited to preliminary screening only.

A. *Equipment:*
   1. Test tubes: Must be uniform composition, wall thickness, and bore. A convenient size for the test is 12 × 144 mm calibrated at 5-, 5.5- and 8.5-mL to top of meniscus; for preparation of cheese samples, use 18 × 150 mm test tubes.
   2. Test tube stoppers: Phenol-free, to fit test tubes.
   3. Pipettes: 1-mL and 5-mL, graduated in 0.1-mL divisions.
   4. Water bath: Thermostatically controlled to 40 ± 1°C.
   5. Thermometer: 0 to 110°C.
   6. Dropping bottles: Amber glass, size 7.5 mL, with dropper calibrated to deliver approximately 50 drops of CQC solution per milliliter.
   7. Light source and filter: Such as dental X-ray viewer consisting of a single 14- to 22-watt daylight-type fluorescent light and a plastic or glass light filter approximately 10 cm high by 30 cm wide; or equivalent.
   8. Color standards: Prepare phenol standards [17.4C3] to contain 1, 2 and 5 µg of phenol per 5 mL of solution. Alternatively, commercial standards are available. (Applied Research Institute, 40 Brighton Avenue, Perth Amboy, NJ 08861).
   9. Hand homogenizer or mechanical blender.

B. *Reagents:*
   1. Water, laboratory grade (LG).
   2. Buffer: Dissolve 100 g of sodium sesquicarbonate dihydrate ($NaHCO_3$ .Na_2CO_3.$2H_2O$) or 46.89 g of sodium carbonate plus 37.17 g of sodium bicarbonate in water and make up to 1L.
   3. Buffer control: Dilute 25 mL of the above buffer to 500 mL. Alternately, dissolve one Buf-Fax (Applied Research Institute) tablet in 50 mL of water. For determinations on cheese, sour cream, and buttermilk, prepare a double-strength buffer control by diluting 25 mL of buffer to 250 mL; alternatively, dissolve one Buf-Fax tablet in 25 mL of water.
   4. Buffer substrate: Dissolve 0.5 g of phenol-free disodium phenyl phosphate crystals in water, add 25 mL of buffer, and make up to 500 mL with distilled water. If disodium phenyl phosphate is not phenol-free, purify as follows: Place 0.5 g in a 100 mL separatory funnel and dissolve in 10 mL of water. Add 25 mL of buffer and 2–3 drops of CQC reagent or Indo-Phax solution. Mix and let stand for 5 min. Extract color with 5 mL of n-butyl alcohol and let stand

until alcohol separates. Remove bottom aqueous layer into 500 mL volumetric flask and dilute volume with distilled water. Optionally, prepare solution by dissolving one Phos-Phax tablet in 50 mL of water. If Phos-Phax is not phenol-free, dissolve tablet in 5 mL of water. Add 2 drops of CQC, shake well, and allow 5 min for color development. Extract color with 2 mL of n-butyl alcohol. After separating, remove alcohol layer with pipette, and dilute the aqueous solution to 50 mL with LG water.
5. CQC reagent: Dissolve 30 mg of crystalline CQC in 10 mL of methyl or ethyl alcohol. Optionally, prepare the solution by dissolving one Indo-Phax tablet in 5 mL of methyl alcohol. Store the solution under refrigeration and discard after one week, or if it turns brown.
6. Catalyst: Dissolve 200 mg of copper sulfate ($CuSO_4 \cdot 5H_2O$) in 100 mL of water. Where Indo-Phax tablets are used, the catalyst can be omitted since the tablets contain the copper salt as well as the CQC.
7. *n*-Butyl alcohol (neutralized): Neutralize the free acid in the alcohol as follows: Add 0.1 N NaOH to alcohol until a small portion tested with bromthymol blue gives a green or light blue color. To test, shake a sample of the alcohol with an equal quantity of neutral distilled water, allow to separate, and add a few drops of indicator solution to the water layer. Alternatively, dissolve four Buf-Fax tablets in 50 mL LG water, add to 3.785 L butyl alcohol, mix thoroughly.
8. *n*-Butyl alcohol, 7.5%: Add 7.5 mL of neutralized *n*-butyl alcohol to 92.5 mL of LG water.

C. Standards:[17]
1. Stock phenol solution: Weigh 1.000 g of reagent-grade anhydrous phenol, transfer to a liter volumetric flask, make up to the mark with 0.1 N HCl, and mix (1 mL contains 1 mg of phenol). Store in amber colored bottle. This stock solution remains stable for several months in the refrigerator at 4°C. Note: Discard if brownish or purplish color develops.
2. Dilute phenol solutions: a) Dilute 5 mL of a stock phenol solution to 500 mL with water, each milliliter contains 10 µg phenol; b) Dilute 10 mL of Solution (a) just described to 100 mL with water. *Prepare fresh* daily; each milliliter contains 1 µg phenol.
3. Phenol standards: To a series of test tubes similar to those which will be used for the test procedure, add dilute phenol and water in the amounts specified in Table 17.1. *PREPARE FRESH DAILY.*

For the Scharer rapid test, prepare only phenol standards containing 1, 2, and 5 µg phenol equivalent/5 mL of solution. Add 0.5 mL of buffer [17.4B2], mix and add 6 drops (0.1 mL) of Indophax [17.4B5] containing copper catalyst, or 2 drops of catalyst [17.4B6],

17.4 Scharer Rapid Phosphatase Test

Table 17.1 Preparation of Phenol Standards

Phenol Solution (mL)	Water (mL)	Phenol Equivalent (μg/5 mL)
0.0	5.0	0.0
0.5	4.5	1.0
1.0	4.0	2.0
2.5	2.5	5.0
5.0	0.0	10.0

and 6 drops (0.1 mL) of CQC to each. Mix immediately and incubate for 5 min. at 40°C in a water bath.
4. Butyl alcohol-extracted standards: Add 3 mL of neutralized *n*-butyl alcohol to each of the above standards. Extract the indophenol blue.
   Test samples that have been extracted should be compared directly with these extracted standards.

D. *Controls:*
   1. Chocolate milk, cheese, ice cream, or other flavored milk products: Do an interfering-substance control [17.3C] on each flavored product.
   2. Cream, butter, and cheese: Do a microbial-phosphatase control [17.3D] on each positive sample.
   3. Positive and negative controls: Do at least one positive control [17.3B] for each series of samples and one negative control [17.3A] for each kind of product.

E. *Procedure:*
   1. Pipette 0.5-mL portions of samples and controls into suitable test tubes.
      a. Buttermilk, sour cream and creamed cottage cheese are best prepared for sampling by homogenizing in a mechanical blender.
      b. Cheese or dry cottage cheese: To prepare for sampling, transfer 5 g to a test tube (18 × 150 mm) containing 2 mL 7.5% neutralized *n*-butyl alcohol and grind with a glass rod. Add an additional 18 mL of 7.5% neutralized *n*-butyl alcohol. Shake well, macerate further, if necessary, and allow to stand for about 30 min. A 0.5-mL portion for testing is then withdrawn from the relatively clear-middle portion.
      c. Butter: To prepare samples, melt butter at 40°C, mix and pipette a 0.5 mL portion into a test tube using a warm pipette.
      d. Concentrated and dry milk products: Reconstitute product with water to the original concentration of milk solids and test in the manner specified for the original product.
   2. When testing milk, cream, ice cream, butter, chocolate milk, or reconstituted milk products, add 5 mL of buffer substrate [17.4B4].

When testing buttermilk, sour cream, cheese, or cottage cheese, add 5 mL of buffer substrate prepared with double strength buffer as stated in [17.4B3]. For reliable results, the pH of the mixture should be tested and adjusted, if necessary, to pH 9.5 to 10 with solid buffer [17.4B2]. Samples should be kept cold before testing.
3. Stopper tubes and mix by inverting several times.
4. Place tubes in a water bath, warm to 40 ± 1°C, and then incubate at 40 ± 1°C for 15 min. A temperature control should be used to follow come-up time. Approximately 1 min is usually required to reach temperature.
5. Remove tubes from the water bath, add 6 drops (0.1 mL) of CQC reagent containing catalyst (add 2 drops of catalyst if CQC does not contain copper sulfate) to each tube; stopper, mix, and reincubate for 5 min.
6. Remove tubes from water bath, cool in an ice water bath, add 3 mL of neutralized cold *n*-butyl alcohol, restopper, and extract indophenol blue by gently inverting the tubes four times through a half circle. (Take about 1 sec to invert tubes, pause about 1 sec, take another 1 sec to return tubes upright; pause 1 sec and then repeat). Lay tubes on their sides on a flat surface for 2 min to permit separation of butyl alcohol, then repeat the mixing and separation steps.
7. If all the butyl alcohol has been emulsified and no clear butyl alcohol layer remains, cool tubes for 5 min in ice water and then centrifuge them for 5 min.
8. Stand tubes erect and determine the concentration of phenol in the test sample by comparing the color of the butyl alcohol layer with the standards. Comparison of colors shall be made using a standard light [17.4A7].

F. *Interpretations:*

Only 0.5 mL of milk or milk product was tested, but the standards contained only one-half the amount of phenol equivalent. Therefore, the net result is equivalent to multiplying by two or converting to 1 mL of sample. A value of 1 µg or more of phenol per milliliter of milk, milk product, reconstituted milk product, or cheese extract is indicative of improper pasteurization or of contamination with unpasteurized products.

17.5 Rapid Colorimetric Phosphatase Method (Class A2)[17,18]

This is a modification of the Scharer rapid test, and samples of various products should be prepared for testing as described [17.4E]. This procedure is similar to earlier colorimetric procedures except that the incubation time has been reduced.

A. *Apparatus:*
1. Test tubes: Borosilicate glass, 16 × 125 mm glass stoppered.

17.5 Rapid Colorimetric Phosphatase Method (Class A2)

2. Stoppers: Phenol-free rubber, or glass.
3. Pipettes: 1 mL and 5 mL, graduated in 0.1 mL division.
4. Water bath: Thermostatically controlled to 40 ± 1°C.
5. Thermometer: 0 to 110°C.
6. Centrifuge: Safety head, or equivalent, that will hold 15 mL tubes at a fixed angle; or any centrifuge to accommodate 16 × 150 mm glass-stoppered centrifuge tubes.
7. Any photoelectric colorimeter or spectrophotometer capable of 650 nm measurements.
8. Cuvettes: As required by colorimeter.

B. Reagents:
See section 17.4 B for reagents.

C. Standards:
1. Stock phenol solution: [17.4C1].
2. Dilute phenol solution: [17.4C2].
3. Phenol standards: [17.4C3].
   For construction of the standard curve, prepare standards containing 0, 1, 2, 5, and 10 µg phenol equivalents per 5 mL of solution. Add 0.5 mL of buffer [17.4B2], mix and add 0.5 mL of Indophax reagent [17.4B5] containing copper catalyst or 2 drops of catalyst [17.4B6], and 0.5 mL of CQC reagent to each standard. Mix immediately and incubate for 5 min at 40°C in a water bath.
4. Extraction with *n*-butyl alcohol: Add 3 mL of neutralized *n*-butyl alcohol to each of the above standards [17.5C3]. Extract the indophenol blue by the procedure given in [17.4E5 and 6], and centrifuge for 5 min.
5. Construction of standard curve.
   a. Withdraw the extracted butyl alcohol layer from each of the extracted standards [17.4C4] into matched cuvettes.
   b. Zero the photoelectric colorimeter or spectrophotometer on the 0 phenol standard at 0 absorbance (100% transmission) with the wave-length set at 650 nm as per instructions for photoelectric colorimeter. Read the absorbance value of each standard.
   c. Plot absorbance readings versus microgram equivalent of phenol on standard coordinate paper and draw a straight line of best fit through the points of 1,2,5, and 10 µg equivalents of phenol/5 mL of prepared standards [17.4C3].

D. Controls:
See [17.4D] for controls applicable to this method.

E. Procedure:
Samples for testing various products should be prepared as indicated in [17.4E].

1. Pipette 0.5 mL of well-mixed sample into a 16 × 125 mm test tube capable of withstanding centrifugation, using a 1.0-mL graduated (0.01 or 0.1 mL divisions) Mohr-type pipette.
2. Add 5.0 mL of purified, buffered substrate, and mix well.
3. Incubate the test mixture, together with a similarly prepared negative control [17.3A] of the same kind of sample for 15 min at 40 ± 1°C in a water bath equipped with a mechanical agitator or stirrer. (Allow a 1-min warm-up time.)
4. Remove samples from the water bath, add 0.1 mL of Indophax solution or 0.1 mL CQC and 2 drops of the catalyst to the samples. Mix the samples and place them immediately in the water bath for 5 min.
5. Remove the samples from the water bath, cool in an ice-water bath for 5 min, add 3 mL of neutralized *n*-butyl alcohol, and mix by inversion six full turns [17.4E6]. Allow samples to stand in the ice-water bath for another 5 to 10 min for the alcohol layer to separate.
6. Centrifuge samples for 5 min and remove the clear alcohol layer to cuvettes, using Pasteur pipettes.
7. Zero the instrument on the negative control and read the absorbance value of the test sample at a wavelength of 650 nm, or as per instructions for colorimeter.
8. Determine the concentration of phenol in the test sample by comparing the absorbance reading of the sample to the standard curve.

F. Interpretation:

Determine the concentration of phenol equivalent in the test sample by comparing the absorbance reading on the sample to the standard curve. Results thus obtained give the micrograms of phenol per milliliter directly, i.e., only 0.5 mL of sample was used in the test but the standard contained only half the amount as phenol equivalent, and therefore the net result is equivalent to multiplying by two or converting to 1 mL of sample. Any value greater than 1.0 µg of phenol per milliliter is indicative of improper pasteurization.

17.6 Rutgers Phosphatase Test[1,9,11 to 14,15] (Skim milk, milk, light cream; class A1. Cheese; class A2. Heavy cream, plain ice cream mix, butter, concentrated dry milk; class B.)

The principle of the test is that phosphatase can hydrolyze phenolphthalein monophosphate; phenolphthalein is released and can be detected by addition of alkali. The color is compared visually with a standard prepared from the same milk.

A. Equipment:

17.6 Rutgers Phosphatase Test

1. Test tubes: Must be of uniform composition, wall thickness, and bore. Tubes 18 × 150 mm, 15 × 100 mm or 10 × 70 mm are recommended.
2. Pipettes: 1 mL, graduated or volumetric.
3. Water bath: Thermostatically controlled to 37 ± 1°C.

B. *Reagents* (Applied Research Institute, 90 Brighton Avenue, Perth Amboy, NJ 08861):
1. Substrate concentrate: Dissolve 3.9 g of dicyclohexylamine salt of phenolphthalein monophosphate and 73.2 g of 2-amino-2-methyl-1-propanol in 21.9 mL of HCl. The pH is 10.15 at 25°C.
2. Standard concentrate: Dissolve 10 mg of phenolphthalein, 40 mg of tartrazine and 73.2 g of 2-amino-2-methyl-1-propanol in 21.9 mL of HCl. The pH is 10.15 at 25°C. Additional standard concentrates containing 0.02% and 0.05% phenolphthalein may also be prepared by using 20 mg and 50 mg phenolphthalein, respectively. Standards approximate milk containing 0.1, 0.2, and 0.5% raw milk.
3. Color developer: Dissolve 10 g of NaOH in water and dilute to 100 mL.
4. *n*-butyl alcohol, 7.5%: Adjust 75 mL of *n*-butyl alcohol to pH 7 by adding approximately 1N NaOH and make up to 1 L.

All reagents are stable indefinitely if refrigerated.

C. *Procedure:*
1. Pipette to 1.0 mL portions of samples into two test tubes.
   a. Cheese: Place 5.0 g of cheese in a test tube (18 × 150 mm) and add 20 mL of 7.5% neutralized *n*-butyl alcohol. Grind sample with a glass rod for 2 min and filter immediately through Whatman No. 4 filter paper. Use 1.0-mL portions of filtrate.
   b. Butter: Melt sample at 40°C, mix, and use 1.0 mL portions.
   c. Concentrated and dry milk: Reconstitute with water to the original concentration of solids. Mix and use 1.0 mL portions.
2. Place tubes in water bath and warm to 37 ± 1°C.
3. Add 1 drop (0.04 mL) of substrate concentrate to one tube and 1 drop (0.04 mL) of standard concentrate to the other tube.
4. Mix and incubate at 37 ± 1°C for 30 min.
5. Add 1 drop (0.04 mL) of color developer to each tube, mix and compare visually.

D. *Interpretation:*

If the tube containing substrate concentrate shows less pink color than the tube containing standard concentrate, the sample can be considered properly pasteurized. The standard contains 4 μg of phenolphthalein per test which is equal to the amount of phenolphthalein liberated by 1 mL of pasteurized milk contaminated with 0.1% raw milk in 30 min. Preparation of a standard in the same milk as the sample results in colors having the same hue, which is very important in visual comparisons.

17.7 Phosphatase Reactivation in Dairy Products Treated by High-temperature, Short-time Methods

The phosphatase test has been used successfully to detect inadequate pasteurization of milk and milk products processed by conventional methods at 62.8°C for 30 min, or 71.7°C for 15 sec. Dairy products heated higher than the conventional temperature by high-temperature, short-time (HTST) or by high-heat short-time (HHST) methods (88.3°C for 1 sec to 100°C for 0.01 sec) or by ultra-pasteurization (UP) (138°C for 2 sec) may give a negative phosphatase test immediately after heating and cooling, and when refrigerated (4.4°C or below). However, a positive test may be observed in products not adequately stored, particularly in those held at or above 10°C for extended periods.[4,5,30 to 33] Development of restored phosphatase activity has been termed "reactivation".

Reactivation may occur in products processed at temperatures above 100°C. Reactivation is greatest in products processed at 104.4°C, incubated at 34°C, and adjusted to pH 6.5. If magnesium (Mg) salts are added to HHST- or UP-processed products after heat treatment but before incubation or storage, they catalyze reactivation. Products with high fat contents show increased reactivation due to high initial concentration of phosphatase, but an increase in holding time during heating will reduce reactivation. Homogenization of products before heat processing decreases reactivation. Phosphatase reactivation is low or does not occur in products with titratable acidities of 0.40 to 0.45% (calculated as lactic acid) or more. Cow's milk shows seasonal variations in phosphatase concentration and in its reactivation behavior.[16]

The differentiation of reactivated phosphatase from residual phosphatase (in raw or under-pasteurized product) is based upon the fact that the enzyme is capable of reactivation when stored with Mg salts at 34°C, and will show a 4- to 10-fold increase in phosphatase activity over the same product stored without added Mg salts.

A. *Differentiation of reactivated from residual phosphatase:*[16,19]

To differentiate reactivated from residual phosphatase, compare results obtained on testing a 1 + 5 dilution of milk or milk product containing Mg acetate with the results of tests of an undiluted portion of the same sample without Mg acetate after storage at 34°C for 1 h.

1. Reagents:
    a. Magnesium acetate solution: Dissolve 35.4 g of $Mg(C_2H_3O_2)_2 \cdot 4H_2O$ in 25 mL of water. Warm slightly to aid solution. Pour the solution into a 100-mL volumetric flask, rinse the original container several times with 5-mL portions of water and add the rinsings to the volumetric flask. Allow the solution to cool and

17.7 Phosphatase Reactivation in Dairy Products

make up to 100 mL. This solution contains 40.1 mg of Mg per milliliter.

b. Control and diluent: Place 10 mL of each sample to be tested in a boiling water bath for 1 min after the temperature of the sample has reached 95°C. Cool the sample and use a portion for dilutions as required and another for a boiled control.

B. *Procedure:*

1. Place a 5-mL aliquot of the sample to be tested in a screw-cap (phenol-free test tube) and add 0.1 mL of water. To a second 5-mL aliquot of the same sample in an identical tube, add 0.1 mL of Mg $(C_2H_3O_2)_2$ solution with a 0.1 mL Mohr pipette and mix well.
2. Incubate both aliquots for 1 h at 34°C; mix samples occasionally while in the water bath.
3. Remove samples from the water bath and cool in ice water. Dilute 1 mL of the sample containing Mg with 5 mL of the corresponding boiled milk or milk product control (1 + 5 dilution).
4. Using the rapid colorimetric procedure [17.5] test the undiluted sample without Mg and the 1 + 5 diluted sample with Mg added for phosphatase activity. The Scharer Rapid Phosphatase Test [17.4] or the Rutgers Phosphatase Test [17.6] may also be used.

C. *Interpretation:*

If the 1 + 5 diluted sample containing Mg has equal or greater activity than the undiluted sample without Mg, the sample is considered negative for residual phosphatase and indicative of reactivation. If the diluted sample has less activity than the undiluted sample, and if the initial conventional phosphatase test result was positive, the sample is considered positive for residual phosphatase.

The minimum residual phosphatase that can be detected[20] in a given UHT or HTST product that shows a positive, conventional test only, is given by $[E_O] = 1.2 [E_{RS} - E_{RC}]$; where, $[E_O]$ = residual phosphatase, $[E_{RC}]$ = observed reactivated phosphatase of the sample without Mg, and $[E_{RS}]$ = observed reactivated phosphatase of the sample containing Mg, and diluted 1 + 5.

D. *Precautions:*

A false-positive phosphatase test may be obtained when a sample containing reactivatable phosphatase has been allowed to stand longer than 4 h at about 16°C or 2 h at about 20°C.

Reactivated phosphatase in an unrefrigerated product will test like residual phosphatase.

A positive test may be obtained with a sterile milk product because of reactivated phosphatase. However, if the sterile product is allowed to stand at room temperature for extended periods (2 h at 20°C), the differential test may no longer be applicable. To confirm presence or absence of alkaline phosphatase, follow the procedure for microbial phosphatase [17.3D] and use appropriate controls [17.3]. If the phosphatase test is positive, it may indicate presence of microbial phosphatase. It may then be necessary to resort to microbiological tests to assess the quality of the product.

17.8 References

1. BABSON, A.L.; GREELEY, S.J. New substrate for alkaline phosphatase in milk. J. Assoc. Offic. Anal. Chem. 50:555–557; 1967.
2. BARBER, F.W.; FRAZIER, W.C. The development of a positive phosphatase test in refrigerated, pasteurized cream. J. Dairy Sci. 26:343–352; 1943.
3. BURGWALD, L.H. The phosphatase test. A review of the literature on its application for detecting irregularities in the pasteurization of milk and diary products. J. Dairy Sci. 22:853–873; 1939.
4. EDDLEMAN, T.L.; BABEL, F.J. Phosphatase reactivation in dairy products. J. Milk Food Technol. 21:126–130; 1958.
5. FRAM, H. Phosphatase reactivation in high-temperature, short-time pasteurized cream. J. Dairy Sci. 40:1649; 1957.
6. GILCREAS, F.W.; DAVIS, W.S. Investigation of the amylase and phosphatase tests as an indication of pasteurization. Annual Report, Intern. Assoc. of Milk Sanitarians. 25:15–39; 1936.
7. HAMMER, B.W.; OLSON, H.C. Phosphatase production in dairy products by microorganisms. J. Milk Technol. 4:83–85; 1941.
8. HENNINGSON, R.W. Chemical stability of disodium phenyl phosphate in carbonate-bicarbonate buffer. J. Dairy Sci. 43:769–772; 1960.
9. HORWITZ, W., ED. Official methods of analysis. Washington, DC: Assoc. Offic. Anal. Chem., 13th ed.; 1980.
10. KAY, H.D.; GRAHAM, JR., W.R. Phosphorus compounds of milk, VI. The effect of heat on milk phosphatase. A simple method for distinguishing raw from pasteurized milk, raw from pasteurized cream, and butter made from raw cream from that made from pasteurized cream. J. Dairy Res. 5:63–74; 1933.
11. KLEYN, D.H.; LIN, S.H.C. Collaborative study of a new alkaline phosphatase assay system for milk. J. Assoc. Offic. Anal. Chem. 51:802–807; 1968.
12. KLEYN D.H.; YEN, W. Spectrophotometric determination of phosphatase activity in milk utilizing a new alkaline phosphatase assay system. J. Assoc. Offic. Anal. Chem. 53:869–871; 1970.
13. KLEYN, D.H.; GOODMAN, S.T. Rapid determination of alkaline phosphatase activity in cheese. J. Assoc. Offic. Anal. Chem. 60:1397–1399; 1977.
14. KLEYN, D.H. Rapid screening method for alkaline phosphatase activity in cheese: Collaborative study. J. Assoc. Offic. Anal. Chem. 61:1035–1037; 1978.
15. LIN, S.H.C.; KLEYN, D.H. New alkaline phosphatase activity assay system as applied to milk. J. Dairy Sci. 50:942; 1967.
16. MURTHY, G.K.; COX, S.; KAYLOR, L. Reactivation of alkaline phosphatase in ultra high-temperature short-time processes liquid milk products. J. Dairy Sci. 59:1699–1710; 1976.
17. MURTHY, G.K.; MARTIN, R.; RHEA, U.S.; PEELER, J.T. Rapid colorimetric test for alkaline phosphatase in dairy products. J. Food Prot. 42:794–799; 1979.
18. MURTHY, G.K.; PEELER, J.T. Collaborative study of rapid methods for alkaline phosphatase in milk and cream. J. Food Prot. 42:800–803; 1979.
19. MURTHY, G.K.; PEELER, J.T. Rapid methods for differentiating reactivated from residual phosphatase in milk and cream: Collaborative study. J. Assoc. Offic. Anal. Chem. 62:822–827; 1979.
20. MURTHY, G.K.; PEELER, J.T. Method for predicting minimum detectable residual alkaline

17.8 References

phosphatase in high-temperature, short-time processed dairy products. J. Food Prot. **43**:46–48; 1980.

21. MURTHY, G.K. Preparation of a positive control sample for use in the routine analysis of milk and milk products for alkaline phosphatase. J. Food Prot. **45**:108–111; 1982.
22. MURTHY, G.K.; PEELER, J.T. Collaborative study of alkaline phosphatase activity in filter paper disks impregnated with skim milk: positive control sample. J. Food Prot. **45**:112–114; 1982.
23. O'BRIEN, J.E. Automated test for milk phosphatase. J. Dairy Sci. **49**:1482–1487; 1966.
24. REYNOLDS, R.G.; TELFORD, W.J.P. Application of an automated procedure to the milk phosphatase test. J. Milk Food Technol. **30**:21–25; 1967.
25. SCHARER, H. A rapid phosphomonoesterase test for control of dairy pasteurization. J. Dairy Sci. **21**:21–34; 1938.
26. SCHARER, H. Improvements in the rapid phosphatase test for detection of improper pasteurization of milk and its products. J. Milk Technol. **1**(5):35–37; 1938.
27. SCHARER, H. Application of the rapid phosphatase test to dairy by-products. J. Milk Technol **2**:16–20; 1939.
28. SCHARER, H. Application of the phosphatase test to cheese. Amer. J. Pub. Health and the Nation's Health. **35**:358–360; 1945.
29. SCHARER, H. The Scharer modified phosphatase methods. J. Milk Food Technol. **16**:86–88; 1953.
30. WRIGHT, R.C.; TRAMER, J. Reactivation of milk phosphatase following heat treatment. I. J. Dairy Res. **20**:177–188; 1953.
31. WRIGHT, R.C.; TRAMER, J. Reactivation of milk phosphatase following heat treatment. II. J. Dairy Res. **20**:258–273; 1953.
32. WRIGHT, R.C.; TRAMER, J. Reactivation of milk phophatase following heat treatment. III. J. Dairy Res. **21**:37–49; 1954.
33. WRIGHT, R.C.; TRAMER, J. Reactivation of milk phosphatase following heat treatment. IV. The influence of certain metallic ions. J. Dairy Res. **23**:248–256; 1956.

CHAPTER 18
CHEMICAL AND PHYSICAL METHODS

R.A. Case, R.L. Bradley, Jr., and R.R. Williams
(R.A. Case, Tech. Comm.)

18.1 Introduction

This chapter includes methods that were in Chapters 19 and 20 and Appendix B of the 14th edition. Where possible, methods have been combined to reduce the amount of cross referencing. Detailed procedures for radionuclides and pesticides have been omitted. Because of their complex nature, only special laboratories are conducting these analyses. *Methods of Analysis, Association of Official Analytical Chemists*[14] contains radionuclide and pesticide procedures.

The term Laboratory Grade (LG) water has been used to reflect the technical changes that have occurred in ways to purify water for laboratory use [5.2A-D]. The LG water may be produced by distillation, deionization, reverse osmosis or a combination of these procedures that produce water of at least ASTM Type III.

Some laboratory instrumentation is changing rapidly. Because of different instrument designs and their modifications, performance specifications are given wherever possible, to permit comparisons with reference methods. This can eliminate the need to repeat collaborative studies every time there is a change in design of basic equipment. The specifications also give the manufacturers of new instruments the standards they must meet to be approved. The user must show that his/her particular instrument meets the performance specifications. This is accomplished by making frequent comparisons with the reference method [Preface].

The methods included were selected to meet the needs of typical dairy laboratories. This is not a complete listing of all chemical and physical methods in the dairy field.

18.2 Acid Degree Value (Lipolysis)[9,13,33,37] (Class 0)

Lipolysis develops when milkfat has undergone hydrolysis. One measure of this is the acid degree value (ADV). ADV is the amount of 1 N base required to titrate 100 g of fat. Normal raw milk has been reported to have an ADV of 0.25 to 0.40. When the ADV reaches about 1.2, lipolyzed flavor can be tasted.

A. *Apparatus and reagents:*
1. Babcock test bottle, either 18 g, 8% milk; or 9 g, 20% ice cream; or 9 g, 50% cream.
2. Centrifuge that accommodates test bottles at proper Babcock centrifugal force (Table 18.2). An unheated centrifuge is satisfactory.
3. Flask, Erlenmeyer, 50 mL.
4. Microburet, 5 mL capacity, calibrated in 0.01 mL divisions.
5. Pipette, 10 mL, may be automatic type.
6. Syringe, hypodermic, 50 mL, graduated and with a No. 15 needle. Standard 17.6 mL TC milk pipette can be used as an alternative.
7. Syringe, tuberculin type, 1 mL with a 15 cm, No. 19 needle.
8. Water bath, boiling, with tray to hold test bottles.
9. Water bath, controlled 57 ± 3°C.
10. Fat solvent: Four parts of petroleum ether (petroleum ether, 35° to 60°C to 1 part of n-propanol (AR) by volume.
11. Methyl alcohol, aqueous: Mix equal volumes of absolute methanol (CP) and LG water.
12. Phenolphthalein, 1%: Dissolve 1.0 g in 100 mL of absolute methanol (CP).
13. Potassium hydroxide, alcoholic (KOH): Prepare 0.02N KOH in absolute methanol (CP). Standardize solution frequently against standard potassium acid phthalate.
14. BDI reagent: 30 g of Triton X-100 and 70 g of sodium tetraphosphate (or sodium hexametaphosphate if uncaked) made to 1 L with LG water.

B. *Procedure:*
1. Using a 50-mL syringe (or milk pipette), place 35 mL of milk or cream in each Babcock bottle.
2. Add 10 mL of BDI reagent and mix thoroughly.
3. Place bottles in a bath of gently boiling water that covers the bases of the bottles. Agitate contents thoroughly after 5 min and again after 10 min in the bath. Bottles should remain in the boiling water at least 5 min more to insure clean fat separation. Total time 15 to 20 min.
4. Remove bottles and centrifuge for 1 min.
5. Add sufficient aqueous methyl alcohol to bring the fat column well within the graduated portion of each bottle neck.
6. Centrifuge again for 1 min.
7. Place bottles in water bath (57 ± 3°C) for 5 min. Water level should be above fat column.

8. Transfer 1 mL of fat at 57°C from each test bottle to an Erlenmeyer flask, using the 1 mL syringe. For more accurate results weigh the fat transferred.
9. Dissolve the fat in 5 mL of fat solvent and add 5 drops of 1% phenolphthalein.
10. Titrate to the first faint but definite color change with standardized alcoholic KOH solution, using a 5-mL microburet. If turbidity develops in the solvent-fat mixture during titration, 2 or 3 mL of additional solvent should clear the mixture.
11. Make blank titrations on the fat solvent (in the absence of fat), using 5 drops of indicator.

C. *Calculations:*

$$\text{ADV} = \frac{(\text{mL KOH for sample} - \text{mL of KOH for blank}) \times N \times 100}{\text{weight of fat}}$$

N = normality of alcoholic KOH solution.

If the fat is not weighed, calculate its weight by multiplying the milliliters of butterfat by the density of butterfat at 57°C (0.90 g/mL).

D. *Notes:*
1. ADV should be reported only to the second decimal place.
2. Suggested evaluations of ADV are:

   Less than 0.4 = Normal
   0.7 to 1.1 = Borderline (indefinite)
   1.2 = Slightly lipolyzed
   1.5 and greater = Unsatisfactory (extremely lipolyzed)

18.3 Acidity

A. *Acidity: titratable*[7] *(Class 0):*

Titratable acidity is valuable as a guide in manufacturing operations and for measuring the quality of dairy products. Its measurement is affected by any conditions that cause a change in the calcium phosphate of the samples. Some of these factors are: the amount of dilution of the sample, speed of titration, amount of indicator, and temperature of the sample. All samples should be titrated as quickly as possible while they are at room temperature, using exact amounts of sample diluent and reagents.[23]

1. Apparatus and reagents:
   a. Balance, 0.1 g sensitivity.
   b. Beakers, 100 mL and 250 mL.
   c. Bottle, wash.
   d. Buret, 25 mL, graduated in 0.1 mL divisions, preferably a precision-bore type. Automatic-zero buret may be used.
   e. Casserole, porcelain, 70 × 30 mm.
   f. Flask, Erlenmeyer, 125 mL.
   g. Flask, volumetric, 100 mL.

h. Graduated cylinder, 100 mL.
i. Electric mixer, Hamilton Beach #936 or equivalent (replace steel disc with round Teflon stirrer, 3.25 cm diam. × 0.64 cm thick): For use with dry milk and dry milk products.[1]
j. Pipette, standard milk, 17.6 mL TC.
k. Pipette, standard cream, 9 mL TC.
l. Pipette, volumetric, 25 mL
m. Stirring rod with policeman.
n. Phenolphthalein, 1%: Dissolve 1.0 g in 100 mL of 95% ethanol.
o. Sodium hydroxide, 0.100 N.
p. Water (LG).

2. Preparation of sample:
   a. Mix milk thoroughly by shaking or stirring gently. If milk is lumpy or frozen, heat to 37 ± 1°C in a thermostatically controlled water bath before mixing.
   b. Ice cream, sherbet and similar samples should be warmed to room temperature to eliminate their high viscosity and then mixed thoroughly.
   c. The 17.6-mL TC milk pipette will deliver approximately 18.0 g of whole milk or skim milk at 20°C. With rinsing, accuracy will be improved for viscous products. For greatest analytical accuracy, weigh all samples. (Some recently manufactured milk pipettes may show 17.6 mL TD. This is an error and should be read as TC, to contain not to deliver).

3. Procedure:
   a. Milks; fluid wheys; melted ice creams, sherberts, and ices; and yogurts

   Measure a 9- or 18-g sample into a 100-mL beaker or casserole using a pipette. Rinse pipette with 2 volumes of LG water. Add rinse to sample in beaker and mix gently but thoroughly.

   If sample is dark it may be necessary, after addition of LG water, to titrate with a pH meter to pH 8.3 using constant stirring. Be careful to avoid striking electrode.

   Add 0.5 mL of phenolphthalein indicator and titrate with 0.1 N sodium hydroxide to the first permanent (30-sec) color change to pink.

   b. Dry buttermilk, dry whey, and whey concentrate products

   Weigh accurately 7.5 g of dry buttermilk, or 6.5 g of dry whey, or 12 g of whey concentrate into a 100-mL beaker. Add 25 mL of LG water and stir slowly with a stirring rod to dissolve or suspend sample. Transfer with the aid of a wash bottle to a 100-mL volumetric flask. Make to volume and mix well.

   Transfer 18.0 g of the prepared sample of dry buttermilk or dry whey, or pipette 25 mL of prepared sample of whey concen-

18.3 Acidity

trate into an Erlenmeyer flask or casserole. Rinse pipette with an equal amount of LG water and add rinse to sample in flask.

Add 0.5 mL of phenolphthalein indicator and titrate with 0.1 N sodium hydroxide to the first permanent (30-sec) color change to pink.

c. Dry whole milk, nonfat dry milk, and malted milk products.[1]

Weigh 10 g of nonfat dry milk or malted milk, or 13 g of dry whole milk into a 250 mL beaker. Add 100 mL of LG water to the sample and mix, using electric mixer, no more than 1 min to avoid formation of excessive foam.

Allow to stand for 1 h at room temperature. Gently but thoroughly mix sample with a stirring rod. Transfer an 18.0-g sample into an Erlenmeyer flask or casserole.

Add 0.5 mL of phenolphthalein indicator and titrate with 0.1 N sodium hydroxide to the first permanent (30-sec) color change to pink.

4. Calculations:

Acidity is expressed as % lactic acid (1 mL of 0.1 N NaOH = 0.009 g of lactic acid).

a. For undiluted samples and samples reconstituted to original solids content:

$$\% \text{ acidity (reconstituted basis)} = \frac{(\text{mL of NaOH}) \times (\text{N NaOH}) \times 9}{\text{weight of sample}}$$

b. Dry buttermilk (dry basis)

$$\% \text{ acidity (dry basis)} = \frac{\% \text{ acidity (reconstituted basis)}}{0.075}$$

c. Dry whey (dry basis)

$$\% \text{ acidity (dry basis)} = \frac{\% \text{ acidity (reconstituted basis)}}{0.065}$$

d. Whey concentrate (dry basis)

$$\% \text{ acidity (dry basis)} = \frac{\% \text{ acidity (as concentrate)}}{\% \text{ solids in concentrate}}$$

B. *Acidity: potentiometric, pH (Class 0):*
  1. Apparatus and Reagents:
     a. Beakers: 30-, 40-, 50-, and 400-mL.
     b. Centrifuge, for butter or margarine.
     c. Electrodes: one glass and one calomel or a combination electrode.
     d. Funnel, separatory.

- e. pH meter.
- f. Pipette, 10 mL, large-bore tip (for butter or margarine).
- g. Refrigerator or ice water bath (for butter or margarine).
- h. Thermometer, 0 to 100°C, graduated in 1°C divisions.
- i. Tissues.
- j. Wash bottle, plastic.
- k. Water bath, thermostatically controlled:
- l. Buffer solutions: pH 4.00, pH 5.00, pH 7.00, and pH 10.00.
- m. Hexane, ACS grade.
- n. Potassium chloride, saturated or as specified by electrode manufacturer to refill electrodes.
- o. Potassium chloride solution: Dissolve 35 g in 100 mL of LG water.
- p. Soap or detergent solution, mild: for washing electrodes.
- q. Water (LG).

2. Care and use of electrodes:
   a. Handle carefully. The electrode is the most important and sensitive part in a pH measurement, and it affects the accuracy of measurement. Scratches and abrasions on the tip will modify the sensitivity (and thus the accuracy) of the electrode.
   b. Before using a new glass electrode, soak it for at least 2 h in a potassium chloride solution to assure proper functioning.
   c. The electrode must be kept exceptionally clean. Rinse electrode between buffer and samples with LG water and blot with tissue. Do not wipe.
   d. When working with samples containing fatty substances, which may reduce sensitivity of the electrode, it may be necessary to remove the film by swabbing bulb with hexane or mild soap solution. Rinse thoroughly with water and blot with tissue.
   e. When used in measuring or standardizing, the electrode should be sufficiently immersed in the sample to cover the pH-sensitive bulb and reference electrode wick.
   f. The protective sleeve on a glass electrode must be removed from the fill port on the electrode to insure proper operation.
   g. The combination electrode calls for the following additional procedures:

   Vent the reference side of the electrode before use, by removing the rubber cap at the filling tube. This permits a slow leakage of solution from the electrolyte bridge, thereby insuring good electrical contact at the liquid junction and preventing contamination of the reference electrode. Since the potassium chloride solution will diffuse outward, it is necessary to periodically replace the potassium chloride solution through the refill aperture.

18.3 Acidity

    h. Store electrode by immersing its pH-sensitive tip in a potassium chloride solution. This maintains the necessary hydrated condition of the glass electrode and reduces leakage. Storing the electrode in LG water can result in a high-resistance junction, which may cause inefficiency in operation. If electrodes are not to be used for several months, they may be stored dry. In this instance, the glass electrode must be hydrated in water for a few days before reuse.

3. Standardization of pH meter:
   a. The pH meter should be turned on ½ h before use so the instrument stabilizes. It is best if the pH meter is left turned on at all times.
   b. The pH meter must be standardized with uncontaminated buffer solutions before making determinations. Buffer solution bottles should be capped when not in use to prevent evaporation and contamination. Never pour used buffer solutions back into their stock bottles. Discard used solutions at the end of day.
   c. Pour approximately 25 mL of buffer solution into a 50-mL beaker. Standardize the meter, using both pH 7.0 buffer and a buffer solution as close as possible in pH to that of the sample being tested. Follow instructions in the pH meter manual for 2-point, or more, calibration.
   d. The pH meter should be restandardized periodically during each day. When changing types of samples, or when meter has not been used for at least ½ hr, it is advisable to repeat the standardization procedure.

4. Preparation of samples:
   a. Liquid samples may be assayed without further preparation.
   b. Prepare dry products according to the procedure for titratable acidity [18.3A].
   c. Blend or grind cheese to provide a uniform sample. The pH of a piece of cheese will vary throughout. Pack the blended sample in a small container to ensure good electrode contact.
   d. Butter and margarine:
      1. It is important that the pH be determined on the serum phase and that this be separated from the fat phase. Preparation of serum: Heat about 114 g of butter or margarine in a 400-mL beaker to 46 to 52°C in a water bath, so that the fat melts and rises to the top, while the serum goes to the bottom (45 min may be required for this separation to occur). The sample should not be agitated during the heating. When separation is complete, remove the serum under the fat layer by pipette. Transfer the serum to a 30-mL beaker or centrifuge tube and centrifuge for 3 min in a balanced centrifuge.

2. Place the beaker containing the serum in a refrigerator or ice water for approximately 2 h to permit any remaining fat to rise and solidify.
3. After fat has formed a hard layer, plunge a pipette through it and draw off serum. Transfer it to a clean, 30-mL beaker. Warm the serum to 25°C and make the pH determination on the serum.
4. Alternatively, after melting butter or margarine, remove serum under fat layer by pipette and place in separatory funnel. Allow fat to rise. If a centrifuge is available, centrifuging the separatory funnel will speed up the separation. Remove the serum from below the fat in the separatory funnel. Warm to 25°C and make the pH determination on the serum.

5. Procedure:
   a. Immerse the electrodes directly into the sample so that the pH-sensitive bulb and reference junction are in good contact with the sample. Make sure cheese is well packed and the electrode is in solid contact with the cheese.
   b. Take temperature of sample and set temperature compensator to corresponding temperature. For best results, the temperature of the sample should be 25 ± 3°C.
   c. Activate the pH meter. Make sure the electrodes are in good contact with the samples for at least 45 sec and until the reading stabilizes. Read the pH directly.

C. *Acidity, pH, gold electrode/quinhydrone method*[35] *(Class 0):*
(Not suitable for measuring alkaline material)
1. Apparatus and reagents:
   a. Beakers, 50-mL and 100-mL.
   b. Electrodes: Gold tip electrode (S.F. Gray Co., 601 Hine Ave., Waukesha, WI 53186) used in conjunction with calomel reference electrode.
   c. Mortar and pestle.
   d. pH meter.
   e. Straws, plastic soda, cut into 2-cm lengths.
   f. Thermometer, 0 to 100°C, graduated in 1°C divisions.
   g. Buffer solutions: pH 4.00, pH 5.00 and pH 7.00.
   h. Hexane, ACS grade.
   i. Potassium chloride, saturated: Weigh 50 g potassium chloride into a preweighed 250-mL beaker and add 100 mL of distilled water. Heat solution to boiling to dissolve all of the crystals. While the solution is hot, transfer it to a storage bottle. Allow to cool, with crystals forming on the bottom of the bottle, before use.
   j. Quinhydrone, 171 to 173°C m.p., Eastman #217.

18.3 Acidity

k. Water (LG).
2. Care and use of electrodes:
   a. Calomel reference electrode: Handle carefully and follow manufacturer's instructions for care and maintenance.
   b. Gold electrode: The wire tip of the gold electrode is soft and will break if repeatedly bent or flexed. Always use gentle pressure to slowly force the electrode into solid samples.

   Thoroughly clean the gold electrode assembly immediately after completing each calibration or measurement. Thoroughly wash the entire assembly at the end of each day.

   Between measurements, store the clean gold electrode with its tip immersed in LG water that contains a few crystals of quinhydrone.
3. Standardization of pH meter:
   a. Insert shorting strap between electrode and reference connection on pH meter. Set the meter for MV readings and activate. The reading should be 0.0 MV.
   b. Assemble the gold electrode and liquid sample tube by wetting the electrode stem with water and gently pushing the stem into the plastic calibration tube.
   c. Pour 10 to 15 mL of a buffer solution into a 50-mL beaker. Add quinhydrone (the size of a pea) to the solution and stir rapidly to disperse. When sufficient quinhydrone has dispersed, a black solution results.
   d. Determine the temperature of the mixture.
   e. Fill the electrode assembly with sufficient buffer-quinhydrone mixture to completely immerse the gold wire.
      1. Push the gold electrode stem into the plastic tube until the gold wire tip is at the bottom of the open tip of the tube.
      2. Place tip of the assembly in the buffer-quinhydrone mixture and withdraw the plunger until the gold wire is just below the top of the plastic tube.
      3. Remove tip from the mixture and invert the assembly so that the open tip points upward. Gently push the electrode stem into the tube until the air has been expelled and the liquid just begins to overflow from the tip of the plastic tube.
      4. If necessary, repeat the air explusion and filling steps. The tip should now be completely immersed in the mixture and should rest midway in the plastic calibration tube.
      5. During measurement, the tip must be entirely immersed in the mixture. A reading cannot be obtained if there is direct contact between the gold electrode and the potassium chloride solution.
   f. Insert the gold electrode lead into the reference electrode input and the calomel cell into the adapter input.

g. Immerse the filled electrode assembly and the tip of the calomel cell in 25 to 50 mL of saturated potassium chloride solution in a 100-mL beaker.
h. Record the millivolt reading.
i. To determine the millivolt reading of the buffer at the temperature of the buffer-quinhydrone mixture use
   MV = 0.4529 + 0.00002t - pH [0.000198322 (t + 273)]
   Where t = temperature in degrees Celsius
   If actual MV reading is more than ± 0.03 MV from the theoretical value, clean the electrode. Repeat.
4. Procedure:
   a. While preparing the cheese sample, immerse tip of calomel electrode in 25 to 50 mL of saturated potassium chloride solution in a 100-mL beaker.
   b. Place approximately 5 g of cheese at room temperature in a mortar.
   c. Add about 1 g of quinhydrone powder to the mortar and use the pestle to thoroughly mix with the cheese. Record temperature of the mixture.
   d. Repeatedly press a section of plastic straw into the cheese-quinhydrone mixture until the straw is firmly packed.
   e. With the plastic tube removed from the gold electrode, hold a finger over one end of the straw and gently push the gold wire into the sample mixture until the tip is completely imbedded.
   f. Place the straw end of the electrode assembly into the beaker with the calomel electrode and saturated potassium chloride solution. Lower the gold electrode so that only the bottom portion of the sample is in the potassium chloride mixture. The end of the gold wire must be above the potassium chloride solution.
   g. Determine and record the millivolt reading (MV).
5. Calculations:

$$pH = \frac{0.4529 + 0.00002t - MV}{0.000198322 (t + 273)}$$

Where t = temperature in degree Celsius of cheese-quinhydrone mixture
MV = millivolt reading (positive number)
If this method is to be used repeatedly, it is recommended that tables be prepared to convert MV to pH at different temperatures.

18.4 Ash and Alkalinity of Ash

A. *Ash (Class 0):*
   1. Apparatus and reagents:

18.4 Ash and Alkalinity of Ash

    a. Balance, 0.1-mg sensitivity.
    b. Burner, Bunsen.
    c. Crucible, porcelain, Coors #1 or platinum dish.
    d. Desiccator, charged with efficient desiccant.
    e. Furnace, muffle, thermostatically controlled.
    f. Hot plate, thermostatically controlled.
    g. Oven, atmospheric.
    h. Steam bath.
    i. Stirring rod, flat end.
    j. Tongs, muffle.
    k. Tripod, iron.
    l. Wash bottle.
    m. Marking ink, permanent type for crucibles.
    n. Aqua regia solution: Mix three parts of HCl with one part HNO_3 and dilute 1:1 with LG water. (Add acid to water.) CAUTION: Eye protectors should be worn.
    o. Water (LG).

2. Preparation of crucibles:

   Heat side of crucible to be marked over low flame for a few minutes. Apply marking ink while crucible is warm. Allow to dry. Fire over burner at dull red heat and cool.

   Submerge crucible carefully in aqua regia. CAUTION: Eye protectors should be worn. Allow crucible to remain submerged for several hours. Remove, rinse in tap water and then in LG water.

   Dry crucible in a atmospheric oven at 100°C. When dry, ignite in muffle furnace or over burner to dull redness. Cool to room temperature. Place in desiccator until ready to use. Crucibles must be cleaned in aqua regia, dried, and ignited in this manner after each use.

3. Procedure:
   a. Weigh accurately into a prepared, preweighed crucible 2 g of dry products or 10 g of other samples. If alkalinity of ash is to be determined, calculations are simplified if exactly these weights are used.
   b. Evaporate to dryness in an atmospheric oven at 100°C or place on steam bath and dry for 1 h. (Dry powders don't require this step.)
   c. Transfer to hood and place on iron tripod over a Bunsen burner and slowly carbonize. A few drops of olive oil may be added to reduce spattering. Heat cautiously to prevent spattering, remove burner if fat ignites. The flame should be adjusted to a height whereby carbonization is complete in about 30 min. The sample should not be carbonized to such a point that any part becomes white.

d. Place in muffle furnace regulated at 540 to 550°C. Incinerate sample until free from carbon and a light gray or white ash remains.
e. Remove from furnace and cool. Moisten ash with a small amount of LG water to form a paste and gently crush the ash particles with a flat-end stirring rod. To remove any adhering ash particles, rinse the rod with a few milliliters of LG water and allow rinsing to drain into crucible.
f. Evaporate sample solution on steam bath. Remove dish and dry contents on a hot plate regulated at 100 to 120°C. (If hot plate is too hot, ash may spatter due to residual moisture present. After partial drying of the sample, temperature of the hot plate may be increased to assure thorough drying of the ash.)
g. Return dried sample dish to muffle furnace, observing the same precautions as noted above and ash again for 1 h. Remove and cool to room temperature in a desiccator until ambient temperature is reached. Reweigh.

4. Calculations:

$$\% \text{ ash} = \frac{\text{weight of residue} \times 100}{\text{weight of sample}}$$

B. *Alkalinity of ash (Class 0):*
1. Apparatus and reagents:
   a. Beaker, 400 mL.
   b. Buret, 50 mL graduated in 0.1-mL divisions.
   c. Graduated cylinder, 50 mL.
   d. Pipette, volumetric, 50 mL.
   e. Watch glasses, 90 mm.
   f. Calcium chloride solution 40%: Dissolve 885 g of calcium chloride ($CaCl_2 \cdot 2H_2O$) in 783 mL of distilled water. Add a few drops of phenolphthalein indicator and adjust to neutrality with 0.1 N hydrochloric acid. Filter and store in a rubber-stoppered bottle.
   g. Hydrochloric acid, 0.100 N.
   h. Phenolphthalein indicator, 1.0%: Dissolve 1.0 g in 100 mL of 95% ethyl alcohol.
   i. Sodium hydroxide, 0.100 N.
   j. Water (LG).
2. Procedure:
   a. Ash sample by method [18.4A].
   b. Moisten ash with a small amount of LG water. Transfer quantitatively to a 400-mL beaker.
   c. Pipette 50 mL of 0.1 N hydrochloric acid and add to sample in beaker. Cover with a watch glass and boil gently for 5 min. Avoid spattering! Allow beaker to cool and rinse the watch glass with

18.5 Chloride (Salt)

freshly boiled LG water. Allow rinsings to drain back into the beaker.

d. Add 30 mL of 40% calcium chloride solution, gently stir, cover with a watch glass and allow to stand for 10 min.

e. Add 10 drops of phenolphthalein indicator solution and titrate with 0.1 N sodium hydroxide to a faint pink color that lasts 30 sec.

3. Calculations:

All results are expressed as "mL of 0.1 N hydrochloric acid per 100 g of dry milk".

If exactly 2 g of sample are weighed out and the hydrochloric acid and sodium hydroxide solutions are exactly 0.1 N, the calculation is simplified and may be expressed as:

mL 0.1 N HCl/100g = 50 × (50 - mL of NaOH)

If the weight of the sample is not exactly 2 g and the solutions are not exactly 0.1 N, the following formula is used:

mL 0.1 N HCl/100 g =

$$\frac{1000 \times 50 \times (N \text{ of HCl}) - (N \text{ of NaOH} \times \text{mL of NaOH})}{\text{weight of sample}}$$

18.5 Chloride (Salt)

A. Mohr method (Class 0):

This method may be used for milk, cream, cottage cheese, butter, margarine, whey, and similar products whose salt is easy to get into solution. The procedures for butter and margarine are given in [18.9C] (Kohman Method.)

1. Apparatus and reagents:
    a. Balance, 0.01-g sensitivity.
    b. Beaker, 250 mL.
    c. Bottle, wash.
    d. Buret, 50 mL graduated in 0.1 mL divisions.
    e. Magnetic mixer.
    f. Filter paper, Whatman #2 folded, 12.5 cm.
    g. Flask, Erlenmeyer, 125 mL.
    h. Flask, volumetric, 100 mL.
    i. Funnel.
    j. Pipette, Mohr, 10 mL graduated in 0.1 mL.
    k. Pipette, standard cream, 9 mL.
    l. Pipette, volumetric, 25 mL.
    m. Thermometer, 0 to 100°C.
    n. Potassium chromate, K_2CrO_4, 10% solution.
    o. Silver nitrate, $AgNO_3$, 0.1 N.
    p. Water (LG).

2. Procedure:
   a. Transfer 9 g of milk, cream or other liquid into a clean flask. Rinse pipette with an equal volume of water into the same flask. Proceed with Step c.
   b. Accurately weigh 10 g of cottage cheese or similar product into a 250 mL beaker. Add about 15 mL of warm LG water (50 to 55°C) and mix to a thin paste with a magnetic mixer. Add about 25 mL of LG water and mix until sample is dispersed.

   Transfer solution to a 100-mL volumetric flask. Rinse beaker and magnetic bar with distilled water several times and add rinsing to volumetric flask. Make to volume and mix thoroughly. Filter through folded filter paper. Collect approximately 50 mL of filtrate.

   Pipette 25 mL of filtrate into a clean flask.
   c. Add 1 mL of potassium chromate indicator. Titrate with 0.1 N silver nitrate to the first visible pale red-brown color that persists 30 sec. The end point is much more sharply apparent if the titration is conducted under a yellow light.
3. Calculations:

$$\% \text{ chloride} = \frac{\text{mL of AgNO}_3 \times \text{N AgNO}_3 \times 0.0355 \times 100}{\text{weight of sample or aliquant}}$$

% sodium chloride (salt) =

$$\frac{\text{mL of AgNO}_3 \times \text{N AgNO}_3 \times 0.0585 \times 100}{\text{weight of sample or aliquant}}$$

4. NOTES:
   a. The weight of 9 mL of sample is assumed to be 9 g. This is not exact, but the salt content of most dairy products is so low that any error due to this difference is beyond the accuracy of the test.
   b. Aliquant weight is determined by dividing sample weight by dilution and multiplying by volume pipetted.

$$\frac{\text{g of sample} \times \text{mL pipetted}}{\text{mL of dilution}}$$

B. *Volhard method (Class A1):*

This method is for cheese, processed cheese, cheese foods, and cheese spreads, in which salt may be bound within the matrix. It can also be used for most dairy foods including whey and whey products by adjusting the amount of silver nitrate to provide an excess.

Parts of this method are hazardous. Safety precautions must be taken to protect the analyst and to remove fumes.

1. Apparatus and reagents:

18.5 Chloride (Salt)

  a. Balance, 0.1 mg sensitivity.
  b. Beaker, 50 mL.
  c. Boiling chips.
  d. Bottle, wash.
  e. Buret, 50 mL, with 0.1 mL divisions.
  f. Buret, for dispensing nitric acid, 100 mL.
  g. Filter paper, Whatman #2 folded, 12.5 cm.
  h. Flask, Erlenmeyer, 250 mL.
  i. Flask, volumetric, 1000 mL.
  j. Funnel, 75 mm.
  k. Graduate, 25 mL.
  l. Hood or adequate air exhaust area.
  m. Hot plate.
  n. Pipette, 5 mL, automatic.
  o. Pipette, Mohr, 10 mL with 0.1 mL divisions.
  p. Stirring rod with flat end.
  q. Thermometer, 0 to 100°C.
  r. Tongs.
  s. Ferric ammonium sulfate, $FeNH_4(SO_4)_2 \cdot 12H_2O$, saturated, standard indicator.
  t. Nitric acid, ACS, sp gr 1.42.
  u. Potassium permanganate, $KMnO_4$, 5% solution.
  v. Potassium thiocyanate, KSCN, 0.1 N.
  w. Silver nitrate, $AgNO_3$, 0.1 N.
  x. Sucrose.
  y. Water, laboratory grade (LG).
2. Procedure:
  a. Accurately weigh about 2 g of sample into a preweighed, 50-mL beaker. Add about 20 mL of warm (50 to 55°C) LG water and break up particles with the flat end of a stainless steel stirring rod until a uniform slurry is formed. Quantitatively transfer slurry to a 250-mL Erlenmeyer flask. Rinse beaker with about 10 mL of LG water at 50 to 55°C and add to flask.
  b. Add exactly 25.0 mL of 0.1 N silver nitrate to the sample. Immediately add 10 mL of nitric acid and about 50 mL of water. Add one clean boiling chip. Place on hot plate and boil the solution in an area where the exhaust system is adequate, preferably in a hood. CAUTION: Eye protection should be worn.
  c. As the solution boils, add in 5-mL portions, a sufficient amount (usually 15 mL) of potassium permanganate until the solution turns brown and remains brown for a minimum of 5 min while gently boiling. CAUTION: Addition of potassium permanganate to the boiling solution is hazardous. Add the permanganate down the side of the flask in small portions. Direct addition of perman-

ganate into the center of the boiling solution could cause the solution to "pop" and spray the analyst with hot acid.
   d. Continue heating until the brown color disappears, and you have a resulting clear, straw-colored solution. If the solution is not clear, add more 5-mL portions of potassium permanganate until the solution becomes clear, indicating complete oxidation of organic material. If too much potassium permanganate is added to the solution and the brown color does not clear by boiling for 15 to 20 min, add a very small amount of sucrose to clear the solution.
   e. Filter the hot solution through folded filter paper into a clean, 250-mL Erlenmeyer flask. Wash filter paper thoroughly with hot LG water. Cool the solution to room temperature. Add 2 mL of ferric ammonium sulfate indicator and titrate excess silver nitrate with 0.1 N potassium thiocyanate to the first pale red-brown color that lasts 30 sec. Subtract blank value obtained, using 2 mL of H_2O in place of 2 g of sample from final titration obtained.
3. Calculations:

$$\% \text{ chloride} = \frac{[(\text{ml} \times \text{N AgNO}_3) - (\text{ml} \times \text{N KSCN})] \times 0.0355 \times 100}{\text{grams of sample}}$$

% sodium chloride (salt) =
$$\frac{[(\text{mL} \times \text{N AgNO}_3) - (\text{ml} \times \text{N KSCN})] \times 0.0585 \times 100}{\text{grams of sample}}$$

C. *Coulometric titration*[6,18] *(Class B):*

The salt content of a sample is determined by the titration of chloride with silver ions, which are generated coulometrically from the silver anode. The constant DC voltage applied across a pair of silver electrodes immersed in diluted sample causes a release of silver ions into the mixture. The chloride ions present in the sample precipitate out as silver chloride. When all the chloride has reacted, the concentration of silver ions increases, and the conductivity of the mixture rises. Sensing electrodes detect the rise in conductance and stop the titration. Since titration of chloride ions is at a constant rate, the instrument calculates chloride levels directly from elapsed titration time.
   1. Apparatus and reagents:
      a. Balance, analytical.
      b. Beakers: 50 mL, 20 mL, and 1 L.
      c. Blender.
      d. Bottle, glass storge, 1 L.
      e. Chloride meter, Corning #926 or Buchler #4-2500 or equivalent.
      f. Filter paper, Whatman #2 or equivalent.
      g. Flask, volumetric, 100 mL.

18.5 Chloride (Salt)

    h. Funnel.
    i. Hot plate with magnetic stirrer, thermostatically controlled at 70°C.
    j. Pipette, 100 mL.
    k. Pipette, volumetric, 1 mL.
    l. Pipette, 100 microliter, automatic, SMI or equivalent.
    m. Silver electrodes.
    n. Stirring bar, magnetic.
    o. Stirring rod, stainless steel, with flat end.
    p. Wash bottle.
    q. Acetic acid, glacial, ACS.
    r. Chloride standard solution.
    s. Nitric acid, ACS.
    t. Polyvinyl alcohol, 99 to 100% hydrolyzed.
    u. Silver electrode polish.
    v. Sodium chloride, AR.
    w. Water (LG).
2. Solutions:
    a. Acid buffer with chloride:
        1. Add 950 mL of 70°C LG water into a 1-L beaker. Place the filled beaker on the hot plate with the magnetic stirrer and add the stir bar.
        2. Turn the stirrer on, then add 2.25 g polyvinyl alcohol. Stir constantly until the polyvinyl alcohol goes into solution.
        3. Turn off the heat, but continue to stir the solution. Add 48.0 mL of glacial acetic acid, 0.04 g of sodium chloride, and 1.6 mL of nitric acid.
        4. Allow the solution to cool, then transfer to a clear, clean, screw-capped storage bottle.
        5. Commercially prepared acid buffer with chloride may be used.
3. Meter preparation and calibration:
    Follow manufacturer's instructions. The unit must be checked at least daily for calibration using a known chloride standard. If repeated standards do not reproduce, check the following: pipetting accuracy, conditions of buffer, condition of electrodes and stirrer. Inaccurate pipetting and dirty electrodes are the primary sources of poor repeatability.
4. Preparation of sample:
    a. Liquid samples may be used without further preparation providing the chloride level is not too high to be read on the meter. Otherwise make a dilution to bring the sample within usage range.
    b. Cheese requires an addition of nitric acid to help release the chloride ions and a filtering step to remove suspended solids from the mixture.

1. Weigh approximately 10 g of cheese into a blender jar.
2. Add 100 g LG water.
3. Blend for 1 min being careful not to leak water from the container during blending.
4. Pipette exactly 2 mL of nitric acid into solution and stir.
5. Filter the solution by gravity into a clean beaker.

5. Procedure:
   a. Follow manufacturer's instructions. Place acid buffer into beaker and lower electrode into the solution.
   b. Carefully pipette the required sample into the buffer using the same pipette used for calibration.
   c. Start the instrument and record the stabilized reading.
6. Calculations:

$$\% \text{ chloride} = \frac{\text{reading} \times \text{dilution volume*} \times 0.0355}{\text{mL pipetted} \times \text{grams of sample}}$$

$$\% \text{ NaCL} = \frac{\text{reading} \times \text{dilution volume*} \times 0.0585}{\text{mL pipetted} \times \text{grams of sample}}$$

*Cheese has three major components; fat, solids, and moisture. Only the moisture content of the cheese contributes to the dilution volume. The amount of water contributed by the cheese is added to the volume of the dilution water and the nitric acid to give the dilution volume. Example: 10 g of cheese with 35% moisture would have a dilution volume of $(.35 \times 10) + 100 + 2 = 105.5$ mL.

18.6 Chlorine, Available (Class O)

This method detects available chlorine in commercially prepared solutions of sodium and calcium hypochlorite and in chlorine rinse waters.

A. *Apparatus and reagents:*
   1. Balance, 0.1-g sensitivity.
   2. Beaker, 50-mL capacity.
   3. Buret, 50 mL with 0.1 mL divisions.
   4. Flask, Erlenmeyer, 250 mL or 500 mL.
   5. Flask, volumetric, 100 mL or 1,000 mL.
   6. Graduated cylinder, 10 mL.
   7. Pipette, Mohr, 10 mL, graduated in 0.1 mL divisions.
   8. Pipettes, volumetric: 25-, 50-, 100-, and 200-mL.
   9. Acetic acid, glacial, ASC.
   10. Potassium iodide solution (KI) 10%: Weigh 10 g of potassium iodide into a 50-mL beaker and dissolve with LG water. Transfer to a 100-mL, stoppered, volumetric flask and bring to volume with LG water. Invert stoppered flask numerous times to mix. Store in a dark place

18.6 Chlorine, Available (Class 0)

when not in use and discard when the solution develops a yellow color.
11. Starch, 0.5%: Make a thin paste from 2.5 g of soluble potato starch and a little cold LG water. Stir in 500 mL of boiling distilled water and allow to stand overnight. Use the clear supernatant liquid. To preserve the starch solution, add 1.25 g of salicylic acid.
12. Sodium thiosulfate ($Na_2S_2O_3$), 0.01 N or 0.1 N solutions. See Table 18.1.
13. Water (LG).

B. *Procedure:*
1. Into an Erlenmeyer flask of appropriate size, pipette the correct amount of sample, as indicated in Table 18.1.
2. Add 10 mL of potassium iodide solution and 5 mL of glacial acetic acid. Swirl to mix. A yellow color after addition of the acid indicates presence of chlorine.
3. Titrate liberated iodine *immediately upon addition of acid* using the thiosulfate solution of required normality [Table 18.1] until a pale yellow color is obtained.
4. Add 1 mL of starch solution and titrate until the blue color disappears for at least 30 sec. If a solution, after addition of acid, is very pale yellow in color, immediately add the starch indicator and then titrate until the solution remains colorless for 30 sec. (Adding starch too early will allow decomposition of the indicator and indistinct end points.)
5. Determine a blank titration using a volume of distilled water corresponding to that of the sample. If a blue color results, titrate with 0.01 N of thiosulfate to the same end point as the sample.
6. Before calculating the available chlorine, subtract the blank titration from the sample titration results. If the blank requires more than 0.2 mL of 0.01 N solution per 50 mL of water, examine reagents and equipment for source of contamination.

Table 18.1 Volume of Sample Suggested for Available Chlorine Assays.

Expected Chlorine Concentration	Sample Size (mL)	N of Sodium Thiosulfate Titrant
Less than 10 ppm	400	0.01
10 to 50 ppm	200	0.01
50 to 100 ppm	100	0.01
100 to 200 ppm	50	0.01
200 to 500 ppm	25	0.01
greater than 500 ppm	50	0.10

C. *Calculation:*
Available chlorine (ppm) =
$$\frac{\text{ml of thiosulfate} \times \text{N thiosulfate} \times 3.55 \times 10^4}{\text{ml of sample}}$$

18.7 Extraneous Matter: Cheese[34] (Class O)

A. *Apparatus and reagents:*
 1. Balance, 0.1-g sensitivity.
 2. Beaker, 400 mL.
 3. Beaker, 2 L, stainless steel or glass with slotted cover for mechanical mixing or solid cover if magnetic stirrer is used.
 4. Beaker, 2 L.
 5. Blender, Osterizer or equivalent, with 1 quart (1 L) jar.
 6. Bottle, glass-stoppered.
 7. Bottle, wash.
 8. Cellophane envelopes, 4 × 7.6 cm.
 9. Extraneous-matter record cards.
 10. Filter flask, with side arm, 1000 mL or 2000 mL.
 11. Graduates, 1000 mL and 250 mL.
 12. Infrared lamp or atmospheric oven.
 13. Knife or spatula, stainless steel.
 14. Lintine disc, 3.1 cm diam.
 15. Microscope, minimum 12 × magnification.
 16. Pipette, Mohr, 0.5 mL.
 17. Standard test chart: USDA Sediment Standard Chart for Milk and Milk Products, No 7 CFR 58.2728.
 18. Hot plate, thermostatically controlled with stirring apparatus.
 19. Suction filtering apparatus to hold 3.1 cm Lintine discs.
 20. Thermometer, 0 to 150°C.
 21. Water (LG), filtered through lintine disc before use.
 22. Sodium citrate, 10% (for natural cheese). Dissolve 100 g of sodium citrate in LG water in a 1-L volumetric flask and dilute to volume. Filter through lintine disc before use.
 23. Pepsin (for processed cheeses), 1-10000 granular. Refrigerate pepsin. Remove from cooler before use and equilibrate to room temperature before opening container.
 24. Pepsin-phosphoric acid (for processed cheese): To approximately 200 mL of LG water in a 1000-mL volumetric flask, add 20 mL of 85% phosphoric acid and 6.5 g of pepsin. Swirl to dissolve pepsin. Make up to volume with LG water. Filter solution through a lintine disc.
 25. Phosphoric acid, 85% (for processed cheese).

B. *Preparation of sample:*

18.8 Fat

1. Soft natural cheese: cut with a knife into 1.25 cm or smaller cubes.
2. Hard natural cheese: Blend with an electric blender for a few seconds, switching the motor off and on during the mixing.
3. Processed cheese: cut with a knife into 1.25-cm or smaller cubes.

C. *Procedure:*
1. Place a 226-g sample in a 2-L stainless steel beaker containing 1000 mL of 10% sodium citrate solution for natural cheese or 1000 mL of pepsin-phosphoric acid solution for processed cheese.
2. Place on hot plate and mechanically agitate for 1 h at 55 ± 3°C for natural cheese or 52 ± 4°C for processed cheese. (Should sample be heated to more than 60°C, the protein material will coagulate and produce a clot, which can clog the filter pad.) The beaker should be covered during agitation to prevent outside contamination.
3. After 1 h, remove the agitator and rinse with filtered LG water, allowing rinsings to drain back into the rest of sample in the beaker. Filter solution using medium suction through an extraneous matter disc-waffle side up. (Filter within 5 min.)
4. Rinse beaker with about 150 mL of filtered LG water and pour rinsings through the same disc.
5. Remove disc from the filtering apparatus and dry with an infrared lamp or an atmospheric oven at 65°C for 20 min. The disc should be protected against contamination while drying.
6. Enclose the disc in a cellophane envelope and mount on an extraneous-matter record card showing identity.
7. Examine the disc under a microscope using at least 12× magnification. Identify and record on the card the kinds of extraneous matter present, using the appropriate USDA standard (para. A17).

18.8 Fat

A. *Babcock method*[14,20,21,23] *(class O):*

Over the years, much effort has been expended to refine the original Babcock method to increase its accuracy and applicability when used to determine the fat content of various types of milk, cheese, whey, and other related dairy foods. These refinements consist basically of: a) using LG water maintained at specified temperatures; b) controlling and standardizing the technique used in adding acid to samples to evolve the heat necessary to disintegrate completely non-fatty material, while at the same time releasing liquified milkfat for measurement; c) employing a mechanical shaker to aid in digestion of nonfatty materials and in quick release of fat globules; and d) controlling variability in temperature of product/acid mixture. The technique referred to in b), above, involves rapid addition of acid accompanied by a smooth rotary swirling of the bottle, first in a clockwise direction immediately followed by a counterclockwise movement. Time allotted for addition of acid to the sample should not exceed 20 sec as heat loss due to time lag causes incomplete digestion of sample resulting in cloudy fat column.

1. Apparatus and reagents:
    a. Acid measure: Calibrated to deliver the correct amount of acid.
    b. Balance, 0.01-g sensitivity.
    c. Bottle, Babcock: Type will depend upon fat content and size of sample. It is highly recommended that Babcock bottles have etched neck markings and be calibrated for improved accuracy. (Forcoven Products, Inc., P.O. Box 1556, Humble, TX 77338)
        1. 0 to 0.5% skim milk bottle, 18-g, 15.24-cm, double-necked, with 0.01% divisions, meeting AOAC specifications. For use with samples containing less than 0.5% fat.
        2. 0 to 8% milk bottle, 18-g, 15.24-cm, with 0.1% divisions, meeting AOAC specifications. For use with samples containing up to 8% fat.
        3. 0 to 20% ice cream bottle, 9-g, 15.24-cm, with 0.2% divisions, meeting AOAC specifications. For use with samples containing from 5 to 20% fat.
        4. 0 to 50% cream bottle, 9-g, 15.24-cm, with 0.5% divisions, meeting AOAC specifications. For use with samples containing 10 to 50% fat.
    d. Bottle holder, for test bottles.
    e. Centrifuge, electrically heated to maintain air temperature of 60°C.
    f. Centrifuge speed indicator (tachometer): Proper speed of the centrifuge may be determined from Table 18.2 "Diameter of wheel" is the distance between the inside bottom of opposite cups, as measured through the center of rotation of the centrifuge wheel while the cups are horizontally extended.

Table 18.2 Rotational Speed (rpms) Required in a Babcock Test Centrifuge.

Diameter of Wheel (in)	(cm)	Speed (rpm)
10	25.4	1074
12	30.5	980
14	35.6	909
16	40.6	848
18	45.7	800
20	50.8	759
22	55.9	724
24	61.0	693

18.8 Fat

g. Dividers or precision Babcock caliper, for measuring fat column.
h. Thermometer, 0 to 110°C.
i. Pipette, milk, standard 17.6-mL TC milk pipette meeting AOAC specification.
j. Reading light, used as background when measuring fat column: Light should be diffused (soft green color preferred) and provide illumination from angles above and below level of fat column.
k. Shaker, mechanical, for test bottles.
l. Water bath, to temper samples before pipetting, provided with a standardized thermometer and maintained at 39 ± 1°C.
m. Water bath, for test bottles, provided with a standardized thermometer and maintained at 55 to 60°C (preferably 57°C).
n. Sulfuric acid (sp. gr. 20/20°C 1.820–1.830 unless otherwise noted), tempered at 22 ± 1°C: Each new lot should be compared against previous lot.
o. Water (LG), maintained at 60°C.
p. Glymol (red or blue reader), sp. gr. not to exceed 0.85 20/20°C for use with cream and cheese test.

2. Procedure:
   a. Milk
      1. Adjust fresh or preserved composite samples to approximately 38°C and mix until homogeneous. Immediately pipette 17.6 mL of sample into a milk test bottle. Do not release the sample until the bulb of the pipet rests on the neck of the test bottle. After active flow has ceased (about 30 sec) blow out the last drops from the pipette tip into the bottle, remove the pipette and adjust milk to 22 ± 1°C. Hold the bottle at a slight angle and add about 17.5 mL of sulfuric (sp. gr. 20/20°C 1.820–1.830) portion-wise (approximately 8, 5, 4 mL) with hand shaking between additions washing all traces of milk into bulb. Time for complete acid addition should not exceed 20 sec. Immediately swirl until all traces of curd disappear (reaction temperature should be 100 to 105°C). Shake on mechanical shaker for at least 5 min.
      2. Place bottle in heated (60°C) centrifuge, counterbalance and turn on the centrifuge. Centrifuge for 5 min after proper speed has been attained. Stop the centrifuge and add hot (60°C) LG water until bulb of bottle is filled to within .6 cm of the neck. Centrifuge the sample for 2 min. Stop the centrifuge and add hot (60°C) LG water until the liquid column approaches the top graduation of the scale. Centrifuge the sample for 1 min.
      3. Transfer the bottle to a water bath that is maintained at 55 to 60°C (preferably 57°C) and temper for a minimum of 5 min. The water level should be slightly above the level of the fat

column in the bottle. Remove the bottle from the water bath, and quickly dry it. Immediately measure the fat column to nearest 0.05% using dividers or calipers. Measure from the bottom of the lower meniscus to highest point of upper meniscus. The fat column at time of measurement should be translucent, golden yellow, or amber, and free from visible suspended particles.

Reject all tests with a cloudy (milky) fat column or showing the presence of charred matter or curd, or where for any reason the reading is indistinct or uncertain. Repeat test. Corrections can usually be made by adjusting the temperature of sample or acid. Other items that may cause problems are: a) the amount of H_2SO_4 added; b) the rate at which the H_2SO_4 is added; or c) the way the sample is shaken when the acid is first added. This test takes practice to obtain clear fat columns.

b. Cream
  1. Adjust cream samples to approximately 38°C, mix until homogenous. Accurately weigh 9 g of cream into a cream test bottle. Pipette 9 mL of LG water with a clean pipette and mix the water and cream thoroughly by swirling the bottle. Adjust cream to 22 ± 1°C. Hold the bottle at a slight angle and add about 17.5 mL of sulfuric acid (sp. gr. 20/20°C 1.820–1.830) portion-wise (approximately 8, 5, 4mL) with hand shaking between portions, rotating the bottle slowly during addition to rinse down any cream that may cling to the neck. Time for complete acid addition should not exceed 20 sec. Immediately swirl until all lumps completely disappear. Shake on mechanical shaker for at least 5 min.
  2. Place bottle in heated (60°C) centrifuge, counterbalance, and turn on the centrifuge. Centrifuge for 5 min after proper speed has been attained. Stop the centrifuge and add hot (60°C) LG water until bulb of bottle is filled to within .6 cm of neck. Centrifuge the sample for 2 min. Stop the centrifuge and add hot (60°C) LG water until the liquid column approaches the top graduation of the scale. Centrifuge the sample for 1 min.
  3. Transfer the bottle to a water bath that is maintained at 55 to 60°C (preferably 57°C) and warm bottle for a minimum of 5 min. The water level should be slightly above the level of the fat column in the bottle. Remove the bottle from the water bath, and quickly dry it. Add 2 drops of glymol, allowing it to run down the inside neck of the bottle to the top of the fat column. Do not use glymol with 0–20% Ice Cream bottles. Immediately measure the fat column (using dividers or calipers) from the bottom of the lower meniscus to the sharp line of

demarcation between the glymol and fat. Read the percentage of fat to the nearest 0.5% fat when using a 0 to 50% fat bottle or 0.1% fat when using a 0 to 20% fat bottle. The column at time of measurement should be translucent, golden yellow, or amber, and free from visible suspended particles.

Reject all tests with a cloudy (milky) fat column or showing the presence of charred matter or curd, or where for any reason the reading is indistinct or uncertain. Repeat test. Corrections can usually be made by adjusting the temperature of sample or acid. Other items that may cause problems are: a) the amount of H_2SO_4 added; b) the rate at which the H_2SO_4 is added; or c) the way the sample is shaken when the acid is first added. This test takes practice to obtain clear fat columns.

C. Cheese
   1. Shred cheese using a grater or blender. Accurately weigh 9 g of grated cheese into a preweighed cream or Paley test bottle.
   2. Pipette about 10 mL of hot (60°C) LG water into the bottle. Mix to thoroughly suspend cheese before adding acid. Cool to 22 ± 1°C.
   3. Hold the bottle at a slight angle and add about 15 mL of sulfuric acid (sp. gr. 20/20°C 1.835–1.840—stronger than with milk) portion-wise (approximately 8, 4, 3 mL), swirling by hand after each addition. Time for complete acid addition should not exceed 20 sec. Immediately swirl until all lumps completely disappear. Shake on mechanical shaker for at least 5 min.
   4. Proceed as in paragraph [18.8A2b2].

B. Pennsylvania modified Babcock method [20,36] *(Class B)*:

The regular Babcock test for some dairy products does not yield the total amount of milk fat in the product because, under the conditions of the test, it is impossible to centrifuge out all of the fat. The Babcock test for skim milk will customarily read from 0.07 to 0.10% low, and in the case of buttermilk the reading may be 0.30% low. This situation led to the development of modification to the Babcock procedures that uses ammonium hydroxide and/or alcohol to free the fat. This procedure shows better agreement with the ether extraction method of analysis.

   1. Apparatus and reagents:
      a. The usual Babcock equipment [18.8A1].
      b. Bottle, Paley, 0.20%, 9 g, 6 in with 0.2% divisions and with side opening for addition of sample.
      c. Ammonium hydroxide (NH_4OH, 28 to 29% NH_3).
      d. Normal butyl Alcohol, bp. 117°C.

e. Sulfuric acid, diluted (sp. gr. 20/20°C 1.720–1.740): mix 3.5 parts H$_2$SO$_4$ (sp. gr. 20/20°C 1.820–1.830) to one part of water. (CAUTION: Add acid to water.)
2. Procedure:
   a. Chocolate milk, ice cream, ice cream mix and frozen desserts
      1. Melt ice cream, ice cream mix and frozen desserts, at room temperature and, if necessary, heat to about 20°C to eliminate foam. Reduce any large particles in fruit and nut ice cream to a finely divided state with a blender.
      2. Accurately weigh 18.0 g of well mixed chocolate milk or 9.0 g of well mixed ice cream into a preweighed test bottle of the correct size. Precaution must be taken to keep the sample from settling.
      3. Hold the bottle at a slight angle, add 2 mL of NH$_4$OH and rotate the bottle slowly to thoroughly mix reagent with sample.
      4. Add 3 mL of normal butyl alcohol and mix thoroughly by swirling the bottle.
      5. Hold the bottle at a slight angle and add about 17.5 mL of dilute (sp. gr. 20/20°C 1.720–1.740) sulfuric acid portion-wise (approximately 8, 5, 4 mL) rotating the bottle to wash traces of milk from the neck. Hand shake between portions to mix acid and sample. Time for complete acid addition should not exceed 20 sec. Immediately shake until all lumps completely disappear.
      6. Proceed as in paragraph [18.8A2b2] however, no glymol is added.
   b. Skim milk, low fat milk, buttermilk or whey
      1. Measure 2 mL of normal butyl alcohol into the 18 g skim milk test bottle. Add exactly 9 mL of skim milk, low fat milk, whey sample, or 9 g buttermilk to the bottle through the large neck slanting the bottle in a position to facilitate escape of air, thus preventing any tendency for the large neck to clog or overflow.
      2. Add 7 to 9 mL of sulfuric acid (sp. gr. 20/20°C 1.720–1.740) to the test bottle. It may be necessary to vary the amount of acid to obtain a fat column of golden yellow to light amber color. Mix contents of the bottle thoroughly and centrifuge for 6 min.
      3. Add sufficient hot (60°C) LG water until bulb of bottle is filled. Centrifuge the sample for 2 min. Stop the centrifuge and add hot (60°C) LG water until the liquid column approaches the top graduation of the scale. Centrifuge the sample for 2 min.
      4. Transfer the bottle to a water bath that is maintained at 55 to 60°C (preferable 57°C) and temper in bath for a minimum of 5

18.8 Fat

min. The water level should be slightly above the level of the bottle.
5. Remove bottle from water bath, dry, and with the aid of a reading light, use dividers or calipers to measure the fat column from lower surface to the highest point of upper meniscus. Fat in the column at time of measurement should be translucent, golden yellow or amber, and free from visible suspended particles. Double the reading to obtain the percentage of milk since a 9-mL sample was used in an 18-g bottle.

c. Cottage Cheese
1. Blend thoroughly a full retail carton of cheese or representative sample of larger containers until the product is uniform and pourable. Final temperature of the mixed sample should not be over 77°F. Accurately weigh 9 g of the well mixed cheese sample into a 0–20% Paley Test Bottle through the side opening. Add 2 ml of concentrated ammonium hydroxide through the side opening and mix thoroughly. Add 3 ml of normal butyl alcohol again through the side opening. Tightly insert the rubber stopper and mix sample and reagents thoroughly by shaking the bottle. Shake on mechanical shaker for 3 min.
2. Add 9 ml dilute sulfuric acid (sp. gr. 15.5/15.5°C 1.720–1.740) through the neck and mix by hand or on mechanical shaker until the curd dissolves.
3. Add 9 ml sulfuric acid (sp. gr. 15.5/15.5°C 1.820–1.830) and mix thoroughly for 3 min on mechanical shaker. Add a small amount of hot (60°C) LG water to the bottle through the side opening to dislodge any fat adhering to the stopper area.
4. Place bottle in heated (60°C) centrifuge, counter balance and turn on centrifuge. After proper speed has been attained, centrifuge for 5 min. Stop the centrifuge and add hot (60°C) LG water to bring the contents to within ¼ in of the neck. Centrifuge the sample 2 min. Stop the centrifuge and add hot (60°C) LG water until the liquid column approaches the top of the graduated scale. Centrifuge for 1 min.
5. Transfer the bottle to a water bath that is maintained at 55–60°C (preferably 57°C) and warm bottle for a minimum of 5 min. The water level should be slightly above the level of the fat column in the bottle. Remove the bottle from the water bath and quickly dry it. Add approximately 2 drops of glymol, allowing it to run down the inside neck of the bottle to the top of the fat column. Immediately measure the fat column, using dividers or calipers, from the bottom of the lower miniscus to the sharp line of demarcation between the glymol and the fat.

Read the percentage of fat to the nearest 0.1%. The column at the time of measurement should be translucent, golden yellow or amber and free from visible suspended particles.

Reject all tests with a cloudy (milky) fat column or those showing the presence of charred matter or curd or where, for any reason, the reading is indistinct or uncertain. Repeat test. Corrections can usually be made by adjusting the temperature of the sample or acid. Other items that may cause problems are 1) the amount of H_2SO_4 added; 2) the rate at which the H_2SO_4 is added; or 3) the way the sample is shaken when the acid is first added. This test takes practice to obtain good, clear reliable fat columns.

C. *Roccal-Babcock method*[20] *(Class B):*

Because of the nature of the nonfat milk solids in chocolate milk and because of the difficulty in obtaining a fat column free of charred matter when using H_2SO_4 alone for testing low fat products, a Roccal solution is used as a wetting agent. This helps in dispersing the protein layer that encloses the fat globule, thus allowing a more complete separation on protein from fat and reducing charring of the sample.

1. Apparatus and reagents:
    a. The usual Babcock equipment [18.8A].
    b. Roccal II—50% concentrate of benzalkonium chloride U.S.P.
    c. Roccal solution: Add 5 mL of 50% Roccal to 200 mL of concentrated (sp. gr. 1.835 to 1.84) sulfuric acid. Mix well and cool before using. Roccal-sulfuric acid solution will keep about 2 wks. When a dark amber color appears, the Roccal has dissipated and will not be effective in testing. Discard the solution when it becomes yellow to brown.
2. Procedure:
    a. Chocolate milk

    Accurately weigh 18 g of well mixed sample into a milk test bottle. Proceed as with the regular Babcock test of milk [18.8A2a] except add 12 mL of Roccal-sulfuric acid solution in small portions instead of sulfuric acid alone.
    b. Skim milk, buttermilk, or whey

    Pipette 17.6 mL of sample into the skim milk test bottle through the large neck, slanting the bottle to facilitate the escape of air, thus preventing any tendency for the large neck to clog or overflow. Blow the last drop out of the pipette. Proceed as with the regular Babcock test of milk [18.8A2a], except add 12 mL of Roccal-sulfuric acid solution in small portions instead of sulfuric acid alone.

18.8 Fat

D. Gerber method[17] (Class O):

This method is applicable to raw, pasteurized, homogenized, and composite (preserved) milks. Frozen desserts may be tested by this method for in-plant production control. When used for this purpose, the method is Class B.

The description of the Butyrometer has been changed to match the international standard.

1. Apparatus and reagents:
   a. Gerber Butyrometer for milk to meet ISO/DIS 488 standards with lock stopper and key.[16] Cream and frozen desserts butyrometer to meet requirements as described in the 14th edition of *Standard Methods for the Evaluation of Dairy Products*.[20] (NJDL Supplies, 658 Etra Road, East Windsor, NJ 08520 and/or Forcoven Products, Inc., P.O. Box 1556, Humble, TX 77338).
   b. Balance for weighing cream sensitivity at 0.01 g and with a pan bottle support on the balance to provide firm vertical support for the bottles.
   c. Bottle support rack to hold bottles in a secure, vertical position. Stainless steel racks consisting of two perforated shelves and a solid bottom, supported by solid sides, provide convenience and durability. Optionally use racks, with stainless steel "rapid clamp-on" covers to lock bottles securely when using shaking machines.
   d. Centrifuge for bottles, standard style, with disc and trunnion head to rotate at 1,100 ± 100 rpm: or an angle head to rotate at 1,350 ± 100 rpm. Trunnion head models are supplied with heaters to maintain internal temperature at approximately 60°C. Check the operating speed of all Gerber centrifuges under full load at monthly intervals, or more frequently if necessary. Bottles should receive an acceleration of 350 ± 50 times gravity. See manufacturer's instruction manual.

   Optionally use Babcock centrifuges for 22.8-cm bottles, equipped with heaters and cup adaptors for Gerber bottles, and operated at 1,100 ± 100 rpms (350 ± 50 times gravity).
   e. Desk reader (optional) that is improvised to expedite reading of fat columns. Models with a shielded tubular electric bulb standardize background illumination for the bottles.
   f. Milk test pipette, Gerber, TC 11.07 mL at 20°C to meet requirements as described in the 14th edition of *Standard Methods for the Evaluation of Dairy Products*.[20]
   g. Reagent dispensers, 10 mL for H_2SO_4, 1 mL for isoamyl alcohol
   h. Shaking machine (optional), standardizes conditions and saves time.

i. Water bath depth that permits vertical immersion of bottles to the terminal bulb, and is equipped to provide intermittent or constant agitation and to control temperature at 60 to 63°C.
j. Washing device (optional): Consisting of a rack and perforated cover, expedites washing. After reading, return bottles to the rack, remove the stoppers, lock the rack cover; empty, wash, rinse, and drain bottles as a unit.
k. Sulfuric acid: (sp. gr. 15.5/15.5°C 1.820–1.825) clean, colorless, free from fat
l. Isoamyl alcohol (CP) sp. gr. 15.5/15.5°C 0.814 to 0.816; bp 128 to 133°C, free from water, acid, fat and furfural; store in tightly closed containers in the dark.

Test each batch of isoamyl alcohol identified by the manufacturer as suitable for Gerber milkfat tests by transferring 10 mL of H_2SO_4, 11 mL of water, and 1 mL of the isoamyl alcohol to a test bottle. Shake and centrifuge bottles as required in the test for fat in milk. If oily particles are recovered, discard the batch. Determine the fat in a single sample of milk in duplicate using both the old and new isoamyl alcohol. Determinations should agree within ± 0.05%.

m. Sulfuric acid, diluted: (Used only for chocolate milk and frozen desserts.) Slowly add 94 parts by volume of cold sulfuric acid (sp. gr. 15.5/15.5°C 1.820–1.825) to six parts volume of cold water.

2. Procedure:
a. Milk

Measure 10 mL of H_2SO_4 ± 0.2 mL at 15 to 21°C into a test bottle. From a well-mixed milk sample at not more than 24°C, withdraw a sample, using an 11-mL pipette, with the top of the meniscus resting on the graduation line of the suction tube, and touch off any part-drop at the tip. Allow the milk to drain into the bottle slowly, at first, to prevent violent reaction with the acid, then permit the pipette to empty normally. Wait a full 3 sec after free flow stops and blow out the last drop. Add 1 mL isoamyl alcohol and insert the lock stopper securely. With stoppered end up, grasp the test bottle at the graduated column and shake until the curd is completely digested.

Holding the bottle (CAUTION: HOT) at stopper and neck, invert it at least four times to mix the acid remaining in the bulb with the contents. Machine shaking of filled, sealed test bottles secured in racks equipped with quick-locking covers insures safe, rapid, uniform treatment and optional temperatures for centrifuging.

Prepare composite (preserved) and aged milks for assay by heating the milk to 35 to 38°C and thoroughly disperse all fat

before transferring the test charge. To prevent charring of warm milk by acid:
1. Cool the acid to 10°C or below just before adding the warmed sample; or
2. Cool contents of the test bottle after adding the acid; or
3. Transfer the warmed milk to the bottle before adding the cooled acid. Chill the acid to below 5°C and hold the test bottle at a 45° angle, with the plane of its flat glass column vertical, to displace milk from bulb and column with acid. (Use this method when test bottles are filled at one location and tests are completed at another.)

Before contents of the bottles have cooled below 60°C, place them in a centrifuge and counterbalance as needed. Before operating the machine, close the cover and lock. After the required speed is attained, centrifuge for 4 min. Stop the machine and place the bottles in a water bath at 60 to 63°C, leaving only the bulbs exposed. After at least 5 min in the bath, lock stopper, optionally using a desk reader. Gently push the straight line at the bottom of the fat column upward so that it coincides with the nearest whole percent graduation mark; then promptly read the scale at the bottom of meniscus at the top of fat column to the nearest 0.05% graduation. Subtract the lower from the upper reading and record the difference as the fat content in each sample before removing the next bottle from the bath.

Repeat tests where fat columns are imperfect. As fat cools in the column, the slope of the meniscus increases, tending to lower the percent fat reading. Soft, glareless background lighting and a desk reader stand, which is readily adjusted to bring the column to tester eye level, expedite rapid, accurate readings.

Proper fat columns contain no light or dark particles or zones. They are pale to strongly yellow in color. Duplicate determinations should be within 0.10%.

b. Cream

Measure 10 ± 0.2 mL of H_2SO_4 at 15 to 21°C into a cream test bottle. Place the bottle in the support on the balance and preweigh. Weigh 5.00 g of prepared cream sample into the bottle. Remove the bottle from the balance and add 5 mL of water at 15 to 21°C and 1 mL of isoamyl alcohol. Insert the lock stopper securely and proceed as with milk. Read the scale to the nearest 0.5% graduation mark and record results.

c. Flavored milks

This method is applicable to sweetened, flavored milks.

Measure 10 ± 0.2 mL of diluted H_2SO_4 at 15 to 21°C into a milk test bottle. From a properly prepared sample at not more than 24°C, withdraw a sample, using an 11-mL pipette, with the top of the meniscus resting on the graduation line of the suction tube, and touch off any part drop of the tip. Allow the sample to drain into the bottle slowly, at first, to prevent violent reaction with the acid then permit the pipette to empty normally. Wait a full 3 sec after free flow stops and blow out the last drop.

If sample is viscous, preweigh the bottle and weigh 11.125 g of properly prepared sample into the bottle. Add 2 mL of isoamyl alcohol. Insert the lock stopper securely and proceed as with cream. Read the scale to the nearest 0.1% graduation mark.

d. Frozen desserts: in-plant production control (Class B)

High concentrations of nonfat milk solids, sugars, stabilizers, emulsifiers, and homogenizing treatment make vigorous shaking necessary to liberate fat after addition of reagents. Prolonged action of H_2SO_4 and isoamyl alcohol appears to increase the solubility of alcohol in fat, thus often resulting in higher fat determinations than those by the ether extraction method. For this reason, use of the Gerber test for frozen desserts and mixes should be restricted to processing control tests. Where critical data are required, use of an ether extraction method is necessary.

Measure 10 ± 0.2 mL of diluted H_2SO_4 at 15 to 21°C into an ice cream test bottle. Place bottle in support on balance and preweigh. Weigh 5.00 g of a well-mixed sample into the bottle. Remove bottle from balance and add 5 mL of water and 1 mL of isoamyl alcohol. Insert lock stopper securely and proceed as with milk except that the bottle should be centrifuged for 5 min. Read the scale to the nearest 0.2% graduation and record the results. Recentrifuge without inverting the bottle, retemper, and reread the fat column. If the second reading differs by more than 0.2% from the first, the initial shaking was inadequate and the test must be repeated. When the second reading is less than 0.2% above the first, report the result of the first test.

E. Roese-Gottlieb method[14] (Class O):

The Roese-Gottlieb ether extraction procedure and the Mojonnier modification of this procedure are the methods of choice because of their inherent analytical capacity for total milkfat. The Roese-Gottlieb procedure is the reference method for milkfat.

1. Apparatus and reagents:
    a. Fat extraction flask, Mojonnier type (FMC—Beverage Equipment Division, 4601 West Ohio Street, Chicago, IL 60644; or Lurex Manufacturing Company, 1298 Northwest Boulevard, Vineland, NJ 08360; or Forcoven Products, Inc., P.O. Box 1556, Humble, TX 77338)

18.8 Fat

b. Aluminum dishes (FMC Beverage Equipment Division, 4601 West Ohio Street, Chicago, IL 60644) or disposable pans (Central States Can Company, 700-16th Avenue, SE., Massillon, OH 44646): can body, white with clear coating, 309 x 307 x 207 mm.
c. Cork or synthetic rubber stoppers that fit flasks and are not affected by the solvents.
d. Crucible tongs.
e. Balance, sensitivity of 0.1 mg.
f. Interval timer.
g. Mohr pipette, or pipetting device (10-, 5-, 2- or 1-mL depending on product analyzed).
h. Hot plate.
i. Water bath maintained at 100°C.
j. Beakers, 150 mL.
k. Litmus paper, red.
l. Watch glass, 50 mm.
m. Glass beads or anti-bumping chips.
n. Stirring rods.
o. Hood, or volatile solvent exhaust apparatus.
p. Mechanical shaker (optional).
q. Desiccator, filled with Drierite (calcium sulfate).
r. Suspension apparatus for holding flasks, or support to hold pipette while full.
s. Wash bottle.
t. Ammonium hydroxide, ACS grade, sp. gr. 0.90.
u. Ethanol, specially denatured, U.S. formula #1, #30, or 3-A.
v. Ethyl ether, ACS grade, peroxide free.
w. Petroleum ether, ACS grade, boiling range 30 to 60°C.
x. Hydrochloric acid, sp. gr. 1.1.
y. Water (LG).

2. Preparation of sample and dishes:
   a. Adjust samples to approximately 20°C. Mix until homogeneous. If lumps of cream do not disperse, warm sample in water bath to approximately 38°C and keep mixing, using a policeman as necessary.
   b. To prepare fat dishes, place clean dishes in an atmospheric oven at 100 to 105°C for 1 h or in a vacuum oven at 100°C for 10 min. Transfer the dishes to a desiccator for 10 min. Weigh on an analytical balance to obtain preweight and keep in the desiccator until ready to use. The fat pans should be weighed as near as possible to the time of analysis.

3. Procedure:
   a. Milk
      Accurately weigh approximately 10 g of well-mixed sample

into an extraction flask. Add 1.25 mL of ammonium hydroxide or 2 mL if sample is sour, stopper and mix thoroughly. Add 10 mL of ethanol, stopper and mix well. Follow the procedure for ether extraction [18.8E4].

b. Cream

Accurately weigh approximately 5 g of well-mixed sample into an extraction flask and dilute to approximately 10 mL with LG water and mix by shaking. Add 1.25 mL of ammonium hydroxide or 2 mL if sample is sour, stopper and mix thoroughly. Add 10 mL of ethanol stopper and mix well. Follow the procedure for ether extraction [18.8E4].

c. Ice cream and frozen desserts

Accurately weigh 4 to 5 g of a thoroughly mixed sample into an extraction flask. Dilute to approximately 10 mL with LG water and mix by shaking. Add 2 mL of ammonium hydroxide, mix thoroughly, and heat in water bath for 20 min at 60°C with occasional shaking. Cool and add 10 mL of ethanol to samples and mix. Follow the procedure for ether extraction [18.8E4].

d. Nonfat dry milk, dry whole milk, whey powder, and malted milk

Accurately weigh approximately 1 g of well-mixed sample into small beaker. Add 1 mL of LG water and rub to smooth paste. Add 9 mL of additional water and 1 to 1.25 mL of ammonium hydroxide. Warm on steam bath. Transfer to fat extraction flask with 10 mL of ethanol and mix thoroughly. Cool and follow the procedure for ether extraction [18.8E4].

e. Cheese (acid hydrolysis)

Accurately weigh approximately 1 g of prepared sample into 150-mL beaker. Add 9 mL of water and 1 mL of ammonium hydroxide. Mix until smooth. Warm mixture at low heat until casein is well softened. Remove from heat. Neutralize the ammonium hydroxide with HCl.

With samples in fume hood, cautiously add while stirring, 10 mL of HCl and a few glass beads or other inert material previously digested with HCl to prevent bumping, cover with watch glass and boil gently 5 min, or place beaker in a boiling water or steam bath for 20 min. Cool solution and transfer mixture to fat extraction flask. Rinse beaker successively with 10 mL of ethanol, 25 mL of ethyl ether, and 25 mL of petroleum ether, and transfer rinsings to flask; shake flask vigorously for 1 min after adding each reagent. Follow procedure for ether extraction following the first petroleum ether addition.

4. Procedure for ether extraction:

Add 25 mL of ethyl ether to sample, stopper and shake vigorously 1 min. Cool if necessary, add 25 mL of petroleum ether and repeat vigorous shaking. (Caution is necessary since pressure builds up in the flask during shaking. Hold flask upright and twist to remove stopper to release pressure after shaking with each organic solvent or when necessary.) Centrifuge flask at approximately 600 rpm or let it stand until upper liquid is clear. Hold flask by small bulb. Raise to eye level height and carefully decant the ether layer into a preweighed fat dish. (In this step, avoid tilting the flask beyond a straight horizontal position so that none of the aqueous layer will pour off with ether layer.) When stopper is removed, wash "inserted portion" with equal parts of the two ethers and add washings to dish. Repeat extraction of liquid remaining in flask twice, using 15 mL of ether each time. The dividing line between the ether and aqueous layer must not be above the top neck of the lower bulb of the flask. One drop of phenolphthalein indicator may aid in observation of interface. If on the last extraction the dividing line is below the neck, add LG water dropwise to raise the dividing line to within 0.6 cm of the top of the neck of the lower bulb of the flask. Omit rinsings of stopper with mixed solvents after final extraction. (Third extraction is not necessary with nonfat milk product.)

Evaporate solvents completely on hot plate or steam bath at temperature that does not cause spattering or bumping (boiling chips may be added when the pan is initially weighed). Evaporate solvents in adequate fume hood and prevent any sparks or flames in the test area.

Dry fat to constant weight in an atmospheric oven at 102 ± 2°C or in a vacuum oven at 70 to 75°C under pressure below 50 mm Hg (71-cm vacuum). Weigh cooled beaker or dish. Remove fat from container with 15 to 25 mL of warm petroleum ether, dry, and weigh as before. Loss in weight of container is equal to weight of fat.

5. Calculation:

$$\% \text{ fat} = \frac{\text{weight of fat}}{\text{weight of sample}} \times 100$$

6. NOTES:
   a. Correct fat weight by subtracting a value obtained by a blank determination on all reagents used. If blank is greater than 0.5 mg, reagents must be replaced or purified.
   b. Difference between duplicate determinations obtained by analyst should not be more than 0.03%.
7. Shaking of extraction flasks:
   a. When adding LG water to the flask be careful to wash down any sample still adhering to the sides of the flask. Stopper the flask. Keep it in an upright (vertical) position and shake vigorously for a short time to uniformly suspend the sample. Shake with a

swirling motion to keep the contents in the lower bulb. Care should be taken to make sure the entire sample is suspended before adding the ammonium hydroxide.
   b. This procedure is a liquid-liquid extraction wherein intimate contact of the aqueous phase (sample) with the organic phase is mandatory to obtain complete and efficient extraction. Thus, shaking must be vigorous.

   If only 1 or 2 samples are being tested, hold each flask with a finger on the stopper and shake vigorously in a horizontal plane with the large bulb held horizontally and the small bulb turned up. After 5 or 6 cycles, turn flask upright and carefully release pressure within the flask. (With ethers in flask, initial release of pressure should occur after one or two excursions in the horizontal position. Failure to equalize pressure in the flask will promote leakage of solvents around the stopper and thus anomalous results.) Repeat this procedure a minimum of four times.

   With four samples, use the centrifuge flask holder and repeat above. Operator's fingers must be postioned to hold stoppers in place.

F. *Mojonnier method*[22] *(Class O):*
The Mojonnier ether extraction procedure is widely used in the United States. The extra additions of alcohol reduce the problem of gel formation.
   1. Apparatus and reagents: Same as [18.8E1].
   2. Preparation of sample and dishes: Same as [18.8E2].
   3. Procedure:
      Details of sample sizes and volume of reagents may be found in Table 18.3.
      a. Accurately weigh well-mixed, homogeneous sample directly into an extraction flask.
      b. Add the required amount of 40 to 60°C LG water if needed, stopper and mix thoroughly. Add ammonium hydroxide, stopper and mix thoroughly. Cool if necessary.
      c. Add the required amount of ethyl alcohol, stopper and shake vigorously 1 min. Add petroleum ether, stopper and shake vigorously 1 min.
      d. Centrifuge flask at approximately 600 rpm for 45 sec, or let stand until upper liquid is clear.
      e. Hold flask by small bulb. Raise to eye level height and carefully decant the ether layer into a preweighed fat dish. Avoid tilting the flask beyond a straight horizontal position as the aqueous layer will pour off with the ether layer. The dividing line between the ether and aqueous layer must not be above the top of the connecting neck of the lower bulb of the flask. One drop of phenolphthalein indicator may aid in observation of interface. If

18.8 Fat

Table 18.3 Sample Weights and Extraction Conditions for Mojonnier Test Procedure. (Conduct a third extraction if the amount of fat exceeds 0.6g. Use no alcohol and same volumes of ether used in second extraction).

Product	Sample Size (g)	Extraction	Water	NH$_4$OH	Ethyl Alcohol	Ethyl Ether	Pet Ether
				(mL)			
Milk	10	1st	0	1.5	10	25	
		2nd	0	0	5	15	
Skim Milk	10	1st	0	1.5	10	25	25
		2nd	0	0	5	15	15
Buttermilk	10	1st	0	1.5	10	25	25
		2nd	0	0	5	15	15
Whey	10	1st	0	1.5	10	25	25
		2nd	0	0	5	15	15
Evaporated Milk	5	1st	4	1.5	10	25	25
		2nd	0	0	5	25	25
Cond. Buttermilk	3	1st	6	1.5	10	25	25
		2nd	0	0	5	15	15
Unsweetened Cond. Milks	5	1st	4	1.5	10	25	25
		2nd	0	0	5	25	25
Cond. Skim Milk	5	1st	4	1.5	10	25	25
		2nd	0	0	5	15	15
Sweetened Cond. Milks	5	1st	8	1.5	10	25	25
		2nd	0	0	5	25	25
Ice Cream Mix	5	1st	5	1.5	10	25	25
		2nd	0	0	5	25	25
Cream Under 25% Fat	2	1st	5	1.5	10	25	25
Over 25% Fat	1	2nd	0	0	5	25	25
Malted Milk	0.5	1st	8	1.5	10	25	25
		2nd	0	0	5	25	25
Milk Chocolate	0.5	1st	8	1.5	10	25	25
		2nd	0	0	5	25	25
Cottage cheese	10	1st	0	1.5	10	25	25
		2nd	0	0	5	15	15
Cocoa	0.5	1st	8	1.5	10	25	25
		2nd	0	0	5	25	25
Cheese	1	1st	8	3	10	25	25
		2nd	0	0	5	25	25
Butter	1	1st	8	1.5	10	25	25
		2nd	0	0	5	25	25
Skim Milk Powder	1	1st	8.5	1.5	10	25	25
		2nd	0	0	5	15	15
Buttermilk Powder	1	1st	8.5	3.0	10	25	25
		2nd	0	0	5	15	15
Whole Milk Powder	1	1st	8.5	1.5	10	25	25
		2nd	0	0	5	25	25

on the last extraction the dividing line is below the neck, add LG water dropwise to raise the dividing line to within about 0.6 cm of the top of the neck of the lower bulb of the flask. As stoppers are removed, wash "inserted portion" with a mixture of equal parts of the 2 ethers and add washings to dish.
- f. Repeat steps c, d, and e for the second extraction. Volumes for second extraction may be found in Table 18.3.
- g. Evaporate solvents completely on a hot plate or steam bath at a temperature that does not cause spattering or bumping (boiling chips may be added when pan is initially weighed).
- h. After evaporating ether, dry the dishes containing the fat in an oven at 70 to 75°C under pressure below 50 mm Hg (71-cm vacuum) until constant weight is reached.
- i. Cool the dishes and weigh.
4. Calculation:

$$\% \text{ fat} = \frac{\text{gain in weight of preweighed pan}}{\text{weight of sample}} \times 100$$

5. NOTES:
   - a. Correct fat weight by subtracting a value obtained by a blank determination on all reagents used. If blank is greater than 0.5 mg, reagents must be replaced or purified.
   - b. Difference between duplicate determinations obtained by analyst should not be more than 0.03%.
   - c. For method of handshaking extraction flask, see [18.8E7].
   - d. If the amount of fat to be extracted is greater than 0.6 g, a third extraction is required. Omit the alcohol and use the same amount of the ethers as in the second extraction.

G. *Automated turbidimetry*[10,29,30,31,32] *(Class A1):*

Milkfat can be quantitated by measuring the turbidity or light scattering caused by the fat globules. The sample is diluted quantitatively with a solution of tetrasodium ethylenediaminetetraacetate (EDTA) to eliminate turbidity caused by the colloidal (casein) particles, and is homogenized to produce fat globules of uniform size. Finally, it passes through a photocell where the light scattering is measured. The light scattering, which is proportional to the fat content of the milk, is registered on a readout system calibrated to read the percentage of fat directly. All operations are done sequentially in the instrument.
1. Apparatus and reagents:
   - a. Milko-tester (DICKEY-john/FOSS, P.O. Box 10, Auburn, IL 62615) or equivalent
   - b. Water bath, thermostatically controlled to maintain 37 ± 1°C
   - c. EDTA solution: Dissolve 45.0 g of tetrasodium EDTA, 10 mL of polysorbate 20 (Tween 20), and 7.6 g of NaOH in 10 L of distilled

18.8 Fat

water. (The disodium EDTA salt, Na_2 EDTA, can be used in place of the tetrasodium EDTA, 0.5 mL of Triton X-100 and Antifoam Y-30 may be used with Mark III model.)

2. Calibration of instrument:

   It is imperative that individual instruments be calibrated and meet performance requirements on the type of milk to be analyzed.

   a. Each instrument must be calibrated, using 20 representative milk samples ranging from 3% to 6% fat. Test the samples in triplicate by both the instrumental method and by preferably the ether extraction method (Roese-Gottlieb or Mojonnier) or a volumetric method (Babcock or Gerber). Fat samples shall be prepared from raw milk less than 48 h old. Use the type of milk to be analyzed (i.e., raw, pasteurized, unhomogenized, homogenized, etc.). For fat standardization purposes, do not use cream that has gone through a separator because physical changes occur in the fat.

   b. Calculate the mean for the triplicate of each sample tested to the nearest 0.01%.

   c. Calculate the mean difference ($\bar{D}$) between the instrument and the reference method.

   $$\bar{D} = \frac{\Sigma D}{N}$$

   d. Calculate the standard deviation of the differences.

   $$S_d = \sqrt{\frac{N\Sigma(D^2) - \Sigma D^2}{N(N-1)}}$$

   Where S_d = Standard deviation of difference

   N = Number of samples tested. If the specification of 20 samples is exceeded, all samples tested must be included in the calculations except for those for which an error can be proven.

   ΣD = Sum of the differences of the mean reference method results on each sample minus the mean instrument results on the same sample.

   $\Sigma(D_2)$ = Sums of the squares of the differences of the mean reference method results on each minus the mean instrument results on the same sample.

   e. The Milko-tester is properly calibrated when the mean difference is 0.05% or less; or the standard deviation of difference is less than or equal to 0.06% for herd or composite samples, or 0.10% for individual cow samples. Should the standard deviation of difference exceed these values, the instrument must be adjusted in accordance with the manufacturer's instructions and the calibration procedure repeated.

f. During any calendar day of use, make performance checks by comparing results obtained on 6 to 10 milk samples using both Milko-tester and Babcock, Gerber or preferably Roese-Gottlieb or Mojonnier results. The checks should have a range of fat similar to the milk that will be tested. If the difference is greater than 0.05% fat, repeat the determination on three additional samples. If the mean differences of the three samples is greater than 0.05% fat, then adjust the instrument and repeat the calibration.
3. Collection and preparation of sample:
   a. Before testing, samples shall be stored at 0 to 4.4°C.
   b. Samples shall be prepared for testing by tempering at 37 ± 1°C in a thermostatically controlled water bath for a maximum of 15 min after temperature is reached. CAUTION: Samples should not be held in water bath longer than 15 to 20 min.
   c. Gently mix samples by inverting the sample about six times, being careful to avoid over mixing or churning action resulting in sample instability. (Must have air space in container, otherwise sample must be mixed by passing from one container to another three times.)
   d. Samples must be tested no later than 30 sec after mixing.
4. Procedure:
   Obtain readings using manufacturer's instructions for instrument operation and maintenance.

18.9 Multi-component Methods

A. Infrared milk analysis—fat, protein, lactose, total solids[2,3,4,39] *(Class A1):*

Analysis of milk by infrared (IR) is based on absorption of IR energy at specific wavelengths by carbonyl groups in ester linkages of fat molecules (5.723 μm), by peptide linkages between amino acids of protein molecules (6.465 μm), and by OH groups in lactose molecules (9.610 μm). The wavelength of the carbon hydrogen bond (3.47 μm) in the fatty acid has been used to help refine the fat results. It has been used both by itself and in conjunction with the ester linkage wavelength. Total solids may be determined by adding an experimentally determined factor to the percentage of fat, protein and lactose.

The cross effects between fat, protein, and lactose may be reduced by making interrelationship corrections.

The IR beam is passed through an optical filter which transmits energy at the wavelength of maxium absorption for the component being measured. The filtered beam then passes through the milk-filled sample cell and then to a detector. The IR beam is also passed through an optical filter, which

18.9 Multi-component Methods

transmits energy at the wavelength where there is a minimum of absorption by the component and then through the milk-filled sample cell to the detector. Signals from the detector are then used to compare the absorption at the two wavelengths and hence to the concentration of the component. Interference effects result from differences in scattering of energy at the sample and reference wavelengths, and from differences in the absorptivity of water at these two wavelengths. Since the sample and reference wavelengths are close to each other, these effects are not large. Electronic corrections, nevertheless, are made to compensate for them. Differences in the degree of homogenization achieved by the instrument homogenizer, whether because a particular sample is high in fat content or because the sample has been homogenized before analysis, may or may not cause differences in results. Scattering tends to shift the wavelength of apparent maximum absorption to a slightly higher wavelength, a phenomenon resulting from the fact that the index of refraction of milk fat is considerably higher on the high wavelength side of the fat absorption band. Analysis by IR is dependent on calibration against suitable reference methods. Analysis with the equipment now available takes about 20 sec per analysis.

1. Apparatus and reagents:
   a. Individual infrared instruments (DICKEY-john/FOSS, P.O. Box 10, Auburn, IL 62615 or Multispec Inc., 23560 Lyons Avenue, Suite 209, Newhall, CA 91321) must meet following performance specifications based upon analysis of eight samples:

	Maximum Acceptable Differences in %			
	Fat	Protein	Lactose	Total Solids
Std deviation between duplicates	± 0.02	± 0.02	± 0.02	± 0.04
Mean difference between duplicates	± 0.02	± 0.02	± 0.02	± 0.03
Std deviation between instrument and reference method	± 0.06	± 0.06	± 0.06	± 0.12
Mean difference between instrument and reference method	± 0.05	± 0.05	± 0.05	± 0.09

   b. Calcium propionate solution: Dissolve 15 g of calcium propionate. H_2O in LG water and dilute to 1 L.
   c. Lactose solution: Dissolve 50 g of lactose. H_2O in LG water and dilute to 1 L.
   d. Water (LG)
2. Precaution:
   The adequacy of the analysis will depend upon the following:

a. Presentation of good quality samples to the instrument at a temperature of 40°C.
b. Homogenizing efficiency remaining constant. It may decrease with time due to wear on the ball and seat of the homogenizer valve or loss of tension in the spring supporting the ball. The result is loss of linearity at the high end of the fat calibration.
c. Adequate control of calibration with milk standards of known composition.
d. Maintaining a low water vapor content in the console of the instrument to prevent changes in optical zero and a shift in calibration level. Replace desiccant frequently, preferably at end of the operating day, as 3 to 4 h are required to restore equilibrium conditions.
e. Calibration with the type of milk to be analyzed (herd, individual cow, homogenized, unhomogenized, market milk, etc). Do not use mixtures of cream and milk for calibration. Avoid abnormal milks for calibration.
f. Maintenance of purging of previous samples at a level of 99% or better. This can be tested as follows:
  1. Fill ten sample vials with water and ten vials with portions of a pasteurized homogenized milk.
  2. Analyze the 20 samples in order of alternate pairs, first of milk and then of water (i.e., $M_1, M_2, W_1 W_2, M_1 M_2, \ldots$).
  3. Calculate the percentage purging efficiency as follows: If M_1, M_2, W_1, W_2 represent the fat readings for first and second results of alternate pairs of milk and water, respectively.

$$\% \text{ purging efficiency, water to milk} = \frac{(\Sigma M_1 - \Sigma W_2) \times 100}{(\Sigma M_2 - \Sigma W_2)}$$

$$\text{milk to water} = \frac{(\Sigma M_2 - \Sigma W_1) \times 100}{(\Sigma M_2 - \Sigma W_2)}$$

g. Proper adjustment of linearity of output signals relates to component concentration. Linearity adjustments, if necessary, should be made before attempting to calibrate the instrument.
  1. Mix accurately measured portions of LG water with measured portions of cream and a solution of calcium propionate or solution of lactose, for checking linearity of fat, protein and lactose, respectively. Use unhomogenized or homogenized cream, depending on the type of sample to be analyzed. The dilutions should be made so that the relative concentrations are known and so that the readings obtained adequately cover the anticipated concentration range of analysis to be done. Approximations of the dilution factors required can be estimated by obtaining readings on the stock solutions. About

18.9 Multi-component Methods

eight samples should suffice for the linearity check on each component.
2. Obtain readings on the diluted creams and diluted stock solutions. Pump each solution into the instrument twice to avoid a carry-over effect and use only the second reading. Plot these against the relative concentrations of the samples used. If the resulting plot is not linear, make a linearity adjustment as indicated in the instrument operating manual.
3. Preparation of samples:

Milk samples should be of good quality. They should not be sour and the fat should not be oiled off. Potassium dichromate may be used as a preservative. Samples should be preheated to 40°C to melt the fat before analysis. Gently mix samples by inverting the sample about six times, being careful to avoid churning action resulting in sample instability. (Must have air space in container otherwise sample must be mixed by passing from one container to another three times).
4. Calibration:

Each instrument must be calibrated using at least 20 representative milk samples of the type (homogenized or unhomogenized) to be analyzed and with as wide a range of values as possible. Herd milks are preferable if they can be obtained with the required compositional range. For calibration purposes, avoid milk with abnormal composition.

Analyze all calibration samples in duplicate, using accepted standard reference methods. (Ether extraction is preferred for fat.)

Following directions in the operating manual, analyze these samples in duplicate with the instrument. Determine the standard deviation between instrument and reference method and the mean difference between instrument and reference method (see [18.8G2] for calculations). If any of these exceed the performance specification given in this section, the instrument must be adjusted.

Repeat the calibration procedure and readjust if necessary.

Check instrument performance by comparing results obtained on six to ten milk samples using both infrared and reference methods. The check milk should have a range of values similar to the milk being tested. If the results exceed the maximum acceptable differences [18.9A1a], adjust the instrument and repeat.
5. Operation of the instrument:

Adequate operating and maintenance instructions are provided in operating manuals. However, particular attention should be given to maintenance of a low moisture vapor content in the optical section of the instrument. The silica gel desiccant used for this purpose should be changed frequently, particularly if the relative humidity is high. Adjustments to optical zero are made necessary by changes in the moisture vapor content in the instrument. This change should be made at the end of an operating day to allow the instrument to stabilize overnight. Checking the magnitude of the optical zero readings the next morning will indicate the total amount of change in optical zero since the previous change in desiccant. Frequency of change should be adjusted so that this value is not greater than 0.2

B. Near infrared analysis: fat, protein, total solids in milk (Class A2):

The near infrared measurment system makes use of wavebands in the overtone region (1400 to 2400 nanometers) of the mid-infrared spectrum. Absorption in this region approximates integral multipliers of the frequencies of absorption in the mid-infrared spectrum. Because of the many overtones for each fundamental absorption waveband, there is considerable overlapping of wavebands of different molecular groups. It is thus difficult to assign exclusively the representative absorption for individual filters. Four or more filters are used for each component estimated, the combination used being that which gives the best results. Estimates of component percentage are calculated with multiple regression equations that are derived when the instrument is calibrated using samples of known composition. All constituent analyses are done simultaneously on the same milk sample and results are displayed in less than 90 sec.

1. Apparatus:
    a. Near infrared milk analyzer with necessary pumps and sample cells to test liquids. (Technicon Instrument Corp., 511 Benedict Ave., Tarrytown, NY 10591) Individual, near infrared instruments utilized must meet the same performance specifications for fat, protein, and total solids as in [18.9A1a], based upon analyses of eight samples.
    b. Wash solutions as required by manufacturer.
    c. Water bath capable of holding samples at 45 ± 1°C.
2. Preparation of samples:
    Milk samples should be of good quality. They should not be sour and the fat should not be oiled off. Samples must be heated to 45 ± 1°C before analysis to melt the fat and match the instrument temperature. Gently mix sample by inverting it about six times, being careful to avoid churning action that would result in sample instability. (Must have air space in container, otherwise sample must be mixed by passing from one container to another three times.)
3. Calibration:
    a. Homogenized and unhomogenized milk will require different calibrations. Skim and diluted milks require separate calibration for total solids and protein. Preserved and pasteurized milks require bias offsets for total solids and protein calibrations.

18.9 Multi-component Methods

   b. Each instrument must be calibrated using at least 20 representative milk samples of the type to be analyzed by the instrument and with as wide a range of values as possible. Herd milks are preferable if they can be obtained with the required compositional range. For calibration purposes, avoid milk with abnormal composition.
   c. Analyze all calibration samples in duplicate using accepted standard reference methods. (Ether extraction is preferred for fat.)
   d. Analyze these samples in duplicate with the instrument, following directions in the operating manual. Determine the wavelengths and coefficients to be used for each component, following the manufacturers instructions.

   Technicon has calculated a combined calibration from more than one instrument using over 100 pre-analyzed samples obtained from many geographic areas. This calibration can be transferred to other instruments by slope and bias adjustment to match results of the above 20 samples.
   e. After this calibration, the instrument must meet the performance specification given in [18.9A1].
   f. If the sample drawer is moved, check bias and adjust calibration if required before proceeding.
   g. If cell cover is removed or if cover position is changed, adjust cell cover in accordance with manufacturer's recommendations and repeat bias calibration.
4. Operation of instrument:
   a. During any calendar day of use, make performance checks consisting of comparison of results obtained on six to ten milk samples using both near infrared and a reference method. If the results are greater than the maximum acceptable differences [18.9A1], adjust the instrument and repeat.
   b. The instrument must be cleaned following manufacturer's instructions.

C. *Modified Kohman method*[14] *for fat, moisture and salt in butter and margarine (Class O):*

This method tests for moisture, fat and salt in a sample of high-fat spread. A weighed sample is heated to evaporate the moisture, then reweighed to determine moisture lost. The fat is extracted with ether and the remaining solids are weighed to determine fat. The solids are then dissolved and salt determined by titration.

1. Apparatus and reagents:
   a. Balance, 0.1 mg sensitivity.
   b. Beaker, 200 mL, tall, aluminum, with pouring spout.
   c. Buret, 25 or 50 mL, with 0.1 mL divisions.
   d. Crucible, Gooch.

e. Desiccator with efficient desiccant.
f. Flask, Erlenmeyer, 125 mL.
g. Flask, filter, 500 mL.
h. Flask, volumetric, 250 mL.
i. Filter paper, to fit Gooch crucible.
j. Hot plate, thermostatically controlled at 232°C ± 3°C.
k. Oven, atmospheric, 100°C.
l. Stirring rods, glass, with rubber policeman.
m. Volumetric pipette, 25 mL.
n. Wash bottle for petroleum ether.
o. Petroleum ether, boiling range 30 to 60°C.
p. Potassium chromate indicator (K_2CrO_4), 10% in LG water.
q. Silver nitrate, 0.1 N.
r. Water (LG).
2. Procedure for moisture:
   a. Using the analytical balance, weigh a clean, dry aluminum beaker.
   b. Accurately weigh about 10 g of sample into the aluminum beaker.
   c. Using crucible tongs, place the aluminum beaker on the hot plate at 232 ± 3°C. Swirl constantly during heating to avoid spattering and burning of milk solids. A mechanical shaker may be used.
   d. Continue evaporation until all foaming and bubbling has ceased. The milk solids take on a light brown color. Care should be taken to obtain the same degree of color in each test.
   e. Cool beaker to room temperature in a desiccator.
   f. Reweigh beaker.
3. Procedure for fat:
   CAUTION: This procedure should take place in an exhaust hood. Do not use petroleum ether near open flame or sparks.
   a. Prepare crucible by pressing the filter paper into the crucible. Wash twice with suction using 10 to 15 mL petroleum ether per wash. Dry with suction for 15 sec and then dry in oven at 100°C for a minimum of 15 min. Cool in desiccator. Weigh prepared crucible on the analytical balance.
   b. Add about 125 mL of petroleum ether to the moisture free residue in the aluminum beaker.
   c. Use a glass rod to break up the residue in the beaker to insure complete fat extraction. Rinse the rod with a few millimeters of petroleum ether before removing it from the beaker.
   d. Transfer residue into a preweighed, prepared crucible while applying suction. Use a wash bottle with petroleum ether to aid in the transfer.
   e. Rinse the beaker twice with 25 mL of petroleum ether per rinse. Add rinse to crucible in such a way as to wash the solids into the crucible.

18.10 Moisture and Solids

f. Dry with suction for 15 to 30 sec after last ether has passed through.
g. Dry crucible and contents at 100°C for 15 min.
h. Cool to room temperature in a desiccator.
i. Reweigh crucible and contents on the analytical balance.
4. Procedure for salt:
   a. Transfer filter paper and residue to a 250-mL volumetric flask with hot (65 to 70°C) LG water. Rinse the crucible three times and add washings to flask.
   b. Cool to room temperature and dilute to volume with LG water. Mix thoroughly. Because of the low levels of salt in the solution the volume of the filter paper is not significant.
   c. The salt will dissolve and milk solids will settle in the flask.
   d. Pipette 25 mL of this solution into a 125-mL Erlenmeyer flask.
   e. Add 1 mL of potassium chromate indicator. Titrate with 0.1 N silver nitrate solution to the first visible pale red-brown color lasting 30 sec. The end point is much sharper if the titration is conducted under a yellow light.
5. Calculations:

$$\% \text{ moisture} = \frac{(B - D \times 100)}{C}$$

$$\% \text{ fat} = \frac{[(D - A) - (F - E)] \times 100}{C}$$

$$\% \text{ NaCl} = \frac{\text{mL AgNO}_3 \times \text{N AgNO}_3 \times \text{Dilution vol} \times 5.85}{C \times \text{mL titrated}}$$

A = Weight of beaker
B = Weight of beaker and sample
C = Weight of sample (B − A)
D = Weight of beaker and moisture-free residue (fat, salt, milk solids)
E = Weight of crucible
F = Weight of crucible and solids (salt, milk solids)

18.10 Moisture and Solids

A. *Vacuum oven*[16,20] *(Class O):*
   1. Apparatus and reagents:
      a. Aluminum moisture dishes: AOAC specifications, round, flat-bottomed, at least 5 cm in diameter, provided with close fitting slip-in cover, Cenco #12715 or equivalent.
      b. Balance, 0.1-mg sensitivity.
      c. Blender, Oster or Waring or equivalent.
      d. Crucible tongs.
      e. Desiccator with efficient desiccant.
      f. Spatula.

    g. Steam bath.
    h. Thermometer, 110°C.
    i. Sulfuric acid, technical or CP.
    j. Vacuum oven equipped with gas washing bottles containing sulfuric acid. These bottles should be arranged so that moisture is removed from air entering the oven from the outside. Use Corning glassware #31760 or equivalent.
    k. Vacuum pump of Welch-distillation type, capable of maintaining a minimum vacuum of 66 cm, preferably 74 cm in the oven.
2. Collection and preparation of sample:
    a. Natural cheese: When natural cheese can be cut, take a narrow, wedge-shaped segment reaching from outer edge to center and place in container with a tight fitting lid. When cheese cannot be cut, take sample with cheese trier. Three plugs of cheese should be taken, one from the center, one from a point near outer edge, and one from a point halfway between the other two. If barreled cheese is to be sampled, take a 25- to 30-cm plug about 7 cm from edge of barrel and angled at 11° to the outside [4.2D]. Use the portion of the plug 10 cm below the surface for test. Cheese plugs should be placed immediately in container with tight fitting lid. Before taking sample for analysis, the wedge-shaped segment should be cut very finely with the aid of a grater, knife or spatula and mixed well. If plugs are taken, mix plugs with the aid of a blender for a total of about 15 sec. Blender should be turned off and samples and jar shaken manually to help move larger particles toward blade-side of mixer. Avoid prolonged mixing, as generated heat will cause fat separation and mashing of product.
    b. Process cheese, cheese food and spreads: With the aid of a spatula, cut cheese into small segments and place in a 120- or 240-mL sample jar. (Avoid taking outer surface of cheese for sample use.)
    c. Regular cheese spreads: With the aid of a spatula, thoroughly mix approximately 150 g of sample of cheese in 240-mL sample jar.
    d. Condiment cheese spreads: To obtain a more homogeneous sample, it is necessary to mix at least a 150-g sample of cheese in a blender for 4, 15-sec periods.
    e. Cheese powders, grated and flavored: With the aid of a spatula, thoroughly mix samples before sampling.
    f. Ice milk, ice cream, milk shake mixes, and malted milks: With the aid of a spatula, thoroughly mix samples before sampling or pour between two vessels.
3. Preparation of moisture dishes:
    The aluminum moisture dishes used shall be washed thoroughly,

18.10 Moisture and Solids

rinsed, and dried in an oven for several hours at 100°C. Store in a clean desiccator until used.
4. Procedure:
   a. Weigh 3 ± 0.5 g of sample into a preweighed dish on an analytical balance. Where possible, distribute sample evenly over the bottom of the dish. Some samples require predrying (Note 5). Place dish with cover placed beneath it, on metal shelf in the vacuum oven for 5 h. The oven should be at the temperature specified for the particular sample [18.10A6] being tested, with a minimum vacuum of 66 cm, 74 cm preferred. During drying, admit into oven a slow current of air (about two bubbles per sec) dried by passage through sulfuric acid washing bottles.
   b. After 5 h shut off the vacuum pump and slowly readmit dry air into the oven. Using tongs press cover lightly into dish and remove dish from the oven. Cool in desiccator until the dish has reached room temperature. Weigh dish.
5. Calculations:
$$\% \text{ moisture} = \frac{\text{loss in weight} \times 100}{\text{weight of sample}}$$
$$\% \text{ solids} = 100 - \% \text{ moisture}$$
6. NOTES:
   a. Temperature should be measured under vacuum by a thermometer in contact with the shelf or wall of oven. Temperature in the open space of the oven will be different.
   b. The vacuum oven moisture method using 100 ± 2°C can be applied to the following dairy type products and ingredients:
   Casein powders
   Cheese: natural, process, food, spread.
   Cheese: powders, grated, flavored.
   Dairy animal feed.
   Dried egg products.
   Dried ice cream and ice milk mixes.
   Dried milks: whole and nonfat, and buttermilk.
   Gelatin: granules, crystals, powders.
   Malted and chocolate milks.
   Nuts and nut products.
   c. The vacuum oven moisture method using 70 ± 2°C applies to the following products:
   Foods with a large amount of sugar.
   Fruit products.
   Sugars.
   d. The vacuum oven moisture method using 65 ± 2°C for 16 h can be applied to the following products:

Dried whey and modified whey.
Dried whey protein concentrate.
  e. The following product types require samples to be either partially dried on a steam bath before placing in the oven; or to be predried in a vacuum oven with door closed for 15 min at atmospheric pressure before proceeding to turn on vacuum normally used in test procedure.
  Fluid milks (whole, skim, lowfat) and buttermilk.
  Liquid egg products.
  Soft natural type and high-moisture-type process foods and cheese spread.

B. *Vacuum oven: sand pan (Class O):*

The following products can be tested for moisture using the vacuum oven, sand pan method: butter, margarine, condensed whey, whey protein concentrate, ice cream toppings and syrups, sweetened condensate, milk, and yogurt.

1. Apparatus and reagents:
   Same as with Regular Vacuum Oven Method [18.10A] plus:
   a. Atmospheric oven (optional).
   b. Clean, fine, white sand.
   c. Small glass rods with flattened tips, 3 × 60 mm.
2. Preparation of sand pan:
   a. Prepare aluminum moisture dishes by washing thoroughly, rinsing, and drying.
   b. Place about 25 g of clean dry sand and a small stirring rod into a moisture dish. Place in vacuum oven at 100°C for at least 5 h or in atmospheric oven overnight to dry.
   c. Remove dish from oven and store in a desiccator until used.
3. Procedure:
   a. Weigh 3 to 5 g of sample into a preweighed sand pan on an analytical balance.
   b. Add sample directly in the center of the sand pan to keep the sample from touching the sides of the pan. Weigh as rapidly as possible.
   c. Remove from balance and mix sand and sample with the stirring rod as thoroughly as possible. Leave stirring rod in pan throughout the determination. Do not spill sand.
   d. Place dish in the vacuum oven for 5 h. The oven should be at a temperature of 100 ± 2°C and at a minimum vacuum of 66 cm. During drying, admit into the oven a slow current of air (about two bubbles per sec) dried by passage through sulfuric acid washing bottles.
   e. After 5 h shut off the vacuum pump and slowly readmit air into the oven. Remove dish from the oven and cool in dessicator at

18.10 Moisture and Solids

least 30 min or until the dish has reached room temperature. Weigh dish on analytical balance.

4. Calculations:

$$\% \text{ moisture} = \frac{\text{loss in weight} \times 100}{\text{weight of sample}}$$

$$\% \text{ solids} = 100 - \% \text{ moisture}$$

5. Note:
   a. Temperature should be measured under vacuum by a thermometer in contact with the shelf or wall of oven. Temperature in the open space of the oven will be different.

C. Atmospheric oven[20] (Class O):

The following products can be tested for moisture using the atmospheric oven method: liquid buttermilk, natural cheese, chocolate and cocoa, cottage cheese, liquid creams, egg albumen liquid and powders, liquid milks, condensed whole and skim milk, sweetened condensed skim milk, nuts, and toppings.

1. Apparatus and reagents:
   a. Aluminum moisture dishes: AOAC specifications, round, flat-bottomed, at least 5 cm in diam, provided with close fitting slip-in cover, Cenco #12715 or equivalent.
   b. Balance, 0.1-mg sensitivity.
   c. Blender, Oster, Waring, or equivalent.
   d. Crucible tongs.
   e. Desiccator filled with efficient desiccant such as Drierite, indicating, 4-mesh (calcium sulfate), calcium chloride, etc.
   f. Spatula.
   g. Steam bath.
   h. Thermometer, 0 to 110°C.
   i. Atmospheric moisture oven.

2. Collection and preparation of sample:
   a. Natural cheese: Same as [18.10A2a].
   b. Cottage cheese curd and nuts: To get a more homogeneous sample, it is necessary to mix at least a 150-g sample in a blender for four, 15-sec periods, turning off and starting blender at the end of each 15-sec period.

3. Preparation of moisture dishes:

   The aluminum moisture dishes should be washed thoroughly, rinsed, and dried in an oven for several hours at 100°C. Store in a clean desiccator until used.

4. Procedure:
   a. Weigh about 2 to 3 g of sample into a preweighed dish on an analytical balance. Where possible, distribute sample evenly over the bottom of the dish. Place dish with cover under it on the

metal shelf in the atmospheric oven. Consult Table 18.4 for steam bath requirement, oven temperature, and length of time in oven for various products.

b. After specified time in oven, use tongs to press cover tightly into dish and remove dish from the oven. Cool in desiccator for at least 30 min or until the dish has reached room temperature. Weigh dish on analytical balance.

5. Calculations:

$$\% \text{ moisture} = \frac{\text{loss in weight} \times 100}{\text{weight in sample}}$$

$$\% \text{ solids} = 100 - \% \text{ moisture}$$

D. *Moisture, microwave oven*[25] *(Class A1, A2, B):*

This method is Class A1 for Cheddar, Colby, and Monterey Jack cheese, milk and cream.[25,10a] This method is Class A2[1a] for Cottage Cheese.[xl] It may also be used for other products, but the method then will be Class B. Microwave energy is used to heat samples to remove moisture. Factors affecting the drying are time, sample size, placement of sample in oven, and energy of microwaves. Microwave ovens vary from unit to unit. Power settings and time may vary from unit to unit and with the age of the unit[1a].

1. Apparatus and reagents:
    a. Balance, may be built into the oven.
    b. Blender, Oster, Waring, or equivalent.
    c. Filter pads, glass fiber.

Table 18.4 Atmospheric Oven Temperatures and Times

Product	Dry on Steam Bath	Oven Temp (°C ± 2)	Time in Oven (h)
Buttermilk, liquid	X	100	3
Cheese, natural type only		105	16 to 18
Chocolate and cocoa		100	3
Cottage cheese		100	3
Cream, liquid and frozen	X	100	3
Egg albumin, liquid	X	130	0.75
Egg albumin, dried	X	130	0.75
Ice cream and frozen desserts	X	100	3.5
Milk-whole, low fat and skim	X	100	3
condensed skim		100	3
evaporated milk		100	3
Nuts almonds, peanuts, walnuts, etc.		130	3

(X-Samples must be partially dried on steam bath before placing in oven).

18.10 Moisture and Solids

      d. Microwave oven, (CEM Corporation, P.O. Box 9, Indian Trail, NC 28079, or equivalent).
      e. Petri dish cover, pyrex, 10-cm diameter.
  2. Preparation of sample:
      a. Natural cheese: to obtain a more homogeneous sample, it is necessary to mix cheese in blender for a total of about 15 sec. (Blender should be turned off and sample and jar shaken manually to help move larger particles toward bladeside of mixer. Prolonged mixing should be avoided as heat generated causes fat separation to occur.)
      b. Process cheese: It is not necessary to mix this type of cheese in a blender, as product is relatively uniform. Strips or large pieces of cheese may be used to make up sample.
  3. Procedure: Take a Pyrex petri dish cover (not necessary but does add stability to the filter papers) and two glass filter pads. Weigh the prepared sample on filter pad in petri dish and cover with a second filter pad to avoid spattering. Make certain that the sample is spread evenly before covering. Sample size may vary due to the oven and balance used. Subject the sample to microwave energy for a time and at a power setting that gives results equivalent to those of an atmospheric or vacuum oven. The time, power, and sample size to be used must be determined on each oven by comparing to reference methods.
  4. Calculation: At the end of the test cycle, reweigh the filter pads. Some units allow the results to be read directly in percent moisture or solids.

$$\% \text{ moisture} = \frac{\text{loss in weight} \times 100}{\text{weight of sample}}$$

% solids = 100 − % moisture

  5. NOTE: It is important that this method be periodically checked against a reference method (atmospheric or vacuum oven) to make certain that instrument calibration is being maintained. Adjustment may be made in time, energy, and/or sample size to produce results similar to reference method.

E. *Moisture balance: dry milk products (Class B):*
  1. Apparatus and reagents:
      a. Moisture balance with heating device.
      b. Moisture pans to fit balance.
  2. Procedure: Follow manufacturer's instructions to zero the balance. Distribute the sample evenly on the moisture pan and weigh. Adjust the power control and times for the specific materials being tested. At the end of the set time turn off the heat and reweigh.

3. With some instruments the moisture may be read directly from the scale. Otherwise:

$$\% \text{ moisture} = \frac{\text{loss in weight} \times 100}{\text{weight of sample}}$$

4. Note: The length of time in the moisture balance will vary depending on amount of moisture in different types of samples. Once correct drying time has been established for a product type, it is desirable to maintain this same drying time for all like samples tested. A setting of 80% power for 8 min has been suggested as a starting point but may vary with different types of samples and instruments.

F. *Refractometer: Whey and Whey Products (Class B):*
   1. Apparatus and Reagents:
      a. Hand refractometer, Bausch and Lomb scale range 40 to 85% or Zeiss Model # 0-85, with double scale and complete with stand or equivalent.
      b. Water, (LG) for checking refractometer.
      c. Light source (illuminator) for refractometer.
      d. Plastic spoons for transferrring sample to prism.
      e. Soft paper tissue for wiping refractometer prisms.
      f. Solvent for cleaning prisms, ethanol or xylol are suitable. If the material being examined is water soluble, use LG water. Do not use acetone or similar solvents because such solvents attack prism cement.
   2. Procedure: Place about 1 to 2 drops of warm sample on prism of refractometer. Close the cover and immediately raise the refractometer to eye level and point toward a good source of light.

      Observe the intersection of field-dividing line and scale. Make the necessary temperature correction indicated on thermometer on the side of the refractometer.
   3. Calculation: % solids = Reading on refractometer scale minus temperature correction factor.
   4. NOTES:
      a. immediately after taking refractometer reading above, wash prism thoroughly with LG water to remove whey product. Allowing product to dry on prism harms the prism surface.
      b. Care should be taken to rinse refractometer prisms with LG water and wipe dry with soft paper tissue. Do not touch prism with glass stirring rod, metal spoon, or other abrasive equipment.
      c. Test must be read quickly, as crystals form upon cooling and inaccurate results are obtained.
      d. Any air incorporated in sample will cause erroneous results. Samples must be thoroughly but carefully mixed.

18.10 Moisture and Solids

e. Accuracy of test is about ± 0.5 to 0.6% if all the steps in the procedure are followed explicitly. Results should be compared to a reference method and appropriate corrections made.

f. This method is applicable to whey concentrate (in the 40 to 60% solids range). A correction factor must be applied to a whey concentrate with a higher solids content.

G. *Solids in milk—lactometric method*[11,19,40] *(Class 0):*
  1. Apparatus and reagents:
     a. Butterfat test equipment: See Babcock Method [18.8A].
     b. Lactometer cylinder: Inside diameter must be sufficiently greater than the largest diameter or lactometer to permit free movement of the lactometer and having the capacity necessary to float the lactometer.
     c. Lactometer, large type.

Total length	28.7 ± 1.1 cm
Stem length	8.9 ± 0.2
Ballast length	1.7 ± 0.2
Stem diameter	0.60 ± 0.05
Bulb diameter	3.40 ± 0.05
Scale length	6.1 ± 1
Scale width	1.2 ± .2

     d. Lactometer, small type: Specifications in centimeters are approximate since certain tolerances are allowable for glass hydrometers (total length, 15; stem length, 4.5; stem diam, 0.5; bulb diam, 3.5; scale length, 2.3). The design for a whole milk lactometer has a scale with a range of 24.5 to 35.0 degrees, the smallest division being 0.5 degree.
     e. Water bath: Maintain bath at 39 ± 1°C. Have it equipped with overflow tube or some device to hold water at the proper level for heating milk samples, and deep enough to bring the water to within 5 cm of the top of the cylinder.
  2. Procedure:
     a. Lactometer test: Rapidly heat about 150 ml of milk sample for a small lactometer or 300 ml for a large lactometer to approximately 39°C by immersing the sample in a clean Erlenmeyer flask in water at about 46°C. Gently shake sample at intervals to prevent overheating. The flask should be loosely stoppered to reduce evaporation.
     b. Transfer the sample with minimal agitation to a lactometer cylinder held in a water bath with a temperature thermostatically maintained at 39°C. The water should be to within 5 cm of the top of the cylinder. The lactometer itself should be preheated for not less than 3 min before use by immersion in a cylinder of water

held in the same constant temperature bath. Remove the lactometer and wipe dry immediately before use.
  c. Slowly immerse lactometer in milk sample. The reading is taken at the top of the meniscus after the lactometer comes to rest. It is important that the stem of the lactometer be clean and dry above the milk surface. Repeat readings can be made by withdrawing the lactometer just enough to enable the operator to wipe the stem with a tissue or soft cloth before slowly immersing it again to the reading point.
  d. Each lactometer degree on the scale is divided into 0.2 degrees for the larger type, and each reading should be estimated to 0.1 degree. Successive readings should agree by ± 0.1 degree if the temperature remains constant. Temperature of the milk should be checked with an accurate thermometer at the time of reading.
  e. Fat test: Conducted by the standard Babcock method [18.8A] or its equivalent. Fat should be read to the nearest 0.05%.
3. Calculations:

$$\text{Milk, \% total solids} = 0.33F - 0.40 + \frac{273L}{L + 1000}$$

$$\text{Skim milk, \% total solids} = 1.33F - 0.40 + \frac{273L}{L + 1000}$$

where F = % fat
       L = lactometer reading in degrees.
4. Notes:
  a. Before use and as often as needed, test lactometers for accuracy by comparison with instruments of known accuracy at 15.5°C at two different sp. gr. levels. Use levels that are near the extremes that are expected. Discard those with readings that exceed 0.0001 sp. gr. from that on the reference standard.
  b. General suggestions: A large number of determinations can be done rapidly by using several lactometer cylinders in a constant-temperature bath. The cylinder of milk can be held at this temperature for about 1 h with only negligible change in specific gravity, provided that the milk is gently stirred before the lactometer is immersed. This long-holding method is not suitable for use with milk that has oiled off or become rancid.

18.11 Protein

Since 1883, the Kjeldahl principle has been used as the reference method for measuring total nitrogen in food and feed. With greater emphasis being placed on the importance of protein in foods today, newer and more accurate, specific, and rapid methods have been developed. These methods must be calibrated against the Kjeldahl.

18.11 Protein

A. Kjeldahl[14] (Class A1):

Two variations of the Kjeldahl test are in common use in the dairy industry. The difference between the two is in the way the NH_3 produced is titrated. Micro and semi-micro Kjeldahl apparatus is now in common use. If one of these methods is used, make proportional reductions in the amounts of reagents.

1. Apparatus and reagents:
   a. Balance, 0.1-mg sensitivity.
   b. Beakers: 2000 mL and 250 mL.
   c. Bottle, rubber-stoppered.
   d. Buret, acid dispensing, 100 mL.
   e. Buret, 50 mL, with 0.1 mL divisions.
   f. Dropping pipette with rubber bulb.
   g. Filter paper, Whatman #1.
   h. Flask, Erlenmeyer, 300 mL.
   i. Flask, volumetric, 2000 mL.
   j. Funnel, ribbed.
   k. Glass beads.
   l. Gloves, heavy canvas.
   m. Graduates, 1000 mL, 250 mL, and 50 mL.
   n. Kjeldahl connecting bulb.
   o. Kjeldahl digestion and distillation apparatus with either a fume hood or fume pipes for efficient exhaust.
   p. Kjeldahl digestion flask, long neck, 500 mL.
   q. Rubber stoppers, Kjeldahl long form #6, 1 hole.
   r. Acetanilide powder.
   s. Boric acid, saturated solution.
   t. Hydrochloric acid, 0.0200 N.
   u. Indicator solutions
      1. Bromocresol green—Methyl Red: to be used with H_2SO_4 in receiving flask. Dissolve 1 g bromocresol green and 0.20 g methyl red in 600 mL 95% ethanol.
      2. Tashiro indicator: to be used with boric acid in receiving flask. Dissolve 0.25 g methylene blue and 0.375 g methyl red in 300 mL 95% ethanol.
   v. Kjeldahl catalyst with potassium sulfate, commercially pre-weighed packets are recommended. Use one of the following with or without pumice to inhibit bumping.
      1. 0.7 g mercuric oxide and 15 g potassium sulfate.
      2. 0.65 g metallic mercury and 15 g potassium sulfate.
      3. 0.19 g cupric sulfate and 9.7 g potassium sulfate.
   w. Paraffin
   x. Potassium sulfate, anhydrous, granular, low nitrogen: order small quantities, as product deteriorates after several months.

- y. Potassium sulfide: Dissolve 85 g potassium sulfide in hot LG water. Transfer to 2000-ml volumetric flask when cool, and make to volume. Filter through Whatman #1 filter paper.
- z. Sodium hydroxide, 0.1000 N
- aa. Sulfuric acid, low in nitrogen, ACS, sp. gr. 1.84
- bb. Sulfuric acid, 0.1 N.
- cc. Water (LG).

2. Procedure
    a. Place weighed sample (0.7 to 2.2 g) in digestion flask. Sample should contain 0.2 to 0.4 g protein.
    b. As a standard control, place a weighed amount (0.5 g) acetanilide in a second digestion flask.
    c. To each flask, add 1 packet of Kjeldahl catalyst and 25 to 30 mL of sulfuric acid (sp. gr. 1.84).
    d. Place flasks in an inclined position on digestion apparatus. Turn on exhaust system for fume removal.
    e. Heat at low heat until frothing ceases. Add small pieces of paraffin to prevent excessive foaming if necessary.
    f. Continue heating, with increased heat, until mixture boils briskly and mixture is colorless. Then heat an additional 30 min. If the sample seems to become solid, turn off heat, allow flask to cool on the rack then add 15 mL additional sulfuric acid (sp. gr. 1.84) to the flask.
    g. The digestion time varies with the type of organic nitrogen compound and the temperature of the boiling mixture. If digestion takes more than an h, add 5 to 10 g potassium sulfate to raise boiling temperature.
    h. Turn off heaters and allow flasks to cool on the rack until fuming stops.
    i. While flasks are cooling, prepare receiving flasks.
        1. Method 1:
            a. From a buret, add a measured amount (50 mL) of 0.1 N sulfuric acid to each receiving flask (300-mL Erlenmeyer). The amount of acid added must be in excess of the amount of ammonia expected to be produced.
            b. Add 15 drops of bromocresol green-methyl red indicator solution to each flask.
        2. Method 2:
            a. Add 50 to 75 mL of saturated boric acid solution to each receiving flask (300-mL Erlenmeyer).
            b. Add 15 drops of Tashiro indicator solution to each flask.
    j. Place the receiving flask on the distilling rack so the tip of the condenser extends below the surface of the acid in the flask. This prevents escape of ammonia fumes during distillation.

18.11 Protein

k. Remove flasks from digestion rack and cool by immersing flasks in cold water. Sample will solidify with cooling.
l. Add 150 mL of distilled water and swirl flask until the sample is *completely* in solution. Immerse flask in cold water to cool again. Add 2 or 3 glass beads to flask. While swirling flask, add 25 mL of potassium sulfide solution.
m. Turn on water valve so that cold water flows around the condensers.
n. Put on protective gloves and face shield. Hold flask at an angle and in place on heater of distilling unit. Rapidly, but carefully, pour 100 mL of 1:2 sodium hydroxide down the side of the flask so that it does not mix at once with the acid solution.
o. Immediately insert stopper firmly in the flask, to prevent escape of ammonia.
p. Mix the contents of the flask by *shaking* it vigorously. It is imperative that the contents be well mixed, to minimize superheating and "bumping."
q. Turn on heater and distill 150 to 200 mL into receiving flask.
r. If Method 1 is being used, be sure that there is excess acid in the receiving flask during distillation as indicated by indicator.
s. When volume in flask is reduced, "bumping" occurs. Turn heat off immediately and disconnect flask. CAUTION: Wear protective gloves and face shield.
t. Remove the receiving flask, letting the tip of the condenser drain into the flask momentarily. Rinse the tip with distilled water, catching rinsings in flask.
u. Titrate in one of two ways.
  1. Method 1:
    Titrate the excess acid in the receiving flask with 0.1000 N sodium hydroxide. The solution changes color from red through grey tones to a final green.
  2. Method 2:
    Titrate with 0.0200 N HCl. The solution changes color from purple through gray tones to a final green.
3. Calculations:
  a. Total nitrogen:
  Method 1:
  1 mL of 0.1 N H_2SO_4 is equivalent to 0.0014 g nitrogen
  % N =

$$\frac{[(mL\ H_2SO_4 \times N\ H_2SO_4) - (mL\ NaOH \times N\ NaOH)] \times 0.14}{grams\ sample}$$

Method 2:
$$\%N = \frac{(mL\ HCl \times N\ HCl) \times 0.14}{grams\ sample}$$
   b. Protein, for dairy products: 1 part nitrogen = 6.38 parts protein.
   % protein = % nitrogen × 6.38
4. NOTES:
   a. Run a standard (acetanilide) sample with every series of samples. The total nitrogen should be 10.33% ± 0.15%, indicating that reagents, equipment, and technique are acceptable.
   b. If the color of the indicator changes before the distillation is completed, repeat the determination with a smaller sample.
   c. CAUTION: If mercury catalyst is used, take precautions to limit exposure of laboratory personnel. Mercury and mercury compounds are very toxic. Waste from use of mercury may require special disposal.
   d. Ratio of salt to acid (wt:vol) should be about 1:1 at end of digestion for proper temperature control. Digestion may be incomplete at lower ratio; nitrogen may be lost at higher ratio.
   e. Each gram of fat consumes 10 mL H_2SO_4 and each gram of carbohydrate consumes 4 mL H_2SO_4 during digestion.
   f. Normal solutions need to be prepared precisely to the fourth decimal place but not necessarily all zeros (i.e., 0.1246 will work as well as 0.1000).

B. *Dye binding: amido black method[26,27] (Class A2):*
   1. Apparatus and reagents:
      a. Pro-milk MK II (DICKEY-john/FOSS, P.O. Box 10, Auburn, IL 62615) or equivalent.
      b. Beakers, stainless steel (2), 3-L capacity.
      c. Bottles, plastic, 2-L capacity.
      d. Bottles, plastic (2), 10-L capacity. (Supplied with Pro-Milk unit. Replacements may be ordered from DICKEY-john/FOSS. Chemicals are in measured quantities to prepare 10 L of dye).
      e. Squeeze bottles, plastic (3), 500 mL.
      f. Filters, glass fiber.
      g. Flask, Erylenmeyer, 3-L.
      h. Forceps.
      i. Funnel, plastic—powder type.
      j. Graduate, 1000-mL capacity.
      k. Hot plate.
      l. Syringe for dispensing dye.
      m. Syringe, plastic, 1 mL.
      n. Ammonium hydroxide, ACS, sp gr 0.90.

18.11 Protein

o. Chemicals for working liquid—Foss # PRM504 (This solution can be made by the user).
p. Triton X-100.
q. Water (LG).

2. Solutions:
   a. Working "0-liquid" (buffered amino dye solution): Carefully measure 10 L of tap water at room temperature into the plastic jug. Mark the volume on the container. Tap water is used for calibrating only. Remove water and rinse jug with LG water Invert to drain for a short period.

   The two components, dye and buffer, are each dispensed into separate 3-L flasks. That individually contain approximately 2 L of LG water. Bottles that hold the dye and buffer should be thoroughly rinsed and the rinsings added to their respective solutions. For complete dispersion of chemicals, the solutions are heated to 70°C. This should be done in a well ventilated area, preferably a fume hood.

   When dilution is completed, transfer solution to the 10-L bottle, rinsing the container in which the chemicals were heated with approximately 1 L of water for each container, adding rinsings to combined solutions. Fill to the 10-L mark with LG water and add by means of a plastic syringe, 1 mL of Triton X-100. Mix the solution thoroughly and hold for 24 h before using. Record date when solution was made in log book, and also on bottle.

   b. Standard liquid for checking "45" reading should be made in sufficient quantity for use with the entire dye lot. Prepare by volume: 400 mL of working "0-liquid," plus 1000 mL of LG water.

   c. Ammonia solution, 2%: to 2 mL of ammonium hydroxide add 98 mL of LG water.

3. Initial calibrations:
   a. Syringe calibration: The capacity of the syringe will change with continual use. To assure that delivery by the milk syringe is constant, check it before use, and daily for an initial period of two weeks. The syringe must be adjusted to consistently deliver 1.00 g of water at 20°C.

   For testing, use LG water at 20°C. Insert the syringe tip a minimum of 1.25 cm into the water and pump the syringe five times to expel air before filling syringe. Invert the syringe and wipe the cannula with tissue to remove any water adhering to the surface. Expel the sample into a covered container of sufficient size to hold ten samples from the syringe. Depress syringe plunger only once per delivery. It is important that the sample be taken in the same manner each time for a maximum uniformity of sample size. When a total of ten deliveries has been made, weigh the container to obtain net weight. Calculate weight of a single delivery. Record of syringe delivery weight must be maintained. It should not vary more than ± 0.0002 g. When records show that this tolerance is maintained for a 2-week period, checking can be done on a weekly interval. If a new syringe is used or another employee is using the syringe, the daily calibration procedure shall again be initiated. An additional clean, calibrated syringe should be available at all times.

b. Dye delivery: It is important that the amount of dye delivered (about 20 g at 20°C) be consistent. The average value is calculated from the weight of ten deliveries made into a covered container. Give a firm, uniform stroke on the dye dispenser handle with the dye opening over the container. Repeat for a total of ten discharges. A record is maintained to assure consistency of delivery. Whenever dye delivery is altered and each time a new dye solution is made, delivery should be checked to calibrate unit.

c. Repeatability of operator's technique: Using a sample of homogenized milk, each operator is to make 25 duplicate tests. The mean of each operator's sample should check within ± 0.02% protein of each other with an individual std. dev. of ± 0.008% protein or less.

d. Calibration of instrument: Proteins shall be determined in duplicate by the instrument and by the Kjeldahl method of analysis on a minimum of 10 samples. Save a duplicate set of samples in freezer until Kjeldahl results are received. The average differences in mean values shall not be more than ± 0.02% protein if unit is in calibration.

If the Pro-Milk is giving protein values higher than Kjeldahl, the amount of dye should be increased. If the instrument protein results are lower than those of the Kjeldahl, the amount of dye should be decreased. This is accomplished by following the manufacturer's operational manual.

Continue the adjustment and repeating analysis until a 10-sample mean is within the ± 0.02% protein tolerance.

Ten samples should be submitted monthly for duplicate tests by the Kjeldahl method. The same tolerances should be maintained. Calibration should be repeated for each new batch of dye.

4. Measuring procedure:
   a. Turn on instrument at least 1 h before use. Instrument can remain on if used daily.
   b. Using a clean filter and tube, check reading on the scale using the standard liquid for checking "45". The needle should rest on

18.11 Protein

"45". Bring needle to correct position using the "45" adjustment knob on the readout unit.
   c. Replace filter and clean tube and check the "0-liquid" as described in the operational manual. The needle should rest on 0. Bring needle to correct position using the 0 adjustment knob.

   If an adjustment is made on the "0" setting then the "45" setting must be checked gain and readjusted if necessary. Continue checking the two solutions alternately until both points are in correct adjustment. The "0-liquid" normally needs checking once a day, but the 45 standard should be repeated every 50 samples. If a major adjustment is required, the the "0-liquid" test should be repeated.
   d. Determine five replicate protein tests on a homogenized milk sample. Carefully fill the syringe with milk, discarding the first fill after distilled water. The syringe should be handled in the same manner to fill with milk by pumping five times and wiping tip.

   The pumping action will assure that the sample is homogeneous when changing milk samples. Milk samples are tested at 20 to 37°C. Samples should be tested at the same temperature as the calibration samples. Samples are held in a bath in the refrigerator and removed in small lots. Temper and mix samples thoroughly before testing (as is done for fat testing).
   e. Place a clean filter and tube on platform.
   f. Move the dye dispenser over to the tube, and with a gentle, firm, uniform stroke, dispense dye into the tube. Release the dispenser to the upright position.
   g. Dispense milk into dye with a single stroke of the milk syringe plunger. Place stopper in position and fasten by turning the operation knob.
   h. Mix the sample and dye by tilting the mixing assembly until it touches the rim of the discharge tray and then return to upright position. Repeat for a total of five full arcs. Apply pressure to rubber bulb until it is firm in the netting. While sample is filtering, clean and fill syringe with next sample. Depress the release button on top of measuring unit, to bleed pressure from mixing tube and expell air. Read the percent protein on the upper scale when the pointer has stabilized. Release the operation knob and turn the stopper away. Turn the mixing assembly until the lock catches, then press lever to release tube. Remove and replace filter paper, clean mixing tube by wiping with tissue placed on the tube stand. Use same tissue to wipe mixing assembly.
5. Rinsing and cleaning:

a. The instrument is flushed with ammonia solution after each group of 50 samples has been tested, and when measurements are terminated. Place a filter on the lower support for the mixing tube and place tube on the filter. Fill ⅓ of the tube with ammonia solution and turn stopper in position over the mixing tube. Press the rubber bulb and flush the cuvette with the entire amount of ammonia solution.

   Release pressure and turn stopper away from the mixing tube. Fill tube with LG water and turn stopper in position over the mixing tube. Apply pressure to the rubber bulb until clear water comes out at the discharge tray. Release pressure and let water remain in the instrument until the equipment is again used.

b. At termination of daily tests, rinse the milk syringe with ammonia solution then distilled water.

   Allow water to remain in syringe until the next series of measurements are made. The milk syringe should be dismantled at least monthly and the individual components washed with a mild detergent solution, followed by a thorough rinsing with tap and LG water.

6. NOTES:
   a. The dye solution may become reduced in activity, with results lower in value as the amount of dye in the bottle is lowered and/or the dye is over 45 da old.

      The time for a dye change must be anticipated in advance and a new lot prepared and allowed to stabilize. Determine the protein in duplicate, on the current homogenized milk samples using the "old" dye solution. Discard the remaining dye solution. Rinse bottle with LG water and allow to drain. Introduce new dye. Check zero and "45" using freshly prepared solution. Make necessary adjustments. Determine five replicate protein values using the same current homogenized milk sample. Results should not vary from those obtained with the old dye by more than ± 0.2% protein.

      If greater variance does occur, recheck "0" and "45" and retest the milk sample. Also rechecks should be made on the volumes delivered by milk and dye syringes. If this does not rectify the difference, the dye should be discarded and a new lot made.

      If variation occurs from one lot of prepared dye solution to the next, it might be advantageous to maintain a log on the dye weight and the lot number.

   b. This method is suitable for determining protein contents of milk, cream, and other dairy products such as: flavored milk, cultured buttermilk, half and half, frozen dessert mixes, and nonfat dry

18.11 Protein

milk. For these products, the sample must be weighed, and sample size must be sufficient to provide about 60 to 100 mg of protein.

The dye binding method is not applicable for testing samples preserved with potassium dichromate.

c. If calibration samples are not within the prescribed tolerances, check the milk syringe, dye syringe, and age of dye, and recheck the 45 and "0-liquids".

C. *Dye binding: Acid orange-12*[38] *(Class A2)*
1. Apparatus and reagents:
   a. Udy protein analyzer: Model L, Part NO. L-1000 (Udy Analyzer Company, 201 Rome Court, Ft. Collins, CO 80524.)
   b. Filter paper, 1.8 cm discs.
   c. Drain tube of the cuvette.
   d. Lint-free tissues.
   e. Dispensing bottle: polyethylene, 19-L capacity.
   f. Automatic pipette, 40 mL.
   g. Syringe: 0 to 5 mL, calibrated to deliver 2.24 mL of water at 20°C.
   h. Polyethylene bottles: 240 mL with dropper cap, 60 mL with dropper cap, and 480 mL, to be used as drain receptacles.
   i. Closures, gum rubber.
   j. Tripod support: for 19-L dispensing bottle.
   k. Calibration test kit: with instructions (used for preparing standard curve).
   l. Graduate cylinder: 1000-mL (used for preparing reagent dye solution from a concentrate).
   m. Hot plate.
   n. Erlenmeyer flask, 2000 mL.
   o. Lint-free cloth or paper for wiping cuvette.
   p. Beaker, 10 mL.
   q. Reference dye solution, equivalent to 0.600 g of dye/L (used to standardize the colorimeter to 42.0).
   r. Cleaning solution for cuvette.
   s. Water (LG).
   t. Dye solution: As supplied with the instrument, it is ready to use without further preparation. It contains 1.3 mg of dye per milliliter. Additional orders for this solution may be placed, either for the completely prepared solution or for a concentrate that must be properly diluted before use. Concentrates are available for preparing 20 L and 50 L. Instructions for dilution follow:
      1. Fill the dilution container (20-L or 50-L) about three-fourths full of LG water.

2. Completely dissolve the buffering salts, enclosed with the dye concentrate, in about 1,500 mL of boiling LG water and add to the dilution container, mixing well by swirling.
3. Add 1,200 mL of glacial acetic acid to the 20 L container, or 3,000 mL to the 50 L container, and mix again.
4. Warm the dye concentrate at 50°C for about 1 h, or until the solution is no longer turbid; then transfer to the dilution container.
5. Complete the dilution with LG water and mix thoroughly.

2. Preparation of color (protein) analyzer.
   a. The instrument should be allowed to warm up for 3 to 4 h before use. In actual routine use, the instrument is best left on at all times. A supply of color analyzer lamps should be kept on hand for replacement as required.
   b. Before placing cuvette in color analyzer, the cuvette should be wiped clean and dry with a soft cloth or lintless paper. The index mark on the funnel end should be facing to the front when the cuvette is in place with drain tube attached and ready for calibrating.
   c. It is important that the discharge end of the drain tube be at an elevation slightly above the midpoint of the main body or short-light-path section of the cuvette proper. At this elevation the cuvette body retains the correct amount of liquid by capillary action.
   d. Fill the cuvette with LG water and remove all air bubbles by exerting a momentary pressure by mouth on the funnel opening.
   e. Whe the analyzer is not in use, the cuvette should be kept filled with water. Periodic cleaning with the liquid cleanser should be done as often as required based on equipment usage and experience.

3. Preparation of calibration curve:
   a. Warm up the analyzer for 3 to 4 h before use.
   b. Fill the cuvette with reference dye solution by adding solution to the funnel top cuvette until steady reading is registered on the meter.
   c. Adjust the meter to a scale reading of 42.0.
   d. Fill the cuvette, in turn, with each of the calibration test kit solutions. Record the reading obtained for each.
   e. Plot the concentration of the dye solutions versus the meter reading obtained for that concentration. Connect the points obtained with a smooth curve. The curve will not be a straight line unless plotted on the logarithmic axis of semi-log paper with the concentration on the arithmetic axis.

4. Preparation of dye dispenser and syringe:

18.11 Protein

   a. Dye dispenser
      1. Place the dispensing vessel on the tripod stand.
      2. Attach the automatic 40-mL pipette to the stand and dispensing vessel, using the tubing and clamps provided.
      3. Fill the dispensing vessel with dye solutions and fill each dispensing tip with dye solution. The automatic pipette is now ready for use. The reservoir is automatically refilled as it is being drained. The pipette is calibrated at the factory to deliver 40.0 mL of water at 20°C, which is equivalent to 40.44 g of dye solution.
      4. To check calibration of the dispenser, add a pipette full of dye from each tip into separate preweighed 100-mL beakers, and weigh on an analytical balance. Deliveries should be 40.44 ± 0.05 g of dye solution.
   b. Syringe
      1. Using trial and error technique, adjust the syringe to deliver 2.234 ± 0.002 g of LG water at 20°C into a preweighed 10-mL beaker.
      2. Weigh beaker and water on an analytical balance. Adjust the syringe as often as necessary to arrive at this weight and retain this tolerance.
      3. Each person using the syringe must become familiar with the equipment and develop a technique for effecting reproducible deliveries.
5. Procedure:
   a. Add, from a calibrated syringe, 2.24 mL of milk at 26°C into a 60-mL polyethylene bottle.
   b. Add 40 mL of dye solution at 25°C to the bottle.
   c. Place a filter paper disc under the support disc in the tip cap of the dropper and screw tightly on the bottle containing sample and dye.
   d. Shake the bottle vigorously for 30 sec.
   e. Standardize the color analyzer to read 42.0, using the reference dye solution in the cuvette.
   f. Slowly squeeze a sufficient amount of the filtrate from the 60-mL plastic sample bottle, through the filter paper in the cap, into the cuvette so that a steady reading is obtained on the meter. Usually 15 to 20 drops will be sufficient.
   g. Record the meter reading to the nearest 0.1 unit. If there is any doubt regarding the reading, restandardize the instrument with the reference dye solution and pass another portion of the sample solution through the cuvette.

6. Calculation:
Convert the meter reading to percent protein by using the calibration curve:

$$\% \text{ Protein} = \frac{52 - (42.24 \times \text{mg of dye/mL})}{7.0512}$$

Where 52 = total milligrams of dye available in 40.0 mL of dye solution;
42.24 = total volume of 40 mL of dye and 2.24 mL of milk; and
7.0512 = derived from multiplying the sample weight of milk, 2.26 g, by 3.12 the conversion factor used to convert results to percent protein. (For samples other than milk, the sample weight used should be multiplied by 3.12 to obtain correct factor to be used in calculation of protein.)

7. NOTES:
   a. The method is suitable for determining protein content of other products such as:
   Chocolate milk.
   Creams.
   Cultured buttermilk.
   Half and half.
   Frozen dessert mixes.
   Nonfat dry milk.
   On the above products (products other than whole fluid and liquid skim milk) the sample must be weighed and sample size must be sufficient to provide about 60 to 100 mg of protein.
   b. The dye binding method is not applicable for testing milk samples preserved with potassium dichromate.
   c. Upgraded models of the UDY Protein Analyzer have a computerized readout system which does not require making a standard curve calibration. If one of these units is being used, follow the instruction manual.

18.12 Water: Added to Milk[12]

Solutes, or materials dissolved in a liquid solvent, generally lower the vapor pressure and freezing point of the solvent. Such lowering usually is proportional to the concentration of solutes in the solvent. As applied to milk when water is added, the concentration of salts and lactose dissolved in the serum is diluted. Dilution of milk with water progressively gives a vapor pressure and freezing point approaching pure water.

Use of the vapor pressure or freezing point in surveillance of milk for presence of added water should be approached with caution and with an understanding of causes for variations found in the vapor pressure and freezing point values.

18.12 Water: Added to Milk

Lactose and chlorides account for 75% of the vapor pressure and freezing point depression and changes in the concentration of one tends to compensate for the changes in concentration of the other. Most of the variation among normal milk samples has been attributed to changes in the nonchloride ash fractions of the milk. The freezing point value of $-0.522°C$ or $-0.540°$ H has been established in some areas as normal for milk. Milk that freezes at or below the value is presumed to be free of added water. However, determination of the freezing points of authentic raw and processed milk samples is essential for establishing evidence of the addition of water. Evidence exists that the freezing point varies with such factors as season of the year, animal health, feed, ambient temperature, breed of the animal, time of milking, the animal's access to water, weather, morning and evening milk, and possibly other factors. Many of these factors apparently are interrelated.

The vapor pressure and freezing point also can be affected by fermentation, vacuum treatment of milk, sterilization of milk, and freezing and holding samples before determining the freezing point. Addition of milk solids and ultra filtration will markedly lower the vapor pressure and the freezing point.

A. *Thermistor cryoscope*[8,15,28] *(Class A1):*

The cryoscope has been used for a number of years to measure the freezing point of milk. Temperatures used in the Hortvet (H) scale for such determinations were initially thought to be equivalent to the Celsius (C) scale. However, the freezing point of a standard 7% sucrose solution at $-0.422°$ H is actually $-0.408°C$ and that of a 10% soution of sucrose at $-0.621°$ H is $-0.600°C$. Sucrose solutions have now been replaced by sodium chloride solutions that have the same freezing points. Most cryoscopes are calibrated in terms of Hortvet but results are reported in Celsius. It is apparent that all the results in the literature that have been obtained by the use of cyroscopes based on Hortvet's principle are not true freezing points. Providing that the method of calibration is given, it is possible to correct results from one method of calibration to the other:

°C = 0.96231H − 0.00240
°H = 1.03916C + 0.00250

where H = Hortvet reading (negative value)
C = Celsius reading (negative value)

1. Apparatus and reagents:
    a. Thermistor cryoscope: It consists of a cooling bath to cool the sample below freezing temperature, a thermistor probe that senses temperature changes of less than 0.001°C, a stirrer, a mechanism to cause the super-cooled sample to freeze, and a mechanism to display the temperature.
    b. Sample tube-sized for the instrument.
    c. Sample tube rack.

d. Pipettes, 2 mL.
   e. Sucrose solutions, 7% and 10%; 7.0000 g and 10.0000 g of sucrose, analytical grade, each made up to 100 mL with LG water; or
   f. Salt solutions equivalent to 7% and 10% sucrose; 0.6892 g and 1.0206 g of NaCl, analytical grade, each made up to 100 mL with LG water.
2. Standardization of cryoscope:
   Calibrate the instrument following directions in manufacturers operating manual using 2 standard solutions. The 7% sucrose solution or equivalent freezes at $-0.408°C$ or $-0.422°$ H. The 10% sucrose solution or equivalent freezes at $-0.600°C$ or $-0.621°$ H. Calibration should be checked at least daily. Determine freezing point of standards, using the same test procedure as for unknowns.
3. Precautions:
   a. Check the level of coolant and temperature of cooling bath. Normal temperature range -4 to $-7°C$.
   b. The tip of the thermistor should be near the center of a 2-mL sample.
   c. Adjust the stirrer and freezer-vibrator to give good mixing and strong vibration for freezing.
   d. Sample tubes must be clean, free from lint, and unscratched to prevent prefreezing of sample.
   e. Use the same technique both on standard solutions and on milk samples.
   f. Milk sample should not have a titratable acidity of over 0.18%.
   g. Instruments may have different required manual steps and readout capabilities.
4. Procedure:
   a. Transfer 2 mL of a well mixed sample to a test tube using a 2-mL pipette.
   b. Cool and freeze sample according to manufacturer's instrument instructions. Hold samples in an ice-water (0 to 4.4°C) bath prior to testing.
   c. If sample prefreezes or doesn't freeze, prepare another sample and rerun. When a sample prefreezes, heat in a water bath to 32 to 38°C, mix well and rerun.
   d. First determinations should establish the freezing point range. Normally, second and third replicates are needed to establish more accurate freezing points.
5. Calculations:
   Calculate added water as follows: 1% of water raises the freezing point of milk about B/100. By common usage,

$$\text{Percent added water} = 100 \frac{(B - T) \times 100}{B}$$

18.12 Water: Added to Milk

Where B = base freezing point of authenticated samples T = true freezing point of test sample. For example, if the authenticated sample freezes at $-0.540°$ H and unknown milk freezes at $-0.520°$ H,

$$\% \text{ added water} = \frac{[-0.540 - (-0.520)] \times 100}{-0.540} = 3.7\%$$

6. Notes:
   a. Samples with large amounts of added water should be tested for fat, solids, or protein to confirm a problem.
   b. Samples that will not freeze may have a high solids content caused by high acid, or by a contaminate such as cleaners or chlorine.
   c. Samples that continue to prefreeze may have high bacteria counts or high somatic cell counts.

B. *Vapor pressure osmometer*[24] *(Class A2):*

An osmometer which measures vapor pressure as a function of dewpoint depression can be used to determine added water in milk. A reading of 280 mOs/Kg has been found to be normal for milk.

1. Apparatus and reagents:
   a. Wescor vapor pressure osmometer (Wescor Inc., 459 South Main St., Logan, UT 84321) or equivalent consisting of a small thermocouple detector, located in a sample chamber headspace and a readout device in milliOsmols (mOs) per kilogram.
   b. Pipette, 8 microliter
   c. Sample paper discs (6.4 mm, Whatman #1)
   d. Stainless steel forceps
   e. Thermocouple cleaning solution, blow clean pressure vessels or automatic head cleaning device
   f. Tissues, lint-free
   g. Standard 290 mOs NaCl solution: Add 9.158 g of reagent grade NaCl to 1000 mL with recently boiled and cooled LG water.

2. Standardization:
   a. Install instrument in an area relatively free from large ambient temperature changes.
   b. Calibrate the instrument following directions in manufacturers operating manual.
   c. Repeat to assure the instrument repeatability is 290 ± 3 mOs.
   d. With LG water, the instrument should read 20 to 40 mOs. Higher readings indicate thermocouple contamination. At 70 mOs cleaning is required. Clean according to manufacturer instructions.

3. Procedure:
   a. Pick up a single sample paper disc at the edge with clean, dry forceps and transfer disc to concave depression in sample holder.

b. Add 8 microliters of sample to center of disc being careful not to touch edge of sample holder.
c. Gently push sample slide into instrument and seal the chamber by rotating the sealing knob clockwise.
d. Read the milliosmolality of the sample.
e. Wipe the sample holder clean with a paper tissue between samples.
4. Calculations:
Calculate the percent added water as follows:

$$\% \text{ added } H_2O = \frac{(R - S) \times 100}{R}$$

R = Mean mOs/Kg of reference authenticated milk samples which are known to be free of added water. In some areas, 280 mOs/Kg is considered the normal milliosmolality of milk.
S = mOs/kg of sample
5. NOTE:
a. Standard solutions will evaporate if left open. Preferably use small vials of standard solutions and replace frequently.
b. Large errors may result if the thermocouple is not clean. The need for close observation of this requirement cannot be overemphasized.

18.13 Iodine: Selective Ion Procedure[5] (Class B)

A. *Apparatus and reagents:*
1. Balance, 50-mg sensitivity.
2. Beakers, 150 and 250 mL.
3. Electrode, iodide selective ion, Orion 44–53A or equivalent.
4. Electrode, reference, double function, Graphic Controls PHE 54473 or equivalent.
5. Magnetic stirrer.
6. Magnetic stirring bars, teflon coated, 25 × 10 mm.
7. Millivolt meter, Orion 701A or equivalent.
8. Stoppered volumetric flasks, 50, 100, 500, and 1000 mL.
9. Volumetric Pipettes, 2, 4, 5, 6, 10 and 40 mL.
10. Nickel nitrate, 2 M solution prepared with double deionized water.
11. Water, double deionized of at least 1 megohm resistivity.
12. Potassium iodide solution, 10,000 ppm, .02 M in nickel nitrate: Weigh 13.10 g of KI in a 150 mL beaker. Dissolve in about 50 mL double deionized water and transfer quantitatively to a 1000-mL stoppered, volumetric flask using double deionized water. Add 10 mL 2 M nickel nitrate solution, and then make to volume with double deionized water.
13. Prepare a set of standard potassium iodide solutions according to the following:

18.13 Iodine: Selective Ion Procedure (Class B)

Make to Volume	Final Concebtration	Take Volume	Of Concentration	Plus
50 mL	8000 ppm	40 mL	10000 ppm	0 mL 2M Ni(NO$_3$)$_2$
100 mL	1000 ppm	10 mL	10000 ppm	1 mL 2M Ni(NO$_3$)$_2$
100 mL	100 ppm	10 mL	1000 ppm	1 mL 2M Ni(NO$_3$)$_2$
1000 mL	10 ppm	10 mL	1000 ppm	10 mL 2M Ni(NO$_3$)$_2$
500 mL	1 ppm	5 mL	100 ppm	5 mL 2M Ni(NO$_3$)$_2$
1000 mL	2 ppm	2 mL	1000 ppm	10 mL 2M Ni(NO$_3$)$_2$
1000 mL	4 ppm	4 mL	1000 ppm	10 mL 2M Ni(NO$_3$)$_2$
1000 mL	8 ppm	8 mL	1000 ppm	10 mL 2M Ni(NO$_3$)$_2$
1000 mL	16 ppm	2 mL	8000 ppm	10 mL 2M Ni(NO$_3$)$_2$
500 mL	32 ppm	2 mL	8000 ppm	5 mL 2M Ni(NO$_3$)$_2$
500 mL	64 ppm	4 mL	8000 ppm	5 mL 2M Ni(NO$_3$)$_2$
500 mL	128 ppm	8 mL	8000 ppm	5 mL 2M Ni(NO$_3$)$_2$

B. Procedures:
1. Calibration of electrodes:
    a. After assembling the meter and electrodes according to the manufacturer's instructions, set the electrodes in the 10-ppm standard (100 mL in a 150-mL beaker stirred using a magnetic stirrer), set the meter to read absolute millivolts, and allow the electrodes to reach equilibrium (5 min is sufficient). The reading will be somewhere between −100 and −90 mv and steady (no more than ±0.1 mv variation). Repeat with 1- and 10-ppm standards. The reading for the 1-ppm standard should be about 59.2 mv more than for the 10-ppm standard and for the 100-ppm standard should be about 59.2 mv less. Since a negative ion is being measured, the potential will get smaller as the concentration increases. If the readings for all three standards are displaced by the same amount from what is expected, check the lower half of the reference electrode for electrolyte level and change or add as necessary. If the readings show more than 0.1-mv fluctuation, check the electrical ground on the meter and the electrode connections. If the readings are steady but the change in reading is much less than 59.2 mv for a tenfold change in iodide concentration, polish the iodide selective ion electrode as recommended in the Orion manual. When the electrodes are functioning properly, proceed to measure the potential for the eight standards (1 to 128 ppm). For a twofold change in iodide concentration, the potential should change 17.8 mv (59.2 × log 2) and the potentials for the eight standards should fall on a straight line when plotted on semilog paper (concentration on log scale). Any suspect standard should be discarded, since the concentration reported in the milk will be in error in proportion

to the error in the standard added to the milk during the determination.

b. Measurement of iodide concentration in milk: Weigh duplicate 100-g samples into 250-mL beakers. Read the potential (E_1) of the milk directly while stirring with a magnetic stirrer after the addition of 1 mL of 2M $Ni(NO_3)_2$. Record E_1 and add 10 mL of a standard whose concentration is at least $10\times$ and not more than $20\times$ the concentration of the milk estimated from E_1 (consult the graph prepared during calibration).[1] Read the potential (E_2) after the addition of the standard and record $E_2 - E_1$ as E.

c. Calculation: For measurements made at 25°C consult the table in the Orion Manual to find the Q value that corresponds to the E measured in the sample. You will note that for duplicate samples, E_1 may vary as much as 10 mv, but E will generally agree within 0.5 mv. The concentration of iodide in the milk in μg/Kg is $1000 \times Q \times C$, where C is the concentration of the standard used to obtain E_2 in ppm.

18.14 Pesticide Residues in Milk

Milk may be contaminated with one or more pesticide residues in the same sample. Residues of chlorinated hydrocarbon pesticides have the greatest chance to appear in milk and have received the greatest attention in residue monitoring programs. Organochlorine residues are fat soluble, very slowly metabolized by the bovine, and appear in milk and body fat and excreta after ingestion. Organophosphate pesticides are metabolized by the bovine, and metabolites, usually less toxic, appear in milk and excreta. Pesticide residues that have been found in milk and milk products have been: DDE, dieldrin, lindane, methoxychlor, heptachlor epoxide, DDT, hexachlorobenzene (HCB), and TDE. Since banning nonspecific uses of all organochlorine pesticides in general, and DDT and aldrin/dieldrin in particular, the overall pattern of residues in milk has changed to levels at or below method sensitivity.

Residues of polychlorinated biphenyls (PCBs) have been found in milk. This group of multicomponent and persistent industrial chemicals has been limited in permitted uses and thus virtually eliminated from a need to be considered as contaminants.

The analytical methods for organochlorine chemicals are described in the 13th edition, *Association of Official Analytical Chemists, Official Methods of Analysis*.[14] In these methods, qualitative and quantitative determinations are based on chromatographic principles that make possible simultaneous determination of multiple residues present in the same sample. Extraction and cleanup procedures use solvent partitioning and column chromatography to quantitatively recover the chemicals sought. Subsequent chromatographic determinations require clean samples for anlytical accuracy. These separation methods have been tested in interlaboratory collaborative studies with dairy products. Over 100 organochlorine pesticides and PCBs, certain parent organophosphorus pesticides, and other nonpolar pesticides are completely or partially recovered by these methods. A compilation of the behavior of these chemicals in the methods is given in the *Food and Drug Administration Pesticide Analytical Manual.* The detection of PCBs and organochlorine pesticide residues together in the same sample require special additional procedures to separate them before determination.

Due to the limited uses of these chemicals, which make contamination of dairy foods unlikely, and the complex nature of these determinations, the *Association of Official Analytical Chemists, Offical Methods of Analysis*[14] should be consulted for detailed methods. Moreover, few laboratories performing routine dairy analyses are equipped with the materials necessary to perform these analytical techniques.

18.15 Radionuclides[20]

The importance of radioactive substances in dairy foods as a result of accidental spills, fallout from nuclear power plants, or weapons has been established. In the food chain, milk is potentially a major source of radionuclides for children and a significant contributor to radioisotope ingestion by adult human beings. Although several radionuclides are present in the fallout, most of them either have a short half-life or they are not metabolized by the cow. Only a few of the radionuclides are found in measurable amounts. These are: strontium-89, strontium-90, iodine-131, cesium-137, and barium-140. In the vicinity of a nuclear test site, however, iodine-133 ($t_{1/2}$ = 20.8 hr) and tritium-3 ($t_{1/2}$ = 12.3 years) may be found in milk. Around nuclear processing plants, zinc-65 ($t_{1/2}$ = 245 days) has also been found in milk. The amount of tritium-3 observed in milk is close to the detection limit of 200 pCi per liter.

The concentrations of strontium-89, strontium-90, iodine-131, cesium-137, and barium-140 will reach levels of significance immediately following nuclear weapons use but will supposedly decline thereafter. It is considered advisable to determine potential concentrations of these milk-borne fission materials and to assess their dietary contributions to assist in evaluating the biological effects, if any, from human ingestion of them. The building of nuclear power generating stations in milk production areas renders advisable the radioassay of milk for its environmental contaminants and radiation effects, which are as yet inadequately defined.

Sampling of milk for analysis depends on whether samples are intended for determination of radionuclides: a) as affected by farm practices, b) in the vicinity of nuclear power generating stations, or c) at the consumer level. To study the effect of farm practices on radionuclides in milk, samples are collected from individual cows or a portion of the pooled milk is taken from the storage tank on the farm. In the vicinity of nuclear power generating stations, milk samples are taken from within a discrete geographical area depending on meteorological conditions and the number of farms considered at risk. The sample may be from an individual farm or from groups of farms. To study market milk, samples are composites of subsamples from each milk processing plant, the quantity being proportional to the plant's average sales in the community served.

Frequency of sampling depends on the seriousness of the situation and the radionuclide being considered for anlaysis. With iodine-131 ($t_{1/2}$ of 8.08 days), it is necessary to collect milk samples every day or at least once a week. With strontium-90 ($t_{1/2}$ = 28 years), it is necessary to collect at least once a month.

At least 3.8 L of milk (raw or pasteurized) is collected, preserved with formaldehyde (H_2CO) [10 mL of 40% H_2CO per 3.8 L of milk], and shipped to the laboratory for analysis. H_2CO does not interfere with analysis of ^{131}I, ^{140}Ba, and ^{137}Cs by gamma spectroscopy or ^{89}Sr and ^{90}Sr by ion-exchange and/or by chemical methods. However, addition of H_2CO to raw milk causes transformation of ^{131}I to protein-bound ^{131}I. The enzyme, xanthine oxidase, in raw milk oxidizes H_2CO to H_2O_2 in the presence of oxygen, and H_2O_2 oxidizes ^{131}I to $^{131}I_2$, which binds to protein. In pasteurized milk or dried milk the enzyme is absent, consequently, addition of H_2CO to processed milk has no effect on ^{131}I. Processed milk containing ^{131}I from fallout or labeled in vitro can be preserved with H_2CO and successfully used for determination of ^{131}I by ion exchange methods. However, if raw milk is the sample of choice and ^{131}I is the only radionuclide of concern, no preservative should be added to milk. Analysis can be simplified by passing milk through an anion-exchange, resin column at the sampling station and then shipping the column to the laboratory for further analysis. This presupposes that all ^{131}I in milk is in the inorganic form as iodide.

Due to the complex nature of these determinations, the *Association of Official Analytical Chemists, Offical Methods of Analysis*[14] should be consulted for detailed methods. Moreover, a few laboratories performing routine dairy analysis are equipped with the materials necessary to perform these analytical techniques.

18.16 References

1. AMERICAN DRY MILK INSTITUTE. Standards for grades of dry milk, including methods of analysis. Bul. 916 Rev., Chicago, IL; 1983.
1a. BARBANO, D.M.; DELLAVALLE, M.E. Microwave drying to determine the solids content of milk and cottage and Cheddar cheese. J. Food Prot. 47:272–278; 1984.
2. BIGGS, D.A. Milk analysis with the infrared milk analyzer. J. Dairy Sci. **50**:799–803; 1967.
3. BIGGS, D.A. Infrared estimation of fat, protein and lactose in milk: evaluation of multispec instrument. J. Assoc. Offic. Anal. Chem. **62**:1202–1210; 1979.

18.15 References

4. BIGGS, D.A. Performance specifications for infrared milk analysis. J Assoc. Offic. Anal. Chem. 62:1211–1214; 1979.
5. BRUHN, J.C.; FRANKE, A.A. An indirect method for the estimation of the iodine content in raw milk. J. Dairy Sci. 61:1557–1560; 1978.
6. COTLOVE, E. Determination of the true chloride content of biological fluids and tissues. II. Analysis by simple, nonisotopic methods. J. Anal. Chem. 35:101–105; 1963.
7. DOAN, F.J.; LARSON, B.L.; WINDER, W.C.; LEEDER, J.G. Procedures for making acidity, tests of fluid milk products. J. Dairy Sci. 40:1643–1644; 1957.
8. ENGLAND, C.W.; NEFF, M. The accuracy of cryoscope methods. J. Assoc. Offic. Anal. Chem. 46:1043–1049; 1963.
9. FRANKEL, E.N.; TARASSUK, N.P. Inhibition of lipase and lipolysis in milk. J. Dairy Sci. 42:409–419; 1959.
10. GINN, R.E. The Milko-Tester and problems related to its use. J. Milk Food Technol. 37:218–221; 1974.
10.a GREEN, W.C.; PARK, K.K. Comparison of AOAC, microwave and vacuum oven methods for determining total solids in milk. J. Food Prot. 43:782–783; 1980.
11. HEINEMANN, B.; COSIMINI, J. JACK, E.L.; WILLINGHAM, J.J.; ZAKARIASEN, B.M. Methods of determining the per cent total solids in milk by means of the lactometer. J. Diary Sci. 37:869–876, 1954.
12. HENNINGSON, R.W. The variability of the freezing point of fresh raw milk. J. Assoc. Offic. Anal. Chem. 46:1036–1042; 1963.
13. HERRINGTON, B.L. Lipase: a review. J. Dairy Sci. 37:775–789; 1954.
14. HORWITZ, W., ED. Official methods of analysis, 13th ed. Assoc. Offic. Anal. Chem. Washington, DC; 1980.
15. INTERNATIONAL DAIRY FEDERATION. Milk determination of freezing point—thermistor cryoscope. Provisional intern. IDF standard. 108:1982, Intern. Dairy Fed., Brussels, Belgium; 1982.
16. INTERNATIONAL DAIRY FEDERATION. Determination of fat content—Gerber butyrometers. ISO/Dis 488. IDF, Brussels, Belgium; 1983.
17. LEVOWITZ, D. An appraisal of the Gerber test for milk fat in milk and market milk products. J. Milk Food Technol. 23:69–72; 1960.
18. LINGANE, J.J. Automatic coulometric titration with electrolytically generated silver ion—determination of chloride, bromide, and iodide ions. J. Anal. Chem. 26:622–626, 1954.
19. MADDEN, D.E.; BRUNNER, J.R. Comparative accuracy of small lactometers for determining the total solids of individual cows' milk. J. Dairy Sci. 41:783–788; 1958.
20. MARTH, E.H., ED. Standard methods for the examination of dairy products, 14th ed. Amer. Pub. Health Assoc. Washington, DC; 1978.
21. MCKINLEY, M.R.; ARNOLD, R.E.; ROSENBURY, J.B.; BORN, E.C.; DUNHAM, C.E.; PARK, D.L. Evaluation of the Babcock method for the determination of milk fat. J. Assoc. Offic. Anal. Chem. 58:949–956; 1975.
22. MOJONNIER, T.; TROY, H.C. Technical control of dairy products. Mojonnier Bros. Co., Chicago, IL; 1925.
23. NEWLANDER, J.A.; ATHERTON, H.V. Chemistry and testing of dairy products. AVI Publishing Co., Westport, CT.; 1977.
24. PENSIRIPUN, K.; CAMPBELL, E.C.; RICHARDSON, G.H. A vapor pressure osmometer for determination of added water in milk. J. Milk Food Technol. 38:204–207; 1975.
25. PIEPER, H.; STUART, J.A. JR.; RENWICK, W.R. Microwave technique for rapid determination of moisture in cheese. J. Assoc. Offic. Anal. Chem. 60:1392–1396; 1977.
26. SHERBON, J.W. Pro-Milk method for the determination of protein in milk by dye binding. J. Assoc. Offic. Anal. Chem. 57:1338–1341, 1974.
27. SHERBON, J.W. Collaborative study of the Pro-Milk method for the determination of protein in milk. J. Assoc. Offic. Anal. Chem. 58:770–772; 1975.

28. SHIPE, W.F. Report on cryoscopy of milk. J. Assoc. Offic. Anal. Chem. **41**:262–267; 1958.
29. SHIPE, W.F. Collaborative study of the Babcock and Foss Milko-Tester methods for measuring fat in raw milk. J. Assoc. Offic. Anal. Chem. **52**:131–138; 1969.
30. SHIPE, W.F.; SENYK, G.F. Collaborative study of the Milko-Tester methods for measuring fat in homogenized and unhomogenized milk. J. Assoc. Offic. Anal. Chem. **58**:572–575; 1975.
31. SHIPE, W.F.; SENYK, G.F. Evaluation of Milko-Tester minor for determining fat in milk. J. Assoc. Offic. Anal. Chem. **63**:716–719; 1980.
32. SHIPE, W.F.; SENYK, G.F. Performance specifications for measuring milkfat by turbidometric means. J. Assoc. Offic. Anal. Chem. **63**:713–715; 1980.
33. SPEER, J.F., JR.; WATROUS, G.H., JR.; KESLER, E.M. The relationship of certain factors effecting hyrolytic rancidity in milk. J. Milk Food Technol. **21**:33–37; 1958.
34. SPICER, D.W.; PRICE, W.V. A test for extraneous matter in cheese. J. Dairy Sci. **21**:1–5; 1938.
35. SOMMER, H.H.; WINDER, W.C. The physical chemistry of dairy and food products. Univ. Wisc., Madison, WI; 1978.
36. SWOPE, W.D. The Pennsylvania methods for determining the percentage of fat in dairy products. PA Agr. Expt. Sta. Bul 412, State College, PA; 1941.
37. THOMAS, E.L.; NIELSEN, A.J.; OLSON, J.C., JR. Hydrolytic rancidity of milk—a simplified method for estimating the extent of its development. Amer. Milk Rev. **17**:50–52, 85; 1955.
38. UDY, D.C. Improved dye method of estimating protein. J. Am. Oil Chem. Soc. **48**:29; 1971.
39. VOORT, F.R. VAN DE. Evaluation of Milko-Scan 104 infrared milk analyzer. J. Assoc. Offic. Anal. Chem. **63**:973–980; 1980.
40. WATSON, P.D. Determination of the solids of milk by a lactometric method at 102°F. J. Dairy Sci. **40**:394–402; 1957.

INDEX

A

Abnormal milk, methods, 239–258
  confirmatory tests, 249
    electronic cell count, 249
    direct microscopic somatic cell count, 249
  coulter counter, 250
  fluorescent dye
    automated method, 253
    semiautomated method, 255
  screening tests, 239
    California mastitis test, 239
    conductivity measurements, 248
    modified whiteside test, 246
    rolling ball viscometry, 244
    Wisconsin mastitis test, 242
Accreditation, 5
Achromobacter, 195
Acid degree value, 327
Acid orange-12 method, 391
Acidified potato dextrose agar (molds), 198
Acidity
  potentiometric, 331
  pH, gold electrode/quinhydrone method, 334
  titratable, 329
Acinetobacter, 195
Aerobic bacterial spores, 196
Aflatoxins, 2
Aflatoxins B_1, and G_1, 69
Aflatoxin M_1, 51, 52, 53, 68, 69, 77
Agar contact (RODAC) method, 298

Agar preparation, 124
  sterilization, 129
  storage, 129
Air, microbiological quality tests, 300
Alcaligenes, 195
Alternative microbiological methods, 151–171
  statistical protocol, 152
American Public Health Association
    pipette specifications, 21
    standard discs, 308
Amido black method, 386
Amines, 70
Antibiotic plate count method (molds), 197
Antibiotic residues detection, 265–288
    bacillus stearothermophilus disc assay, 269
    bacillus subtilis disc assay, 266
    charm screening assay, 279
    cylinder plate method
      modified sarcina lutea, 281
      specific anti-microbials, 285
    delvotest, 275
Aseptically packaged, 207
Ash and alkalinity of Ash
  ash, 336
  alkalinity of ash, 338
Aspergillus flavus, 45
Aspergillus parasiticus, 45
Atmospheric oven-moisture, 377
Automated turbidity method, 364

405

INDEX

B

Babcock method, 347
  cheese, 351
  cream, 350
  milk, 349
Bacillus, 189, 194, 195
Bacillus cereus, 194
Bacillus coagulans, 194
Bacillus megaterium, 285, 286
Bacillus stearothermophilus, 265, 269, 272, 279
Bacillus stearothermophilus disc assay, 269
  qualitative, 269
    procedure, 271
    reporting, 271
    stock culture, 270
  quantitative, 272
    procedure, 273
    reporting, 275
Bacillus subtilis, 194, 265, 266
Bacillus subtilis disc assay, 266
  procedure, 267
  reporting, 269
Bacteria, direct microscopic methods for, 219–237
Bacterial population "total", 7, 8
Bacti-disc-cutter, 205
Barium-140, 401
Beta-lactam residues, 279
Botulism, 67
Bulk-tank, farm, 94
Butter, 72, 103, 210

C

California mastitis test, 239
  grading and interpretation, 241
  procedure, 240
Campylobacter jejuni, 50
Carbon dioxide-free distilled water, 120
Casein, 102
Cellulose sponge method, 298
Cesium-137, 401
Charm screening assay, 279
  procedure, 280
Cheese, 54, 58, 104, 211
Chemical and physical methods, 327–404
  acid degree value, 327
  acidity, 327
  ash and alkalinity of ash, 336

Chemical and physical methods (*Cont'd*)
  chloride (salt), 339
  chlorine, available, 344
  extraneous matter-cheese, 346
  fat, 347–366
  iodine-selective ion procedure, 398
  multi-component methods, 366–372
  moisture and solids, 373–382
  pesticide residues, 400
  protein, 382–394
  radionuclides, 401
  water: added to milk, 394–398
Clostridium, 189, 195
Clostridium botulinum, 67, 189
Clostridium perfringens, 44
Coliform bacteria, 173–187
Coliform detection, impedimetric method, 182
Coliform bacteria, 173–187
  definitions, 174
    presumptive test, 174
    confirmed test, 174
    completed test, 174
    plate vs. tube methods, relative values, 176
Coliform test, 177
  completed test, 180
  confirmed test, 179
    solid medium, 179
    impedimetric, 182
    membrane filter, 183
    MPN (table) 178, 182
  reporting results, 180
  solid medium, 177
    VRB agar, 177
    modified VRB agar, 177
  liquid medium, 179
Collaborative studies, 4
Color standards, (Phenol), 316
Color standards, preparation, 317
Coloring materials, 108
Completed test, 174, 180
Compressed gases, 40
Concentrated milk, 203
Condensed milk, 203
Conductivity measurements, 248
Confirmed test, 174, 179
Containers, microbiologist test for, 289
Control tests, 25
Coryneform group, 189
Cottage Cheese, 51
Coulometric titration, 342

Corrosive Materials, 36
Cream, 89
Creamers, 100
  dried, 100
  frozen, 100
Cryoscope, thermister, 395
Culture added products, 211
Cultured milks, 211
Cylinder plate method,
  modified sarcina lutea, 281
  controls, 283
  procedure, 282
Cylinder plate methods-specific antimicrobials, 285

D

Dairy and related products (sampling), 89–117
DDE, 400
DDT, 400
Deionized distilled water, 120
Delvotest, 275
  confirmatory test, 278
  controls, 276
  screening test, 277
Detergent residues, test for, 122
Dieldrin, 400
Disintegration method, 293
Dilution water, preparation, 121
Direct microscopic clump count, 219
Direct microscopic method, 219–237
  direct microscopic clump count, 219
  direct microscopic somatic cell count, 219
  dry milk, 220
  microscopic method, error, 220
  pasteurized milk, 220
  preparing films, 229
  raw milk, 219
  staining films, 228–230
Direct microscopic somatic cell count, 219
Disc assay, 266–275
  bacillus stearothermophilus, 269, 272
  bacillus subtilis, 266
Discs, sediment, 307
Dry dairy products, 208
Dry whip topping, 102
Dye binding-amido black method, 386
Dye binding-acid orange-12, 391
Dye-reduction method, 8, 259–264

Dye solution, preparation, 261
  resasurin reduction method, 262

E

Egg and egg products, 108
Electrical impedance method, 165–170
Electronic somatic cell count, 249
Emulsifiers and stabilizers, 108
Enterobacter, 173, 195
Enterobacter hafniae, 64
Enterococci, 193
Enteropathogenic E. coli, 62
Equipment, microbiological test for, 289
Equipment and supplies, laboratory, 17–25
Escherichia coli, 63, 64, 65, 66, 67, 71, 72, 173
Extraneous matter: cheese, 346

F

Farm bulk tank, 94
Fat, 347–366
  Automated turbidity, 364
  Babcock method, 347
    Pennsylvania modified, 351
  Gerber method, 355
  Majonnier method, 362
  Roccal-babcock method, 354
  Roese-Gotlieb method, 358
Facilities, laboratory, 15
  air quality, 16
  cleanliness and maintenance, 17
  laboratory use only, 16
  lighting, 16
  storage, 16
  utilities, 16
  ventilation, 16
  work benches, 16
  work space, 15
Field method, 231
Field-wide single strip method, 231
Flavobacterium, 193
Flavoring materials, 108
Fluorescent dye method, 253
Frozen products, 106, 214

G

Gerber method, 355
Grade A pasteurized milk ordinance, 7

H
Heptachlor epoxide, 400
Hexachlorobenzene (HCB), 400
Histamine, 70
Hydrophobic grid membrane filter method, 183

I
Ice cream, 106, 216
Impedimetric coliform detection method, 182
Infant formula, 205
Infrared milk analysis—multi component, 366
In line sampling device, 96
Inspections, 6
  farm, 6
  plant, 6
Iodine-131 and -133, 401
Iodine: selective ion procedure, 398

K
Keeping quality test,
  impedimetric, 148
  Mosley, 147
Klebsiella, 173
Kjeldahl-protein, 383
Kohman method, modified, 371

L
Lactobacillus, 189
Lactobacillus bulgaricus, 71
Lactobacillus cremoris, 71
Lactobacillus helviticus, 71
Lactobacillus selection agar, 127
Lactometric method—solids, 381
Lactose broth, 126
Levine's eosin methlylene blue agar, 126
Lindane, 400
Line samples, 99
Lipolytic microorganisms, 190
Listeria monocytogenes, 54
Low acid, 209

M
Margarine, 103, 212
Majonnier method, 362

Mastitis, tests for, 239–254
Media, 119–132
  dehydrated, 123
  formulae, 125–129
  prepared, 124
  preparation, basic steps, 124
  quality control, 130
  sterilization, 129
  storage, 129
Methoxychlor, 400
Methylene blue reduction method, 261
Methylene blue thiocyanate, 259
Microbacterium, 189
Micrococcus, 189
Microbiological methods, dairy products, 203–220
  aseptically packaged, 207
  butter, 210
  cheese, 211
  concentrated milk, 203
  culture added products, 211
  cultured products, 211
  dry dairy products, 208
  evaporated milk, 203
  frozen products, 214
  ice cream, 214
  infant formula, 203
  low acid products, 207
  margarine, 210
  sweetened condensed milk, 203
  ultra high temperature processed, 207
Microbiological tests, equipment, containers, water and air, 289–304
  air, 300–303
    agar plate preparation, 302
    impaction, 302
    liquid impingement, 302
    method selection, 301
    sedimentation, 302
  equipment and containers, 289
    disintegration method, 293
    rinse solution methods, 289
    surface contact methods, 296
      cellulose sponge method, 298
      rodac plate method, 298
      swab contact method, 296
  water, 300
Microbiologically suitable water, 119
Microscope, calibration, 226
Microscopic count, 8, 209, 217
Microscopic factor, 226

INDEX

Microscopic examination, 226
  field method, 231
    counting procedure, 232
  field-wide single strip method, 231
  strip count methods, 233–236
Microwave oven - moisture, 378
Milk, added water, 394
Milk, 89
  aseptically packaged milk, 109, 209
  concentrated, 203
  dry, 102
  dry products, 208
  evaporated, 203
  pasteurized, 99
  raw, 94
  sweetened condensed, 203
  ultra high temperature processed milk, 209
  whey, 102, 380
Milk cans, sampling, 97
Milk, sediment in, 305–310
Milk ordinance, Grade A Pasteurized, 7
Mixed sample method sediment, 306–308
Mobile tank trucks, sampling, 97
Modified Kohman method-multi component, 371
Modified whiteside test, 246
  procedure, 246
Moisture and solids, 373–382
  atmospheric oven, 377
  microwave oven, moisture, 378
  moisture balance, dry products, 379
  refractometer, whey, 380
  vacuum oven, 373
  vacuum oven sand pan, 376
Mohr method - chlorides, 339
Molds, 7, 199–200
Monoamine oxidase inhibitor (MAOI), 70
Mosley keeping quality test, 147
Most probable number, 178, 182
Multi component methods, 366
  infrared milk analysis: fat, protein, lactose, total solids, 366
  near infrared analysis: fat, protein, total solids in milk, 370
  modified kohman method: fat, moisture and salt in butter and margarine, 371

N

National Conference for Interstate Milk Shipments, 5

Near infrared analysis - multi component, 370
New methodology, adopting, 4
Non-fat dry milk
  fat determination, 360
  penicillin detection, 281

O

Off-bottom method - sediment, 306, 309
Osmometer, Vapor pressure, 397
Oval tube count, 151
Oval tube method, 164–165
Oxytetracycline, 285

P

Pasteurized milk, keeping quality, 148
  impedimetric, 148
  Mosley, 147
Pasteurized milk ordinance, Grade A, 2
Pathogens, in milk, 43–87
  butter, 72, 210
    aflatoxin M_1, 51
    bacillus cereus, 44
    campylobacter, 50
    coxiella burnetti, 44
    clostridium perfringens, 44
    escherichia coli, 44
    listeria monocytogenes, 44
    mycobacterium byovis, 44
    salmonellae, 45
    staphylococcus aureus, 49
    streptococcus agalactiae, 51
  cheese, 54
    aflatoxin, 68–70
    amines, 70
    botulism, 67–68
    enteropathogenic escherichia coli, 62–67
    staphylococcus, 54–58
    starter cultures, 55
    salmonellosis, 58–62
  cultured milk, 70–72
  frozen desserts, 78–79
  non-fat dried milk, 73–79
    aflatoxin M_1, 77
    salmonellosis, 74
    staphyloccocal poisoning, 73
  pasteurized milk
    listeria monocytogenes, 53
    yersinia enterocolitica, 53
  raw milk, 44

Penicillium camemberti, 66
Penicillin detection, 269
Pennsylvania modified babcock method, 251
Pesticide residues, 2, 400
pH adjustment - media, 125
pH gold electrode/quinhydrone method, 334
pH meter
    standardization, 331, 333, 335
Phenol color standards, 315–316
Phosphate buffer, 121
Phosphate dilution water 121
Phosphatase methods, 311–326
    controls, 313
    disodium phenyl phosphate - use precautions, 312
    phosphatase reactivation, 322
    rapid colorimetric method, 318
    rutgers phosphatase test, 320
    scharer rapid phosphatase test, 315
Phosphatase reactivation, 320
Plate cylinder method, 281
Plate loop method, 152–156
Polychlorinated biphenyls (PCB), 400
Potato glucose agar, acidified, 196
Potentiometric - acidity, 331
Preliminary incubation, raw milk, 146
Presumptive test, 174, 177
Protein, 382
    dye binding - amido black method, 386
    dye binding - acid orange-12, 391
    Kjeldahl, 383
Proteolytic microorganisms, 193
Pseudomonas, 195, 196
Psychrotophic microorganisms, 6, 7, 190, 194–197

Q

Quality assurance, 9

R

Radionuclides, 2, 115, 401
Rapid colorimetric method, 318
Raw milk, direct microscopic count, 217
Raw milk, pathogens in, 46
Raw milk, sampling, 94
Record keeping, 25
Redistilled water, 120

Reduction methods, 261, 262
    methylene blue reduction method, 261
Resasurin method, 262
Refractometer-whey and whey products, 380
Rinse solution methods, 289
Roccal-babcock method, 354
Rodac plate (agar contact) method, 298
Roese-Gottlich method, 358
Rolling ball viscometry, 244
    procedure, 245
Rutgers phosphatase test, 320

S

Safety, 9, 26–42
    compressed gases, 40
    corrosive materials, 36
    flammable chemicals, 37
    personnel practices, 41
    radioactive materials, 40
    toxic chemicals, 38
Salmonella species and serotypes, 45–48, 58–62, 71, 74–78
Sample collection equipment, 89
Sampling dairy products, 89–117
    aseptically (UHT) processed and packaged milk, 109
    butter, 103
    casein, 102
    cheese, 104
    coloring and flavoring materials, 104
    concentrated milk, 102
    condensed milk, 102
    cultured products, 104
    dry milk, 102
    dry whip topping, 102
    eggs and egg products, 108
    evaporated milk, 102
    fluid milk and cream, 89
        pasteurized products, 99–100
        raw, 94
    frozen products, 106
    ice cream, 106
    margarine, 103
    stabilizers and emulsifiers, 108
    sweetening agents, 108
    whey, 102, 380
Sampling - specific procedures, 110
    microbiological tests for
        air, 110, 114
        closures, 110

INDEX

Sampling - specific procedures (*Cont'd*)
    containers, 110
    equipment, 110
    paper, 110
  phosphatase test, 114
  radionuclides, 115
  reduction methods, 110
  rinse test, 112
  sediment, 114
  single service containers, 111
  swab test, 113
Sanitary quality, measuring, 6
  accuracy of methods, 6
Sarcina lutea, 281
Scharer rapid phosphatase test, 315
sediment, 305–310
  mixed sample method, 306, 307, 308
  off-bottom method, 306–309
  procedures, 308
  standard sediment disc, 308
Solids in milk
  lactometric method, 379
Somatic cell count methods, 239–256
Spiral plate method, 5, 151, 156–163
Split sample test program, 6
Stabilizers and emulsifiers, 108
Stains, Levowitz-Weber modification, 229
Standard methods, 1
Standard methods agar, 127
  physical standards, 128
  productivity tests, 128
  specifications, 127
Standard plate count (SPC), 8, 133–146, 151
  computing counts, 142
  counting colonies, 140
    automated, 141
    manual, 140
  diluting samples, 136
  incubation, 139
  personnel errors, 146
  plating, 138
  preparing samples, 135
  reporting counts, 146
  sterility controls, 139
Staphylococcus aureus, 49, 54, 55, 56, 58, 72, 73
Streptococcus, 187
Streptococcus agalactiae, 51
Streptococcus cremoris, 64, 71
Streptococcus diacetilactis, 71
Streptococcus lactis, 55, 71

Streptococcus thermophilus, 71
Strip count methods, 233–236
Strip reticle factor, 233
Strontium-89 and -90, 401
Sulfa drugs, 285
Swab contact method, 296
Sweetening agents, 108

T

TDE, 398
Tests by title
  acid degree value (ADV) 327
  acid orange-12, 391
  acidity, pH, gold electrode/quinhydrone, 334
  acidity, potentiometric, 331
  acidity, titratable, 329
  alkalinity of ash, 338
  amido black, 386
  ash, 336
  atmospheric oven, 377
  automated fluorescent dye, 253
  automated turbidity, 364
  babcock method, 347
  babcock method, Pennsylvania modified, 351
  bacillus subtilis disc assay, 266
  California mastitis test, 239
  cellulose sponge, 298
  charm screening, 277
  chlorides, 339
  chlorine, available, 344
  coliform, 177
  conductivity measurement 248
  coulometric titration, 342
  coulter counter, 250
  delvotest, 275
  direct microscopic clump, 219, 237
  direct microscopic somatic cell count, 219, 249
  disc assay for penicillin, 266, 269, 272
  disentegration, 293
  dye-binding 386, 391
  dye reduction 261, 262
  electrical impedence, 165
  electronic somatic cell count, 249
  extraneous matter, 346
  field, 231
  field wide single strip, 231
  fluorescent dye, automated, 253

Tests by title (*Cont'd*)
  fluorescent dye, semi automatic, 255
  gerber fat method, 355
  hydrophobic grid membrane filter, 183
  impedimetric, 148
  impedimetric coliform detection, 182
  infrared milk analysis (multi-component), 366
  Kjeldahl, 383
  lactometric, 381
  majonnier, 362
  methylene blue reduction, 261
  microbiological, air, 300
  microbiological, butter and margarine, 210
  microbiological, containers, 289
  microbiological, equipment, 289
  microbiological, ice cream and frozen products, 214
  microwave oven, 378
  milkotester, 365
  mixed sample, sediment, 306, 308
  modified Kohman, 371
  modified sarcina lutea cylinder plate, 281
  modified whiteside, 246
  moisture balance, 379
  Mohr, 339
  Mosley keeping quality, 147
  near infrared analysis (multi-component), 370
  off-bottom, sediment, 306, 309
  oval tube, 164
  Pennsylvania modified babcock, 251
  pH gold electrode/quinlydrone, 334
  phosphatase, 311
  plate loop, 152
  potentiometric, pH, 331
  rapid colorimetric phosphatase, 318
  refractometer, 380
  resasurin reduction, 262
  rinse solution, 289
  roccal-babcock, 354
  RODAC plate, 298
  Roese-Gottlieb, 358
  rolling ball viscometer, 144
  Rutgers phosphatase, 320
  sanitation of equipment, 289
  scharer rapid phosphatase, 315

Tests by title (*Cont'd*)
  selective ion procedure, 398
  spiral plate count, 156
  standard methods, water, 300
  standard plate count, 133
  strip count, 233
  sulfa drugs, 285
  surface contact, 296
  swab contact, 296
  thermister cryoscope, 395
  vacuum oven, 373
  vacuum oven - sand pan, 376
  vapor pressure osmometer, 397
  volhard, chlorides, 340
  Wisconsin mastitis, 242
Thermoduric bacteria, 7, 189–191
Thermister cryoscope - water, 395
Thermophilic, 7, 191
Titratable acidity, 329
Toxicity tests, water, 122
Tritium-3, 403
Tyramine, 70

U - V

Vacuum oven method - moisture, 373
Vacuum oven: sand pan method - moisture, 376
Vapor pressure osmometer - water, 397
Violet red bile agar, 128, 177
Volhard method chlorides, 340

W

Water: added to milk, 394
  thermister cryoscope, 395
  vapor pressure osmometer, 397
Water, tests, 300
Weigh tank, sampling, 97
Whey, 102, 380
Wisconsin mastitis test, 242
  procedure, 243

X - Y - Z

Yeasts, 7, 197–198
Yersinia enterocolitica, 53
Zinc-65, 403